COMBUSTION PHENOMENA

Selected Mechanisms of Flame Formation,
Propagation, and Extinction

COMBUSTION PHENOMENA

Selected Mechanisms of Flame Formation, Propagation, and Extinction

Edited by

JOZEF JAROSINSKI
BERNARD VEYSSIERE

CRC Press
Taylor & Francis Group
Boca Raton   London   New York

CRC Press is an imprint of the
Taylor & Francis Group, an **informa** business

CRC Press
Taylor & Francis Group
6000 Broken Sound Parkway NW, Suite 300
Boca Raton, FL 33487-2742

First issued in paperback 2017

© 2009 by Taylor & Francis Group, LLC
CRC Press is an imprint of Taylor & Francis Group, an Informa business

No claim to original U.S. Government works

ISBN 13: 978-1-138-11388-6 (pbk)
ISBN 13: 978-0-8493-8408-0 (hbk)

Library of Congress Cataloging-in-Publication Data

Combustion phenomena : selected mechanisms of flame formation, propagation, and extinction / editors, Jozef Jarosinski and Bernard Veyssiere.
    p. cm.
Includes bibliographical references and index.
ISBN 978-0-8493-8408-0 (alk. paper)
1. Combustion. I. Jarosinski, Jozef. II. Veyssiere, Bernard.

QD516.C6156 2009
541'.361--dc22

2008035803

Visit the Taylor & Francis Web site at
http://www.taylorandfrancis.com

and the CRC Press Web site at
http://www.crcpress.com

Contents

Preface

This book is a supplementary source of knowledge on combustion, to facilitate the understanding of fundamental processes occurring in flames during their formation, propagation, and extinction. The characteristic feature of the book lies in the presentation of selected types of flame behavior under different initial and boundary conditions. The most important processes controlling combustion are highlighted, elucidated, and clearly illustrated.

The book consists of eight topical chapters, which are subdivided into several problem chapters.

Chapter 1 provides a brief introduction summarizing the main challenges in combustion. It recalls the key events in the progress of combustion science concisely.

Chapter 2 is devoted to combustion diagnostics. The majority of the novel diagnostic techniques have already been presented in *Applied Combustion Diagnostics* (2002) by the same author and in a number of survey papers (e.g., *Proceedings of the Combustion Institute*, 2005; *Progress in Energy and Combustion*, 2006). In this book, K. Kohse-Höinghaus presents work on how to make measurements in combustion chemistry, since this is an area where there have been many recent developments.

Chapter 3 deals with flammability limits, ignition of a flammable mixture, and extinction of limit flames. Both phenomena, ignition and extinction, are dependent on time.

In Chapter 3.1, J. Jarosinski presents the problem of flammability limits and extinction mechanisms of limit flames in the standard flammability tube. Extinction in this particular type of tube is very specific and is presented against the background of other flames, which aroused interest for a very long time.

In Chapter 3.2, M. Kono and M. Tsue examine the mechanism of flame development from a flame kernel produced by an electric spark. They discuss results of numerical simulations performed in their laboratory in confrontation with experimental observations and confirm numerical simulation as a significant tool for elucidating the mechanism of spark ignition.

Very specific types of flames are presented in Chapter 4.

In Chapter 4.1, C-J. Sung deals with the propagation of counterflow premixed flames. Only symmetrical premixed twin flames propagating in an opposed-jet configuration are considered. This technique is used for determining the laminar flame speed, which serves as the reference quantity in studies of various phenomena involving premixed flames. The methods of determining laminar flame speeds and of deducing the overall activation energy on the basis of this technique are discussed.

In Chapter 4.2, S. Ishizuka presents his recent experimental and theoretical results on flame propagation along a vortex core. The validity of the existing models linking flame speed, vortex parameters, and mixture properties is discussed in the light of experimental results.

Some characteristics of edge-flames are identified by S.H. Chung in Chapter 4.3. Flames with edges occur in many forms. A thorough understanding of this subject is essential for turbulent combustion modeling.

Instability phenomena in flames are addressed in Chapter 5.

The fundamentals of combustion instability are presented by G. Searby in Chapter 5.1 and phenomena examined by him fall into two categories: instability of flame fronts and thermo-acoustic instabilities. Each category can be subdivided further, and these are discussed.

In Chapter 5.2, S. Candel, D. Durox, and T. Schuller consider certain aspects of perturbed flame dynamics. The relation between combustion instability and noise generation is described by reference to systematic experiments. The data indicate that acoustic emission is determined by flame dynamics. On this basis, combustion noise can be linked with combustion instability.

In Chapter 5.3, D. Dunn-Rankin discusses the shape of deflagrations in closed tubes and the conditions under which it assumes the form of a tulip. The propagation of a premixed flame in closed vessels has been studied from the nineteenth century. The tulip flame is an interesting example of flame–flow interaction originating from the Landau–Darrieus instability.

Chapter 6 is devoted to different methods of flame quenching.

In Chapter 6.1, A. Gutkowski and J. Jarosinski present results of an experimental and numerical study of flame propagation in narrow channels and the mechanism of quenching due to heat losses. This work takes up again classical studies of the quenching distance. The most characteristic features of limit flames are determined experimentally.

Chapter 6.2, contributed by S.S. Shy, is devoted to the problem of flame quenching by turbulence, which is important from the point of view of combustion fundamentals as well as for practical reasons. Effects of turbulence straining, equivalence ratio, and heat loss on global quenching of premixed turbulent flames are discussed.

In Chapter 6.3, C-J. Sung examines extinction of counterflow premixed flames. He emphasizes flame quenching by stretch and highlights four aspects of counterflow premixed flame extinction limits: effect of nonequidiffusion, part played by differences in boundary conditions, effect of pulsating instability, and relation to the fundamental limit of flammability.

In Chapter 6.4, J. Chomiak and J. Jarosinski discuss the mechanism of flame propagation and quenching in a rotating cylindrical vessel. They explain the observed phenomenon of quenching in terms of the formation of the so-called Ekman layers, which are responsible for the detachment of flames from the walls and the reduction of their width. Reduction of the flame speed with increasing angular velocity of rotation is explained in terms of free convection effects driven by centrifugal acceleration.

Chapter 7 focuses on turbulent combustion.

In Chapter 7.1, R. Borghi, A. Mura, and A.A. Burluka present an up-to-date survey of turbulent premixed flames, stressing on the physical aspects. Areas where additional work are needed are defined and discussed comprehensively.

In Chapter 7.2, J.H. Frank and R.S. Barlow describe the basic characteristics of non-premixed flames with an emphasis on fundamental phenomena relevant to predictive modeling. They show how the development of predictive models for complex combustion systems can be accelerated by combining closely coupled experiments and numerical simulations.

Chapter 7.3 deals with the fine resolution modeling of turbulent combustion. In a synthetic survey of numerical methods for turbulent combustion, L. Selle and T. Poinsot highlight the important progress that has been made in high performance numerical calculations and discuss prospects in the near future. They consider that massively parallel Large Eddy Simulation solvers can soon be expected to be cost-competitive, for development in industrial systems. Understanding and controlling combustion instabilities could be one of the future tasks for effective fine resolution modeling.

Chapter 8 contains other interesting examples of combustion and flame propagation.

In Chapter 8.1, F. Takahashi presents candle and laminar jet diffusion flames highlighting the physical and chemical mechanism of combustion in a candle and similar laminar coflow diffusion flames in normal gravity and in microgravity. This apparently simple system turns out to be very complex, and thereby its study is of great importance for the understanding of diffusion flame fundamentals.

In Chapter 8.2, J.D. Smith and V. Sick give an account of combustion processes in spark-ignition engines. They consider three modes of combustion in such engines: homogeneous-charge spark-ignition (premixed-turbulent combustion), stratified-charge spark-ignition (partially premixed-turbulent combustion), and spark-assisted compression ignition or spark-assisted homogeneous-charge compression-ignition (CI), which is a more recent concept introduced to achieve combustion with ultra-low emissions. The advantages and disadvantages of each method are discussed.

In Chapter 8.3, Z. Filipi and V. Sick review recent progress in CI engines. Apart from the classical CI engine they also examine two other concepts: the high-speed light-duty engine and low-temperature combustion in premixed CI engines. They argue that CI engines will operate by combining both these concepts in the future.

In Chapter 8.4, A. Teodorczyk presents the complex problem of transition from deflagration to detonation (DDT). He reviews the phenomena associated with the DDT process in smooth tubes according to the classical scheme and then describes DDT in obstructed channels.

In Chapter 8.5, B. Veyssiere exposes the state of knowledge in detonations. Particular features of the complex multidimensional structure of detonations are presented in relation with the recent results obtained either by nonintrusive optical diagnostics or numerical simulations from high performance calculations. The role of transverse waves in detonation propagation, the existence of correlations between the characteristic dimension of the cellular structure and the critical conditions for detonation initiation and detonation transmission, and the influence of the nonmonotonous heat release process behind the front are examined. Recent developments in the study of spinning detonations are also discussed.

The idea for writing this book originated in 2006 during the course of discussions that the editors had at the University of Poitiers (Laboratoire de Combustion et de Détonique). It gradually took shape and resulted in Combustion Phenomina: *Selected Mechanisms of Flame Formation, Propagation, and Extinction*. This book has effectively used photographic documentation to present physical mechanisms controlling combustion development. It was decided to collect all illustrations in a CD-ROM, which can be found at www.routledge.com/9780849384080.

We began to assemble a team of authors and soon had 26 prospective contributing authors. The International Symposium on Combustion in Heidelberg and the International Colloquium on the Dynamics of Explosion and Reactive Systems in Poitiers gave us a good opportunity to discuss certain issues concerning the book with some of the authors. We deeply appreciate assistance from many combustion experts who made this book possible. Though most of them were extremely busy, they agreed to participate in the project and managed to adhere to the tight publication schedule.

J. Jarosinski wishes to acknowledge his colleagues in the scientifi c community at the Lodz University, Poland, for their collaboration. Special thanks are due to Dr. A. Gorczakowski and Dr. Y. Shoshin. Support and encouragement were also received from Dr. W. Wisniowski, Institute of Aeronautics.

B. Veyssiere appreciates the contribution made by French scientists and is grateful for the encouragements received from the French combustion community.

We hope that reading the book will prove to be intellectually stimulating and enjoyable for members of our scientific community.

Jozef Jarosinski
Bernard Veyssiere

Editors

Jozef Jarosinski worked at the Institute of Aeronautics as a department head responsible for the development of the combustion chamber jet engines and combustion systems for piston engines produced in the Polish industry. He received his PhD in mechanical engineering in 1969 and his DSc in mechanics in 1988, both at the Warsaw University of Technology. In 1992, he joined the faculty of mechanical engineering of the Technical University of Lodz, Lodz, Poland, where he is currently serving as a professor. His research activity is related to fundamentals of combustion. He won the Fulbright Fellowship and spent an academic year (1977–1978) in Roger Strehlow's laboratory at the University of Illinois at Urbana-Champaign. In 1986–1987, he worked for a year and half at the combustion laboratory of McGill University, Montreal, Canada. He is a member of the Combustion Institute and a member of the board of directors of the Polish Combustion Institute. At present, he is continuing his research activity at the Institute of Aeronautics and at the Technical University of Lodz.

Bernard Veyssiere is a research director at Centre National de la Recherche Scientifique (CNRS). He works at the Laboratoire de Combustion et de Détonique of the Ecole Nationale Supérieure de Mécanique et d'Aérotechnique (ENSMA), University of Poitiers, Poitiers, France. He graduated as an engineer in mechanics from ENSMA in 1974, and acquired a PhD and a diploma of Docteur-es-Sciences Physiques in 1978 and 1985, respectively, from the University of Poitiers. He specializes in detonations and combustion phenomena in two-phase mixtures. Specifically, he investigates the regimes of propagation and the structure of detonations in gaseous reactive mixtures with fine solid particles in suspension, especially in the case of aluminum particles. Concurrently, he focuses on the study of propagation mechanisms of dust flames and also conducts research on the acceleration of premixed homogeneous gaseous flames and transition to detonation.

Contributors

Robert S. Barlow
Combustion Research Facility
Sandia National Laboratories
Livermore, California

Roland Borghi
Laboratoire de Mécanique et d'Acoustique
Centre National de la Recherche
   Scientifique, UPR 288
Ecole Centrale de Marseille
Marseille, France

Alexey A. Burluka
School of Mechanical Engineering
Leeds University
Leeds, United Kingdom

Sébastien Candel
Ecole Centrale Paris
EM2C Laboratory
Châtenay-Malabry, France

Jerzy Chomiak
Department of Applied Mechanics
Chalmers University of Technology
Göteborg, Sweden

Suk Ho Chung
School of Mechanical and
   Aerospace Engineering
Seoul National University
Seoul, Korea

Derek Dunn-Rankin
Department of Mechanical and
   Aerospace Engineering
University of California
Irvine, California

Daniel Durox
Ecole Centrale Paris
EM2C Laboratory
Châtenay-Malabry, France

Zoran Filipi
Department of Mechanical Engineering
University of Michigan
Ann Arbor, Michigan

Jonathan H. Frank
Combustion Research Facility
Sandia National Laboratories
Livermore, California

Artur Gutkowski
Department of Heat Technology
Technical University of Łódź
Łódź, Poland

Satoru Ishizuka
Graduate School of Engineering
Hiroshima University
Higashi-Hiroshima, Japan

Jozef Jarosinski
Department of Heat Technology
Technical University of Łódź
Łódź, Poland

Katharina Kohse-Höinghaus
Department of Chemistry
Bielefeld University
Bielefeld, Germany

Michikata Kono
Department of Aeronautics
   and Astronautics
University of Tokyo
Tokyo, Japan

Arnaud Mura
Laboratoire de Combustion
   et de Détonique
Centre National de la Recherche
   Scientifique, UPR 9028
ENSMA et Université de Poitiers
Futuroscope, Poitiers, France

Thierry Poinsot
Groupe Ecoulements et Combustion
Institut de Mechanique des Fluides de Toulouse
Toulouse, France

Thierry Schuller
Ecole Centrale Paris
EM2C Laboratory
Châtenay-Malabry, France

Geoff Searby
IRPHE Laboratory
Centre National de la Recherche Scientifique
Marseille, France

Laurent Selle
Groupe Ecoulements et Combustion
Institut de Mechanique des Fluides de Toulouse
Toulouse, France

Shenqyang S. Shy
Department of Mechanical Engineering
National Central University
Jhong-Li, Taiwan

Volker Sick
Department of Mechanical Engineering
University of Michigan
Ann Arbor, Michigan

James D. Smith
Department of Mechanical Engineering
University of Michigan
Ann Arbor, Michigan

Chih-Jen Sung
Department of Mechanical and
    Aerospace Engineering
Case Western Reserve University
Cleveland, Ohio

Fumiaki Takahashi
National Center for Space
    Exploration Research on
    Fluids and Combustion
NASA Glenn Research Center
Cleveland, Ohio

Andrzej Teodorczyk
Faculty of Power and Aeronautical Engineering
Warsaw University of Technology
Warsaw, Poland

Mitsuhiro Tsue
Department of Aeronautics
    and Astronautics
University of Tokyo
Tokyo, Japan

Bernard Veyssiere
Laboratoire de Combustion
    et de Détonique
Centre National de la Recherche
    Scientifique, UPR 9028
ENSMA et Université de Poitiers
Futuroscope, Poitiers, France

1

Introduction: Challenges in Combustion

Jozef Jarosinski and Bernard Veyssiere

Combustion is an applied science that is important in transportation, power generation, industrial processes, and chemical engineering. In practice, combustion must simultaneously be safe, efficient, and clean.

Combustion has a very long history. From antiquity up to the middle ages, fire along with earth, water, and air was considered to be one of the four basic elements in the universe. However, with the work of Antoine Lavoisier, one of the initiators of the Chemical Revolution and discoverer of the Law of Conservation of Mass (1785), its importance was reduced. In 1775–1777, Lavoisier was the first to postulate that the key to combustion was oxygen. He realized that the newly isolated constituent of air (Joseph Priestley in England and Carl Scheele in Sweden, 1772–1774) was an element; he then named it and formulated a new definition of combustion, as the process of chemical reactions with oxygen. In precise, quantitative experiments he laid the foundations for the new theory, which gained wide acceptance over a relatively short period.

Initially, the number of scientific publications on combustion was very small. At that time combustion experiments were conducted at chemical laboratories. From the very beginning up to present times, chemistry has contributed a lot to the understanding of combustion at the molecular level.

The chronology of the most remarkable contributions to combustion in the early stages of its development is as follows. In 1815, Sir Humphry Davy developed the miner's safety lamp. In 1826, Michael Faraday gave a series of lectures and wrote *The Chemical History of Candle*. In 1855, Robert Bunsen developed his premixed gas burner and measured flame temperatures and flame speed. Francois-Ernest Mallard and Emile Le Châtelier studied flame propagation and proposed the first flame structure theory in 1883. At the same time, the first evidence of detonation was discovered in 1879–1881 by Marcellin Berthelot and Paul Vieille; this was immediately confirmed in 1881 by Mallard and Le Châtelier. In 1899–1905, David Chapman and Emile Jouguet developed the theory of deflagration and detonation and calculated the speed of detonation. In 1900, Paul Vieille provided the physical explanation of detonation

phenomenon, as a shock wave followed by a reaction zone with heat release. In 1928, Nikolay Semenov published his theory of chain reactions and thermal ignition and was awarded the Nobel Prize for his work in 1956 (together with Ciril Norman Hinshelwood). The theory of chain reactions started the development of chemical gas kinetics and reaction mechanisms. Fundamental work on explosions and explosion limits due to chain reactions emerged. This in turn led to great advancement in gas phase reaction kinetics, realizing the role of free radicals, the nature of elementary reactions, and the elucidation of chemical reaction mechanisms. In more recent times, this has made kinetic modeling possible. In 1940, Yakov Zel'dovich analyzed thermal diffusive instability of two-dimensional flames and in 1944 published his book *Theory of Gas Combustion and Detonation*. The influence of turbulence on flame propagation was investigated by Gerhard Damköhler in 1940 and his work was extended by Kirill Shchelkin in 1943, on the basis of simple geometrical considerations. The development of combustion science was greatly influenced by Theodore von Kármán, the founder of the U.S. Institute of Aeronautical Sciences (1933) and the Jet Propulsion Laboratory (1944), and the cofounder of the Combustion Institute (1954). Around 1950, he organized an international team to compile and spread multidisciplinary knowledge on combustion science. Since then the term "aero-thermo-chemistry" has become synonymous with combustion.

However, a new era in the development of combustion science started with the foundation of the Combustion Institute in 1954, on the initiative of Bernard Lewis. The institute's influence was further strengthened in 1957 with the release of its journal *Combustion and Flame*. The creation of the institute brought about two important factors into research activity: organization of the combustion community and stimulation of international cooperation.

The mission of the Combustion Institute is to promote research in the field of combustion science. This is done through the dissemination of research findings at systematically organized international symposia on combustion and through publications. The institute plays an important role in promoting the specialized scientific disciplines that constitute the broad arena of combustion. From its inception the institute has also helped to promote international research activities.

Since 1967, the International Colloquia on the Dynamics of Explosions and Reactive Systems (ICDERS) were organized in addition to the Combustion Symposia. ICDERS was initiated by a group of visionary combustion scientists (Numa Manson, Antoni K. Oppenheim, and Rem Soloukhin). They considered the subject of these colloquia to be important to the future of combustion technology and control of global environmental emission.

Combustion is largely (but not solely) an application-driven scientific discipline, creating some technology drivers. In its early period of development, safety issues were of primary importance, together with related knowledge on flammability limits and explosions. In the 1950s, combustion research was stimulated by aero-propulsion and then by rocket propulsion (e.g., work on ions in flames was connected with the absorption of microwave radiation by a weak plasma). Interest in combustion-generated pollutants such as CO/NO_x and soot rose in the late 1960s. In the early 1970s an increased number of research projects was devoted to urban and wild-land fires. The energy crisis of the 1970s stimulated research on energy saving and combustion efficiency. In the 1980s and 1990s, interest in supersonic combustion grew and studies were undertaken on the role of combustion in climatic changes. All these technology drivers are still present in combustion research. New drivers, such as the development of micropower generation, catalytic combustion, mild combustion, SHS combustion, or the synthesis of nanoparticles are also emerging.

Progress in combustion science is enhanced by important developments in scientific tools and new methods of analysis. Some of these are

1. Introduction of the rigorous conservation equations for flows that react chemically.

2. Developments in computer techniques making it possible to solve complicated fluid motions in a combustion environment that are affected by diffusion and involve complicated chemistry (large numbers of elementary reactions, which individually are not "complex" but quite simple, i.e., most of them involve two reacting species, sometimes three, and the formation or breaking of just one bond), and with a large number of transient intermediates formed in the course of fuel oxidation and pollutant formation.

3. Application of laser diagnostics to probe elementary reaction processes and the structure of flames.

4. Development of activation energy asymptotics for the mathematical analysis of combustion phenomena.

During the 28th International Symposium on Combustion in 2000, Irvine Glassman spoke about the importance of research in the area of combustion science to modern society in his Hottel Lecture. In his message he appealed to the combustion community to contribute to the solution of real issues and to be more creative in solving economic, social, and environmental problems.

This appeal can only be realized if there is a profound understanding of the fundamental processes that occur during combustion, and if it is treated as a phenomenon that requires multidisciplinary science. This book responds to this challenge by presenting up-to-date information on some fundamental problems of combustion.

2

Diagnostics in Combustion: Measurements to Unravel Combustion Chemistry

Katharina Kohse-Höinghaus

CONTENTS

2.1 Introduction

Flames are fascinating phenomena, and they reveal some of their inherent features directly to man's senses: they radiate heat and light, they may hiss, crackle, and smell. To a combustion scientist or engineer, characterization will, however, involve measurements of quantitative combustion characteristics, including temperature, pressure, heat release, or the amount of gaseous and particulate emission. Since optimization of combustion devices involves computer-based simulation of the entire process, combustion measurements are often used to validate relevant submodels from the fuel delivery to the combustion effluent. With practical combustion devices as different in scale as kilns for waste management, power plant combustors, rocket engines, gas turbines, internal combustion engines, or household burners, measurements of relevant flame parameters need a large arsenal of techniques.

For improving the knowledge on combustion chemistry and physics in a practical device, areas of detailed investigation include, among others, the mixing of fuel and oxidator, which is often a two-phase process; ignition, which may rely on sparks, discharges, plasmas, or autoignition; flame–flowfield interaction, when the combustion process takes place in a partly homogenized or turbulent three-dimensional flow; and pollutant formation, where particularly the formation of NO_x, polycyclic aromatic hydrocarbons (PAHs), particulates, and other regulated air toxics such as aldehydes present open

questions. Novel engine concepts such as homogeneous charge compression ignition (HCCI) or controlled autoignition (CAI), novel fuels or fuel combinations such as bio-derived or Fischer–Tropsch fuels and fuel additives, as well as shifting the boundaries to previously seldom explored combustion conditions at very high pressures, very low temperatures, or very lean stoichiometries demand extension of the present database employed in combustion modeling.

It will come as no surprise that to explore these and other combustion aspects, a direct analysis of the process in question is preferable to any global characterization, e.g., at the exhaust level. For this direct inspection of the combustion problem, a variety of methods exist, which are, in their majority, based on laser spectroscopy. Interesting features of laser techniques include temporal and spatial resolution, imaging capability, generation of coherent signals, multiquantity or multispecies detection, and others, a major aspect being their noninvasive nature. Detailed literature is available on laser measurements in combustion [1–9], and the principles and advantages as well as typical instrumentation and application examples of many of these techniques have been summarized repeatedly.

Here, selected, more established laser diagnostic techniques will only be briefly named with their principal area of application in combustion measurements, but not further described; details on these and other methods can be found in the above-mentioned literature and references cited therein. Workhorse techniques in combustion diagnostics include Raman and Rayleigh measurements of major species concentration and temperature; coherent anti-Stokes Raman spectroscopy (CARS) for the measurement of temperature; laser-induced fluorescence (LIF) for the characterization of mixing processes, for the measurement of temperature and intermediate species concentrations; cavity ring-down spectroscopy (CRDS) for sensitive detection of minor species; laser Doppler velocimetry (LDV) and particle imaging velocimetry (PIV) for flowfield characterization; and scattering techniques as well as laser-induced incandescence (LII) for the measurement of incipient particles and soot. Combinations of these and other techniques have been proven to be immensely useful in the analysis of laboratory flames and large-scale combustion machinery, providing "hard facts" on combustion performance in the investigated environment.

For the present contribution, it is assumed that the available books and reviews [1–9] will be a good starting point for information, and that an additional, updated attempt to summarize important developments is not yet necessary. It is thus not intended to provide a balanced overview of all recent contributions to laser and probe measurements in combustion, which would be far beyond the limitations of this contribution. Instead, this chapter focuses on selected methods that are necessary to further unravel details in combustion chemistry. The motivation for this specific accent is threefold. First, one of the most interesting questions regarding cleaner combustion—the chemistry involved in the formation of soot and its precursors—is still only partly answered. Second, the perceived need for novel fuels and fuel blends warrants study of their specific decomposition and oxidation chemistry. Third, novel combinations of laser and probe measurements, in particular using different mass spectrometric approaches, offer the potential to study the chemistry in unprecedented detail, especially with respect to the role of isomeric compounds. Introduction to this area of work will highlight some recent developments involving our own group and those of close collaboration partners.

2.2 Techniques to Study Combustion Intermediates

In the complicated reaction networks involved in fuel decomposition and oxidation, intermediate species indicate the presence of different pathways that may be important under specific combustion conditions. While the final products of hydrocarbon/air or oxygenate/air combustion, commonly water and carbon dioxide, are of increasing importance with respect to combustion efficiency—with the perception of carbon dioxide as a

pollutant rather than a product for climatic reasons—the prediction of other undesired emissions and by-products is not possible on a global, thermodynamic level, but needs information on the pivotal species, which may influence the importance of different reaction channels. Many of these species are radicals, and they are often present only in rather small concentrations, in the ppm or even ppb range.

Flames to study details of the combustion kinetics should preferably be steady (unless, e.g., ignition is the process in question) and should be simplified regarding the flowfield and geometry. Several burner configurations are established for these investigations, including one-dimensional flat premixed flames, and counterflow or coflow diffusion flames [10–12]. In the present contribution, examples will be presented for premixed, flat flames at low pressure, where the flame front is expanded to permit detailed investigation. The relevant parameters, including temperature and species composition, are then displayed as a function of height h above the burner surface, with the fresh fuel/oxidizer mixture at $h = 0$, the flame front location at a few millimeters distance, and the burnt gases at heights of 15–20 mm or above.

2.2.1 Spectroscopic Techniques

Several spectroscopic techniques for the detection of intermediate species have been extensively reviewed some years ago [13]. Of the many combustion-relevant intermediate species in the H–C–N–O system alone, atoms including H, O, C, and N; diatomic radicals including OH, CH, C_2, CO, NH, CN, and NO; triatomic intermediates including HCO, $^1CH_2$, NH_2; larger molecules including CH_3, CH_2O, C_2H_2; and many others have been detected using, in the majority of studies, laser absorption or fluorescence spectroscopy. Of the well-established techniques, LIF and CRDS in particular, have been described in detail before [1,3,13–16]. Instrumentation for these techniques is moderately involved, requiring commercially available, tunable laser light, especially to excite electronic transitions in the UV and visible domain. For LIF, the more standardly employed configurations use pulsed lasers with nanosecond time resolution, and are frequently based on either excimer or Nd:YAG lasers as pump sources and tunable dye lasers to obtain light of the desired excitation frequency. For quantitative measurements in the context of combustion kinetics, the measurement is often performed in a point-wise fashion, where fluorescence from a single, well-defined location in the flame is focused, passed through a suitable wavelength-selective element, and detected with a photomultiplier. Imaging along a line or of a two-dimensional area is possible with suitable shaping of the exciting laser light and detection with a charge-coupled device (CCD) camera. More advanced variants of this technique may use multiphoton excitation or multispecies detection; also, several combustion parameters such as the local temperature and the concentration of a species may be obtained simultaneously.

The LIF technique is extremely versatile. The determination of absolute intermediate species concentrations, however, needs either an independent calibration or knowledge of the fluorescence quantum yield, i.e., the ratio of radiative events (detectable fluorescence light) over the sum of all decay processes from the excited quantum state—including predissociation, collisional quenching, and energy transfer. This fraction may be quite small (some tenths of a percent, e.g., for the detection of the OH radical in a flame at ambient pressure) and will depend on the local flame composition, pressure, and temperature as well as on the excited electronic state and ro-vibronic level. Short-pulse techniques with picosecond lasers enable direct determination of the quantum yield [14] and permit study of the relevant energy transfer processes [17–20].

LIF has been used in a multitude of studies to measure the concentrations of some important radicals, most frequently of OH, CH, and NO. While OH is an important contributor to the fuel degradation and oxidation pathways and an indicator of hot areas in flames, CH has often been used to trace the flame front location, whereas the direct investigation of NO formation is of importance with regard to NO_x being a regulated air toxic. Species including other elements, such as sulfur, phosphorus, alkali, etc., can be detected by LIF in combustion systems, and often, an indication of their presence may already be a useful result, even if quantification is not possible.

Of the very sensitive laser-based absorption techniques, CRDS has developed into a widely applied technique in combustion diagnostics within only a few years [15,16], with intracavity laser absorption spectroscopy (ICLAS) being increasingly employed by some groups [21,22] as well. Similar experimental infrastructure can be used for CRDS as for LIF, with both techniques being, in principle, compatible in a single apparatus. Tunable laser light is often used in the UV–VIS range with CRDS to detect important flame radicals [13], including many of the species that can also be measured by LIF. In addition, further species are accessible with CRDS, which do not fluoresce because of predissociative states [13,15,16]. Cavity-enhanced absorption techniques are also applied in flames in the near infrared [22–25]. The multiple absorption paths provided with these techniques are the reason for their superb sensitivity in the ppb range. Absolute concentrations can be obtained, provided the absorption coefficient for the respective transition is known. CRDS can be used in conjunction with other laser-based combustion diagnostics for

special purposes, as, e.g., in the combination of CRDS and LII [26] to detect PAHs and soot. A classic combination is that of LIF and CRDS [15,27], where CRDS can be used to place LIF measurements on an absolute scale. With respect to the application to fuel-rich flames, we have recently performed a systematic investigation with CRDS under such conditions [28]. The large number of species involved in pathways toward PAHs and soot and studies in flames of realistic hydrocarbon and oxygenated fuels demand that many intermediates of the type C_xH_y or $C_xH_yO_z$ should be detected quantitatively, if at all possible. For this general approach, *in situ* mass spectrometry (MS) is preferable to optical techniques, which can, however, be relied upon in combination with MS for the measurement of temperature and for the concentrations of some smaller intermediates.

2.2.2 Mass Spectrometric Techniques

Some applications of MS to study the chemistry of flames have been reviewed recently [12], with extensive reference to earlier studies. Several features of typical experimental strategies for the study of premixed flames under reduced pressure will be given specific attention here. All *in situ* combustion MS experiments are invasive, since a sample is drawn from the flame for analysis. Under the conditions at reduced pressure, the sampling probe can resolve the species profiles in the flame front (with a thickness of several millimeters, contrasted to a spatial resolution of <0.5 mm). To enable detection of radicals, a molecular beam (MB) sampling approach is typically used with 2–3 differential pumping stages. Species are ionized and separated according to their mass/charge (*m/z*) ratio, frequently using time-of-flight (TOF) configurations. Ionization can be performed by different means; here, electron ionization (EI), resonance-enhanced multiphoton ionization (REMPI), and photoionization (PI) using tunable vacuum ultraviolet (VUV) radiation from a synchrotron are considered. In all cases, quantification of a species requires unambiguous detection (separation of overlapping signatures, e.g., from isotopic contributions or from different compounds with identical nominal mass); minimization of fragmentation; and calibration, for example, with samples of known concentration. Separation of structural isomers is possible if the ionization energies are sufficiently different to be resolved with the energy resolution of the instrument. Calibration with standard samples may be used for stable species with sufficient vapor pressure. Calibration for radical species is feasible when ionization cross-sections (for photons or electrons) are known or can be reliably estimated.

For the examples given here, three different instruments have been used. While an EI-MBMS setup and a REMPI-MBMS apparatus are available in our own laboratory, experiments using VUV-PI-MBMS have been carried out in collaborations, predominantly at the Advanced Light Source (ALS) in Berkeley, United States, as well as at the Hefei Light Source (HLS), Hefei, China.

For EI-MBMS measurements, we have used an instrument which is similar to those employed in many previous studies [12,29], and which has been described before [30,31]. Sampling is performed through a quartz nozzle of ~150 μm diameter, and the gas sample is expanded from the burner housing (50 mbar) in two pumping stages, first to $<10^{-4}$ mbar, whereupon the center of the resulting beam is expanded through a skimmer of 2 mm diameter to $<10^{-6}$ mbar into the ionization chamber. By this process, the sample is cooled to ~400 K [29]. The REMPI-MBMS instrument has a quite similar configuration [32]. The mass resolution of both setups is sufficient ($m/\Delta m \sim 3000$) for the separation of many combustion species (including e.g., C_3H_8, CO_2, and C_2H_4O near $m/z = 44$). It is significantly higher than that of the spectrometer at the ALS ($m/\Delta m \sim 400$) and at the HLS ($m/\Delta m \sim 1400$), where species are not only separated by mass, but also by their different ionization energies [33–35].

The three different ionization methods have specific advantages. In the EI-MBMS, we use nominal energies in the range of 10.5–17 eV to minimize fragmentation. The energy spread of the ionizing electron beam is broad enough to enable the detection of all species in a mass spectrum at a fixed nominal energy. Argon can be used as an inert standard. Thus, an overview over the species pool is possible from a single spectrum, avoiding ambiguities by changing conditions between runs, and errors from long-term drifts by normalization to the Ar profile. REMPI-MBMS is more selective because of the resonant excitation/ionization process; this technique is very well suited for the detection of small aromatic species [32]. A novel feature provided by the instruments using tunable synchrotron radiation for VUV-PI-MBMS is the superior energy resolution of ~40 meV, which permits for the first time the discrimination between isomeric species [36]. For this purpose, however, several mass spectra are necessary at different fixed ionization energies. Both photoionization techniques are much less sensitive to fragmentation, which is advantageous especially for the detection of oxygenated or other species featuring relatively weak bonds. With regard to these different instrumental features, a combination of all techniques is particularly valuable to compare results. Analysis of differences has led us to further development of measurement and analysis procedures, and provides a more realistic assessment of systematic errors, which is extremely helpful in comparison with models [37]. In any case, calibration is of eminent importance, especially for radicals. Calibration procedures have recently been compared in some detail [38]. Optical and mass spectrometric techniques may be used in combination for species that can be detected both ways to minimize calibration errors.

To probe the chemistry of a particular fuel in low-pressure flames, profiles of quantitative species mole fractions x_i versus height above the burner h are typically obtained. Many of the intermediates are formed in the flame front, and it is infeasible to infer the importance of a specific radical or molecule from the shape of the measured profile, or from its mole fraction. Similarly, it may be misleading to use such data to extract specific reaction rate coefficients, since the influence of a single reaction or reaction sequence cannot be isolated. A comparison with reaction models is useful, and information on the status of chemical kinetics for this purpose as well as on typical associated errors is readily available [39–41]. Caution is also warranted when studying novel conditions, including previously unaddressed fuels and fuel blends, pressure, temperature, and stoichiometry ranges; when using new approaches or variants of the existing techniques; or when detecting species for the first time in a given situation. It may be problematic to infer verification or falsification of a model, which has originally been designed for different conditions, or to change parameters in the model to fit a single measurement. Much can be learned, however, from studying many species in flames of different fuels—preferably using different techniques. Some recent examples from collaborative work highlighting the investigation of fuel-rich combustion in hydrocarbon and oxygenate flames as well as in blends of both types of fuels, are described next.

2.3 Exemplary Results

For the analysis of the chemical structure of flames, laser methods will typically provide temperature measurement and concentration profiles of some readily detectable radicals. The following two examples compare selected LIF and CRDS results. Figure 2.1 presents the temperature profile in a fuel-rich (C/O = 0.6) propene–oxygen–argon flame at 50 mbar [42]. For the LIF measurements, ~1% NO was added. OH-LIF thermometry would also be possible, but regarding the rather low OH concentrations in fuel-rich flames, especially at low temperatures, this approach does not capture the temperature rise in the flame front [43]. The sensitivity of the CRDS technique, however, is superior, and the OH mole fraction is sufficient to follow the entire temperature profile. Both measurements are in excellent agreement. For all flames studied here, the temperature profile has been measured by LIF and/or CRDS.

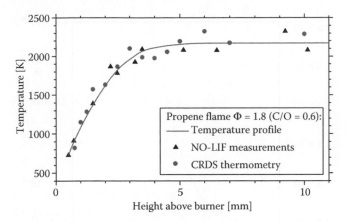

FIGURE 2.1
Temperature profile measurement in a flat, premixed fuel-rich propene low-pressure flame by LIF, using seeded NO, and CRDS, using naturally present OH radicals. (Adapted from Figure 3 in Kohse-Höinghaus, K. et al., *Z. Phys. Chem.*, 219, 583, 2005.)

Both techniques are suitable to determine intermediate species concentration profiles, as seen in Figure 2.2 for the C_2 radical. The shapes of the independently measured profiles from a fuel-rich premixed low-pressure propene flame are in very good agreement. The sensitivity of the CRDS measurement is excellent, and mole fractions ≥ 2 ppb can be determined with high precision under these conditions. The LIF measurements have been placed on an absolute scale using the CRDS results. It should be noted that the LIF measurement shows a slightly better spatial resolution, as evident from the region near the maximum, which is not captured as well by the CRDS measurements. Both techniques permit good resolution of the flame front, which is centered near 3.5 mm. In comparison with Figure 2.1, it is seen that

this corresponds approximately to the beginning of the plateau in the temperature profile, where the maximum temperature is attained near 5 mm.

A similar comparison of several techniques has also been used for the different ionization strategies in the MBMS measurements. Here, three independent instruments have been employed for the first time to measure the same species profile under nominally identical conditions. The benzene mole fraction profile in a fuel-rich propene–oxygen–argon flame is shown in Figure 2.3 for EI-MBMS, REMPI-MBMS, and VUV-PI-MBMS measurements, with the two former set-ups situated in Bielefeld and the latter at the ALS in Berkeley. The three measurements are in good quantitative agreement, differences in the peak mole fraction being of the order of 25%–30%. A rigorous error analysis confirms that an error of about this magnitude is not unexpected—different burners, flow meters, pressure gauges, burner housings, calibration mixtures on the one side, and different sampling nozzles, mass spectrometers, ionization techniques, detectors, and analysis routines on the other, contribute to these uncertainties. Repeated measurements of a "standard" flame condition with the same EI-MBMS instrument over several years have also resulted in errors of about 20% for stable species measurements, for which a calibration with samples of known concentration is possible. A closer inspection of the three profiles shows deviations of ~1 mm in the peak position, which is found at 5.5–6.5 mm. Shifts of this size are also not uncommon, and are a function of different burner and sampling geometries.

The formation of PAHs and soot involves a molecular precursor stage in which hydrocarbon radicals, especially with 2–5 carbon atoms, play a role to build up

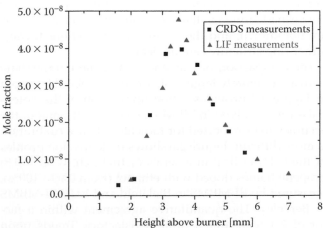

FIGURE 2.2
Comparison of the C_2 radical mole fraction profile measured by LIF and CRDS in a flat, premixed fuel-rich propene flame at 50 mbar.

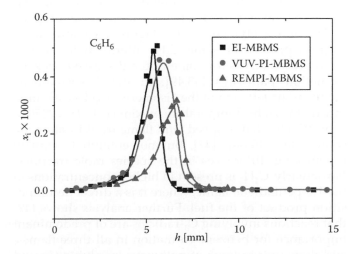

FIGURE 2.3
Benzene profile in a fuel-rich propene flame (C/O = 0.77), measured independently with three different mass spectrometric techniques; x_i: mole fraction, h: height above burner.

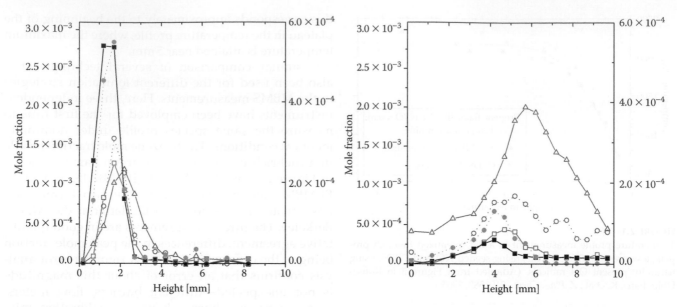

FIGURE 2.4
Mole fraction profiles of several C_4H_x intermediate species in fuel-rich flames of 1,3-pentadiene (left) and an acetylene/propene (1:1) mixture (right) with identical C/O ratio of 0.77 and C/H ratio of 0.625, the measurements were performed with EI-MBMS. $\triangle$: C_4H_2 (left axes), $\square$: C_4H_4 (left axes), $\blacksquare$: C_4H_6 (left axes); $\bullet$: C_4H_5 (right axes), $\bigcirc$: C_4H_3 (right axes). (From Atakan, B., Lamprecht, A., and Kohse-Höinghaus, K., *Combust. Flame*, 133, 431, 2003. With permission.)

benzene as the first aromatic ring, which is regarded as a central step for further growth [12,44–46]. Several pathways have been discussed as important for benzene formation, and it is commonly accepted that their specific importance may depend on the chemical nature of the fuel. An analysis of the influence of the fuel structure on the intermediate species pool is particularly instructive when flames of identical C/O and C/H ratio are studied under the same conditions. An example from previous work [47] compares fuel-rich combustion of the C_5H_8 isomers 1,3-pentadiene and cyclopentene with that of a 1:1 mixture of propene (C_3H_6) and acetylene (C_2H_2). Figure 2.4 shows mole fraction profiles of some C_4H_x intermediates, which are representatives for some of the discussed benzene formation pathways [48,49], in the 1,3-pentadiene flame on the left, and of the "isomeric" C_3H_6/C_2H_2 mixture on the right. Temperature maxima (2100 vs. 2250 K, respectively), flame speed, and flame front location are somewhat different [47], but not enough to explain the striking differences in the species mole fractions. Particularly, C_4H_5 is present in higher concentrations in the 1,3-pentadiene flame, where it is a direct decomposition product of the fuel. Further analysis shows [47] that reactions involving C_3-radicals are of predominant importance for benzene formation in all three flames, and that contributions of pathways involving C_4- and C_2-species are seen preferentially in the 1,3-pentadiene flame, while C_5-species are of some significance in the cyclopentene flame.

The analysis of fuel-rich flames by EI-MBMS, as in the example given above, typically provides profiles of about 35–45 species with $m/z \leq 100$, which can be used to improve the knowledge on reaction pathways and to extend present flame models. With the emergence of isomer-selective VUV-PI-MBMS, however, it seems that the number of species commonly included in such considerations may be by far too small. The cyclopentene flame described in Refs. [47,50] was analyzed again using the flame instrument at the ALS [37], and Figure 2.5 reveals the different chemical structures identified at $m/z = 90$ (C_7H_6) and $m/z = 92$ (C_7H_8) from photoionization efficiency (PIE) curves. Most of these species have not been identified in flames before, their concentrations are not known, and their roles in the reactions toward larger ring structures are unclear.

The comparison of benzene as a stable intermediate with a moderately large mole fraction of about 500 ppm in Figure 2.3 provides a first impression of the potential errors involved in MBMS measurements, and of the accuracy to be expected for radicals, where a calibration is more difficult. Figure 2.6 shows mole fraction profiles of the C_3H_3 radical in a series of fuel-rich (C/O = 0.5) propene flames doped with ethanol (from 0% to 100%), measured by EI-MBMS in Bielefeld and VUV-PI-MBMS in Berkeley. The quantitative agreement within a factor of 2 is considered quite satisfactory. Trends upon ethanol addition, illustrated in the insets, are in good quantitative agreement, with the PI-MBMS experiment exhibiting a superior signal-to-noise ratio. The decrease

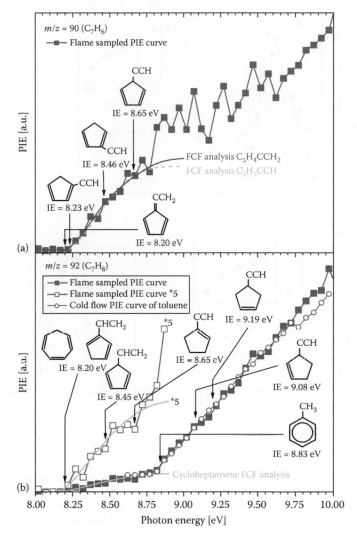

FIGURE 2.5
Comparison between flame-sampled PIE curves for (a) m/z = 90 (C_7H_6) and (b) m/z = 92 (C_7H_8) with the PIE spectra simulated based on a Franck–Condon factor analysis and the cold-flow PIE spectrum of toluene. Calculated ionization energies of some isomers are indicated. (From Hansen, N. et al., *J. Phys. Chem. A*, 2007. With permission.)

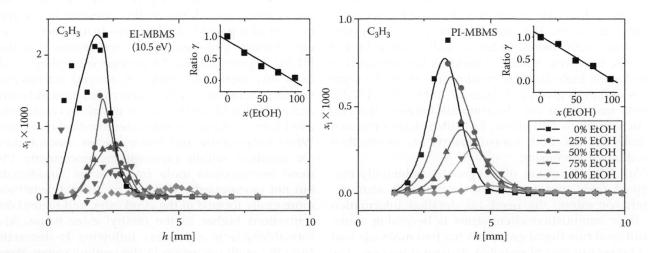

FIGURE 2.6
Mole fraction profiles of C_3H_3 in a series of fuel-rich propene flames (C/O = 0.5) doped with ethanol, measured by EI-MBMS (left) and VUV-PI-MBMS (right). The inset shows the effect of ethanol addition on the peak mole fraction, normalized to the value in the undoped flame; x_i: mole fraction, h: height.

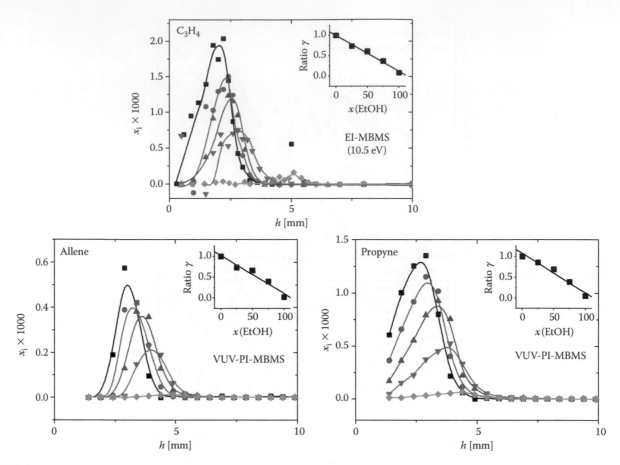

FIGURE 2.7

Mole fraction profiles of C_3H_4 measured with EI-MBMS (top) and VUV-PI-MBMS (bottom) in a series of fuel-rich propene flames (C/O = 0.5) doped with ethanol. While VUV-PI-MBMS can differentiate between propyne and allene, EI-MBMS detects the sum of both isomers. Note the quantitative agreement of the total C_3H_4 mole fractions obtained with both techniques. The inset shows the effect of ethanol addition on the peak mole fraction, normalized to the value in the undoped flame; x_i: mole fraction, h: height.

in the propargyl radical concentration by the addition of oxygenate fuel is expected [31] and is consistent with the assumption that oxygenates may decrease PAHs or particulate emissions. The interaction between the species pool from both fuel constituents will deserve further analysis, however. A similar trend as for propargyl is seen for the mole fraction profiles of C_3H_4 in the same series of flames given in Figure 2.7. While the VUV-PI-MBMS instrument can discriminate between the two isomers propyne and allene, EI-MBMS detects the sum of both species. Both measurements are in excellent quantitative agreement.

With the discussion of oxygenate, potentially bio-derived, fuels and fuel additives such as alcohols, ethers, or esters, the need for detailed information on their combustion chemistries is becoming acute. Additional functional groups in the fuel molecule lead to a larger number of possible structural isomers. The influence of the chemical structure of the fuel molecule

on the dominant decomposition and oxidation pathways is an interesting aspect for further study, and recent work has addressed this question in the combustion of a pair of alkyl esters—methyl acetate and ethyl formate [38]—and of the four isomers of butanol [51,52]. Figures 2.8 and 2.9 provide exemplary results from this work. In Figure 2.8, several intermediate species, including methyl, acetylene, formaldehyde, and acetaldehyde are shown in the methyl acetate and ethyl formate flames burnt under identical conditions. While temperature and major species concentrations are identical within experimental uncertainty [38], these intermediate mole fractions are significantly, but not unexpectedly, different, with acetylene being more easily formed in the ethyl ester, and methyl concentrations higher in the methyl ester flame. Also, formaldehyde is a product following H-abstraction from the methoxy group in the methyl acetate flame, whereas acetaldehyde can result following abstraction

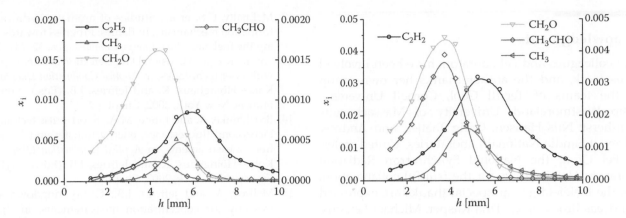

FIGURE 2.8
Intermediate species concentrations in fuel-rich flames (C/O = 0.5) of the two isomeric esters methyl acetate (left) and ethyl formate (right) burnt under identical conditions; x_i: mole fraction, h: height; species named on the left side of each graph correspond to left y-axes, species on the right to right y-axes.

of a secondary hydrogen from the ethoxy group in the ethyl formate flame.

Distinct fuel-specific reaction chemistry is also seen in premixed flat flames of the four butanols. Figure 2.9 shows PIE curves for $m/z = 72$ (C_4H_8O). The species pool is quite different, with butanal present in the 1-butanol flame, 2-methyl propanal in the i-butanol flame, and 2-butanone in both the 2-butanol and the t-butanol

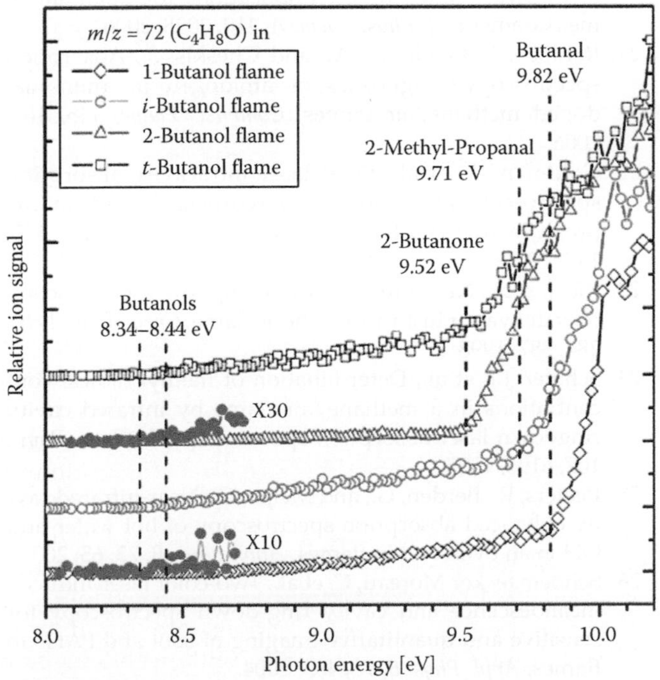

FIGURE 2.9
PIE curves for $m/z = 72$ (C_4H_8O) measured in four butanol flames. The accepted ionization energies for species identified by the observed ionization thresholds are indicated. (From Yang, B. et al., *Combust. Flame*, 148, 198, 2007. With permission.)

flames. This is consistent with the carbonyl function formed at the position of the hydroxyl group in the fuel, with the exception of t-butanol, where fission of the C–C bond is involved [52].

2.4   Summary and Conclusion

The present contribution has shown some new developments and recent examples for the investigation of detailed combustion chemistry, using laser spectroscopic methods, on the one hand, and several *in situ* mass spectrometric techniques, on the other. Different hydrocarbon and oxygenate fuels and fuel blends have been studied using these techniques, with particular emphasis on the detection of species, which are thought to be involved in the formation of small PAHs and soot. In the oxygenate and blended flames, attention has also been devoted to the formation of carbonyl compounds, since several members of this family are regulated as hazardous air pollutants. The study of nominally identical flames with different techniques in fully independent setups has revealed significant details about the reliability of these methods, and has led to considerable development of analysis and calibration strategies. A milestone in combustion diagnostics has been the emergence of isomer-selective techniques, and study of the rich combination of isomeric species shows new details about the combustion chemistry, which warrant further investigation. Also, the study of isomeric fuels under similar conditions demonstrates the importance of the fuel structure for the intermediate species pool, which may lead to a different product composition with regard to undesired emissions. With this powerful arsenal of techniques, chemical features can be studied, which may be of impact for the design of practical combustion devices.

Acknowledgments

Many colleagues and collaborators have been involved in this work, and the author thanks her own group and the teams of Terrill Cool, Cornell University; Phillip Westmoreland, University of Massachusetts at Amherst; Nils Hansen, Craig Taatjes, and Andrew McIlroy at Sandia National Laboratories, United States; and Fei Qi at the National Synchrotron Radiation Laboratory, Hefei, China, for the fruitful collaboration. From the Bielefeld group, special thanks are expressed to Andreas Brockhinke, Tina Kasper, Michael Letzgus, Markus Köhler, Patrick Oßwald, and Ulf Struckmeier. Discussions with Burak Atakan, University Duisburg-Essen; Marina Braun-Unkhoff, DLR Stuttgart; Michael Kamphus, Max-Planck-Institut für Chemie, Mainz, Germany; and Charles McEnally, Yale University, United States, on the subjects of this chapter are gratefully acknowledged.

References

1. Eckbreth, A.C., *Laser Diagnostics for Combustion Temperature and Species*, 2nd ed., Gordon and Breach, United Kingdom, 1996.
2. Kohse-Höinghaus, K. and Jeffries, J.B. (Eds.), *Applied Combustion Diagnostics*, Taylor & Francis, New York, 2002.
3. Kohse-Höinghaus, K. et al., Combustion at the focus: Laser diagnostics and control, *Proc. Combust. Inst.*, 30, 89, 2005.
4. Schulz, C. and Sick, V., Tracer-LIF diagnostics: Quantitative measurement of fuel concentration, temperature and fuel/air ratio in practical combustion systems, *Prog. Energy Combust. Sci.*, 31, 75, 2005.
5. Wolfrum, J., Lasers in combustion: From basic theory to practical devices, *Proc. Combust. Inst.*, 27, 1, 1998.
6. Daily, J.W., Laser induced fluorescence spectroscopy in flames, *Prog. Energy Combust. Sci.*, 23, 133, 1997.
7. Rothe, E.W. and Andresen, P., Application of tunable excimer lasers to combustion diagnostics: A review, *Appl. Opt.*, 36, 3971, 1997.
8. Masri, A.R., Dibble, R.W., and Barlow, R.S., The structure of turbulent nonpremixed flames revealed by Raman-Rayleigh-LIF measurements, *Prog. Energy Combust. Sci.*, 22, 307, 1996.
9. Chigier, N. (Ed.), *Combustion Measurements*, Hemisphere, New York, 1991.
10. Glassman, I., *Combustion*, 2nd ed., Academic Press, San Diego, 1987.
11. Law, C.K., *Combustion Physics*, Cambridge University Press, New York, 2006.
12. McEnally, C.S. et al., Studies of aromatic hydrocarbon formation mechanisms in flames: Progress towards closing the fuel gap, *Prog. Energy Combust. Sci.*, 32, 247, 2006.
13. Smyth, K.C. and Crosley, D.R., Detection of minor species with laser techniques, in *Applied Combustion Diagnostics*, Kohse-Höinghaus, K. and Jeffries, J.B. (Eds.), Taylor & Francis, New York, 2002, Chapter 2.
14. Brockhinke, A. and Linne, M.A., Short-pulse techniques: Picosecond fluorescence, energy transfer and "quench-free" measurements, in *Applied Combustion Diagnostics*, Kohse-Höinghaus, K. and Jeffries, J.B. (Eds.), Taylor & Francis, New York, 2002, Chapter 5.
15. McIlroy, A. and Jeffries, J.B., Cavity ringdown spectroscopy for concentration measurements, in *Applied Combustion Diagnostics*, Kohse-Höinghaus, K. and Jeffries, J.B. (Eds.), Taylor & Francis, New York, 2002, Chapter 4.
16. Scherer, J.J. et al., Cavity ringdown laser absorption spectroscopy: History, development and applications to pulsed molecular beams, *Chem. Rev.*, 97, 25, 1997.
17. Brockhinke, A. and Kohse-Höinghaus, K., Energy transfer in combustion diagnostics: Experiment and modeling, *Faraday Discuss.*, 119, 275, 2001.
18. Brockhinke, A. et al., Energy transfer in the OH $A^2\Sigma^+$ state: The role of polarization and of multi-quantum energy transfer, *Phys. Chem. Chem. Phys.*, 7, 874, 2005.
19. Bülter, A. et al., Study of energy transfer processes in CH as prerequisite for quantitative minor species concentration measurements, *Appl. Phys. B*, 79, 113, 2004.
20. Brockhinke, A. et al., Energy transfer in the $d^3\Pi_g$-$a^3\Pi_u$ (0–0) Swan bands of C_2: Implications for quantitative measurements, *J. Phys. Chem. A*, 110, 3028, 2006.
21. Rahinov, I., Goldman, A., and Cheskis, S., Absorption spectroscopy diagnostics of amidogen in ammonia-doped methane/air flames, *Combust. Flame*, 145, 105, 2006.
22. Goldman, A. et al., Fiber laser intracavity absorption spectroscopy of ammonia and hydrogen cyanide in low pressure hydrocarbon flames, *Chem. Phys. Lett.*, 423, 147, 2006.
23. Xie, J. et al., Near-infrared cavity ringdown spectroscopy of water vapor in an atmospheric flame, *Chem. Phys. Lett.*, 284, 387, 1998.
24. Scherer, J.J. et al., Determination of methyl radical concentrations in a methane/air flame by infrared cavity ringdown laser absorption spectroscopy, *J. Chem. Phys.*, 107, 6196, 1997.
25. Peeters, R., Berden, G., and Meijer, G., Near-infrared cavity enhanced absorption spectroscopy of hot water and OH in an oven and in flames, *Appl. Phys. B*, 73, 65, 2001.
26. Schoemaecker Moreau, C. et al., Two-color laser-induced incandescence and cavity ring-down spectroscopy for sensitive and quantitative imaging of soot and PAHs in flames, *Appl. Phys. B*, 78, 485, 2004.
27. Dreyer, C.B., Spuler, S.M., and Linne, M., Calibration of laser induced fluorescence of the OH radical by cavity ringdown spectroscopy in premixed atmospheric pressure flames, *Combust. Sci. Tech.*, 171, 163, 2001.

28. Schocker, A., Kohse-Höinghaus, K., and Brockhinke, A., Quantitative determination of combustion intermediates with cavity ring-down spectroscopy: Systematic study in propene flames near the soot-formation limit, *Appl. Opt.*, 44, 6660, 2005.

29. Kamphus, M. et al., REMPI temperature measurement in molecular beam sampled low-pressure flames, *Proc. Combust. Inst.*, 29, 2627, 2002.

30. Lamprecht, A., Atakan, B., and Kohse-Höinghaus, K., Fuel-rich propene and acetylene flames: A comparison of their flame chemistries, *Combust. Flame*, 122, 483, 2000.

31. Kohse-Höinghaus, K. et al., The influence of ethanol addition on premixed fuel-rich propene-oxygen-argon flames, *Proc. Combust. Inst.*, 31, 1119, 2007.

32. Kasper, T.S. et al., Ethanol flame structure investigated by molecular beam mass spectrometry, *Combust. Flame*, 150, 220, 2007.

33. Cool, T.A. et al., Selective detection of isomers with photoionization mass spectrometry for studies of hydrocarbon flame chemistry, *J. Chem. Phys.*, 119, 8356, 2003.

34. Cool, T.A. et al., Photoionization mass spectrometer for studies of flame chemistry with a synchrotron light source, *Rev. Sci. Instrum.*, 76, 094102, 2005.

35. Cool, T.A. et al., Photoionization mass spectrometry and modeling studies of the chemistry of fuel-rich dimethyl ether flames, *Proc. Combust. Inst.*, 31, 285, 2007.

36. Taatjes, C.A. et al., Enols are common intermediates in hydrocarbon oxidation, *Science*, 308, 1887, 2005.

37. Hansen, N. et al., Initial steps of aromatic ring formation in a laminar premixed fuel-rich cyclopentene flame, *J. Phys. Chem. A*, 111, 4081, 2007.

38. Oßwald, P. et al., Isomer-specific fuel destruction pathways in rich flames of methyl acetate and ethyl formate and consequences for the combustion chemistry of esters, *J. Phys. Chem. A*, 111, 4093, 2007.

39. Warnatz, J., Maas, U., and Dibble, R.W., *Combustion*, Springer-Verlag, Berlin, 1996.

40. Miller, J.A., Pilling, M.J., and Troe, J., Unravelling combustion mechanisms through a quantitative understanding of elementary reactions, *Proc. Combust. Inst.*, 30, 43, 2005.

41. Smith, G.P., Diagnostics for detailed kinetic modeling, in *Applied Combustion Diagnostics*, Kohse-Höinghaus, K. and Jeffries, J.B. (Eds.), Taylor & Francis, New York, 2002, Chapter 19.

42. Kohse-Höinghaus, K. et al., Combination of laser- and mass-spectroscopic techniques for the investigation of fuel-rich flames, *Z. Phys. Chem.*, 219, 583, 2005.

43. Hartlieb, A.T., Atakan, B., and Kohse-Höinghaus, K., Temperature measurement in fuel-rich non-sooting low-pressure hydrocarbon flames, *Appl. Phys. B*, 70, 435, 2000.

44. Bockhorn, H. (Ed.), *Soot Formation in Combustion*, Springer-Verlag, Berlin, 1994.

45. Frenklach, M., Reaction mechanism of soot formation flames, *Phys. Chem. Chem. Phys.*, 4, 2028, 2002.

46. Richter, H. and Howard, J.B., Formation of polycyclic aromatic hydrocarbons and their growth to soot—a review of chemical reaction pathways, *Prog. Energy Combust. Sci.*, 26, 565, 2000.

47. Atakan, B., Lamprecht, A., and Kohse-Höinghaus, K., An experimental study of fuel-rich 1,3-pentadiene and acetylene/propene flames, *Combust. Flame*, 133, 431, 2003.

48. Frenklach, M. and Warnatz, J., Detailed modeling of PAH profiles in a sooting low-pressure acetylene flame, *Combust. Sci. Tech.*, 51, 265, 1987.

49. Wang, H. and Frenklach, M., A detailed kinetic modeling study of aromatics formation in laminar premixed acetylene and ethylene flames, *Combust. Flame*, 110, 173, 1997.

50. Lamprecht, A., Atakan, B., and Kohse-Höinghaus, K., Fuel-rich flame chemistry in low-pressure cyclopentene flames, *Proc. Combust. Inst.*, 28, 1817, 2000.

51. McEnally, C.S. and Pfefferle, L.D., Fuel decomposition and hydrocarbon growth processes for oxygenated hydrocarbons: Butyl alcohols, *Proc. Combust. Inst.*, 30, 1363, 2005.

52. Yang, B. et al., Identification of combustion intermediates in isomeric fuel-rich premixed butanol-oxygen flames at low pressure, *Combust. Flame*, 148, 198, 2007.

3

Flammability Limits: Ignition of a Flammable
Mixture and Limit Flame Extinction

CONTENTS

3.1   Flammability Limits: History and Mechanism of Flame Extinction

Jozef Jarosinski

3.1.1   Introduction

The concept of flammability limits was first formulated over 200 years ago by Humboldt and Gay Lussac [1]. Later, the theory of flames was gradually developed by Mallard and Le Chatelier [2], Jouguet [3], Daniell [4], Lewis and Elbe [5], Zel'dovich [6], and others. Zel'dovich made an outstanding contribution to the knowledge of flame propagation limits. In his book [6], Zel'dovich concluded that flammability limits were concentration limits, owing to flame cooling by radiative heat losses from the flame and the adjacent hot combustion gases. He showed that at the concentration limits, the laminar burning velocity cannot be zero, but has to take a finite value. He also determined the relation between this velocity and the lowering of the temperature below the adiabatic value at the limit. Later, Spalding [7] arrived at similar conclusions.

To be in parallel with the development of theory, for practical reasons, flammability limits were measured in many laboratories: their knowledge for different mixtures was necessary to bring down the number of accidents in mines and in industry. Most measurements were made in vessels of various shapes. Hence, it was not a surprise that the obtained flammability limits were apparatus-dependent. Coward and Jones [8] gave a very good summary of empirical knowledge on this issue and also proposed a new standard apparatus for determining flammability limits. This was a vertical tube, 51 mm in diameter and 1.8 m long, closed at the upper end and open to the atmosphere at the bottom. If a mixture is ignited at the bottom of the tube and a flame can be formed, to propagate all the way to the top, the mixture is said to be flammable. However, if the flame is extinguished while it is part of its way up the tube, the mixture is said to be nonflammable.

The standard tube was used by Levy [9] in 1965 to study upward and downward flame propagation in lean methane/air and propane/air limit mixtures. He made several new observations, very important at the time:

1. Under the influence of gravity, the shapes of upward and downward propagating flames are very different.

2. Bubble-shaped limit flame propagating up moves with a velocity determined by buoyancy forces, like an air bubble in a column of water (in both cases, the Davies and Taylor formula [10] can be applied).

3. For lean methane/air mixtures, the flammability limits (as a matter of fact, flame extinction limits) appear to be different for upward and downward propagations.

In spite of the growing number of experimental and theoretical studies [11–17], it was difficult to account for observation 3. Progress could only be made based on the knowledge of flame stretch and preferential diffusion, taking into account the effect of these parameters on the burning velocity. Although the concepts of flame stretch and preferential diffusion had been introduced much earlier, their practical application to laminar flames was still in the initial stages at that time. Considerable improvement in the understanding of the effect of these phenomena on the behavior of a laminar flame and its parameters was achieved by Law [18]. Subsequent investigations by Law, Sung, Egolfopoulos, and Dixon-Lewis, both experimental and numerical, contributed to arriving at new, more reliable data on flame parameters as well as on the development of the theory of fundamental flammability limits [19–22].

Empirical flammability limits are only an approximation to the fundamental limits, which can be treated as a characteristic property of each mixture. The fundamental limit represents a concentration or pressure limit beyond which steady propagation of a one-dimensional, planar flame with volumetric heat loss becomes impossible. Such idealized flames cannot actually be maintained, even under laboratory conditions. Real flames are influenced by conductive and convective heat losses, flame stretch, gravity-related effects, etc., and usually their extinction occurs beyond the fundamental limit.

Recent contributions made by our research group to the study of limit flames propagating in the standard flammability tube are presented next.

3.1.2 Limit Flames Propagating Upward

Experiments with such flames are usually conducted in the standard flammability tube; description of typical

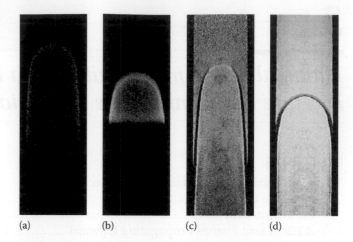

FIGURE 3.1.1
Upward propagating lean limit flames. (a) Methane/air in a standard cylindrical tube—direct photography, (b) propane/air in a standard cylindrical tube—direct photography, (c) methane/air in a square 50 mm × 50 mm tube—schlieren photography, and (d) propane/air in a square 50 mm × 50 mm tube—schlieren photography.

procedure can be found elsewhere [8,9,15,23,24]. For a mixture of limit composition, ignition of a flammable mixture at the open bottom end of the tube initiates propagation of a bubble-shaped flame from the bottom to the top end of the tube (Figure 3.1.1). As mentioned earlier, the flame moves with a velocity determined by the buoyancy forces [9]. The velocity is independent of the mixture properties and can be calculated from the Davies and Taylor formula [10],

$$w = 0.328\sqrt{gD}$$

where
  w is the velocity of the bubble-shaped flame
  g the acceleration due to gravity
  D the tube diameter

It can be seen in Figure 3.1.1 that the total surface area of the propane lean limit flame is much less than that of the methane one. This is because the laminar burning velocity for the limit mixture is much higher for propane than for methane.

3.1.3 Flow Structure

The flow structures of lean limit methane and propane flames are compared in Figures 3.1.2 and 3.1.3. The structure depends on the Lewis number for the deficient reactant. A stretched lean limit methane flame (Le < 1), propagating up is affected by preferential diffusion, giving it a higher burning intensity. Hence, the flame extinction limit is extended. On the other hand, for a stretched lean limit propane flame (Le > 1), the same effect reduces the burning intensity, which can

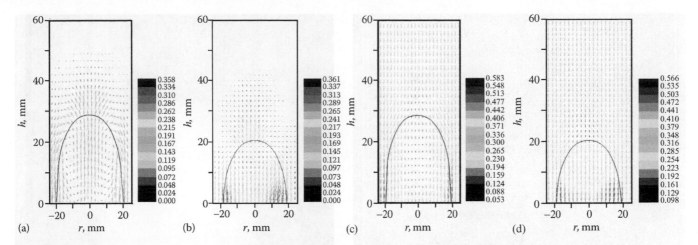

FIGURE 3.1.2
Flow velocity field determined by PIV. Lean limit flames propagating upward in a standard cylindrical tube in methane/air and propane/air mixtures. (a) Methane/air—laboratory coordinates, (b) propane/air—laboratory coordinates, (c) methane/air—flame coordinates, and (d) propane/air—flame coordinates.

lead directly to flame extinction. At the flame leading point, the laminar burning velocity of both limit flames is considerably lower than the buoyant velocity, w.

On comparing the two flames, it is evident that the flow structure of the lean limit methane flame fundamentally differs from that of the limit propane one. In the flame coordinate system, the velocity field shows a stagnation zone in the central region of the methane flame bubble, just behind the flame front. In this region, the combustion products move upward with the flame and are not replaced by the new ones produced in the reaction zone. For methane, at the lean limit an accumulation of particle image velocimetry (PIV) seeding particles can be seen within the stagnation core, in

the combustion products (Figure 3.1.4). A similar flow structure, with a stagnation core, is observed for a limit propane flame, but in a rich propane/air mixture, propagating with Le < 1 (Figure 3.1.5).

The structure of a bubble-shaped lean limit propane flame (Le > 1) is different. At the flame front inflow, the streamlines are much less divergent and soon converge again.

It is also interesting to examine the global gas dynamic structure of upward propagating flames. Figure 3.1.6 gives an example of the global velocity field for the lean limit methane flame in the flame coordinates. The velocity distributions for all near limit flames studied share certain features. The central part of the bubble-shaped flame is

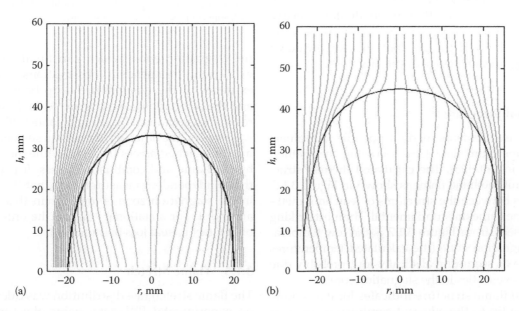

FIGURE 3.1.3
Streamlines of lean limit flames propagating upward in (a) methane/air and (b) propane/air mixtures (flame coordinates).

FIGURE 3.1.4
PIV image of upward propagating lean limit methane flame.

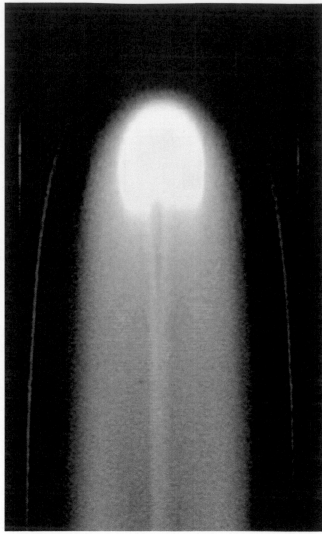

FIGURE 3.1.5
Schlieren picture of upward propagating rich limit propane flame with superimposed direct photography.

occupied by the stagnation zone. In the region of the flame skirt, near the tube wall, the gas experiences a vertical acceleration—the velocity vectors of the combustion products converge. The vertical components of the velocity remain uniform over a distance of about 10 cm. Lower down than this, near the tube centerline, they gradually start to decay, finally forming a secondary stagnation zone. At the same time, near the tube wall, the gas continues to flow down without any significant change of velocity.

3.1.4 Thermal Structure

PIV velocity measurements made it possible to evaluate the flame temperature field [23], following the method demonstrated in Ref. [25]. The calculated thermal structure of lean limit methane flame is shown in Figure 3.1.7. The differences between the structures of lean limit methane and propane flames are fundamental. The most striking phenomenon seen from Figure 3.1.7 is the low temperature in the stagnation zone (the calculated temperatures near the tube axis seem unrealistically low, probably due to very low gas velocities in the stagnation core).

This thermal flame structure indicates local heat flow from the flame tip to the adjacent combustion gases in

the stagnation zone, which can finally lead to flame extinction. Independent observations confirmed the existence of the temperature drop between the flame and the stagnation zone (Figure 3.1.8). One set of these results was obtained by using a contact 10 μm platinum wire sensor, the other one using thin silicone carbide filaments stretched across the tube, while they were heated by the flame and combustion gases. It is evident from these independent measurements that the temperature in the stagnation zone is lower than that in the flame around it. For a stationary flame, the only explanation of this fact is heat loss by radiation.

3.1.5 Flame Stretch

The flame stretch rate distribution was calculated based on experimental PIV data, using the semitheoretical

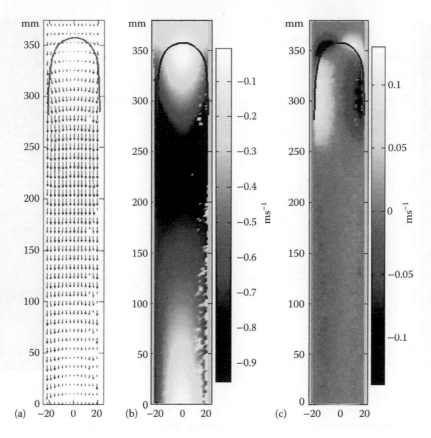

FIGURE 3.1.6
Global velocity distribution behind flame front. Upward propagation in 5.15% methane/air mixture. (a) vector map, (b) and (c) scalar maps of axial and radial velocity components, respectively. Spots are caused by condensation of water vapor on the glass walls.

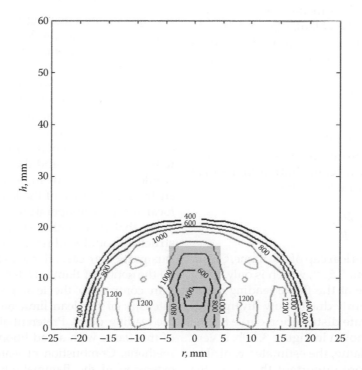

FIGURE 3.1.7
Thermal structure of lean limit methane flame. Isotherms calculated from the measured gas velocity distribution.

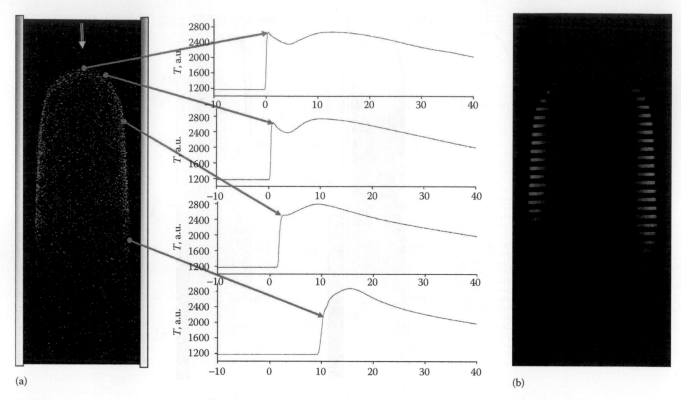

FIGURE 3.1.8
Temperature measurements. (a) Platinum wire sensor measurements and (b) silicon carbide filaments.

method adapted from Ref. [16]. The lean limit methane flame front studied is very thick (its total thickness is ≈6 mm), and there is a problem in the selection of the leading edge, required for the calculation. Finally, the stretch rate was calculated, as is commonly done, at the edge of the flame preheat zone ($T = 400\,K$ isotherm). For the lean limit methane flame, the local stretch rate along the flame surface is shown in Figure 3.1.9 (both methane and propane results are shown in Figure 3.1.9a and b, respectively). It is seen that it reaches maximum at the flame tip and gradually falls in the lower parts of the flame. The flame curvature also passes through a maximum at the flame tip. Its contribution to the total flame stretch rate can be estimated to be

$$k_c = \frac{2u_L}{R}$$

where R is the radius of spherical cap of the flame. Calculations give its value as about $3.5\,s^{-1}$, which is only about 10% of the total stretch rate at the flame leading point. Therefore, the stretch is mainly due to flow divergence.

The shape of the stretch rate distribution curve for the lean limit propane flame, shown in Figure 3.1.9b, is very similar. However, for this flame, the estimated contribution of the curvature is more important than with the methane flame, reaching about 40% of the maximum stretch rate.

3.1.6 Extinction Mechanism of an Upward Propagating Flame

An upward propagating flame is positively stretched. Extinction of both the lean limit propane (Le > 1) and methane (Le < 1) flames starts at the respective leading points. As has been shown, this is where the stretch rate is at a maximum. However, according to the theory [26], the responses of the stretched flames are opposite when the effective Lewis number for the mixture is greater than or less than 1. This means that the burning intensity at the leading point of the limit propane flame would fall to the lowest value and that of the limit methane flame would increase to the highest value for the entire flame surface. In the case of the lean limit propane flame, the local reaction temperature would fall to some critical value that even with a slight disturbance of the flame structure could lead result in flame extinction. Thus, in this particular case, flame extinction is directly caused by the action of flame stretch.

In contrast to the lean propane flame, the burning intensity of the lean limit methane flame increases for the leading point. Preferential diffusion supplies the tip of this flame with an additional amount of the deficient methane. Combustion of leaner mixture leads to some extension of the flammability limits. This is accompanied by reduced laminar burning velocity, increased flame surface area (compare surface of limit methane

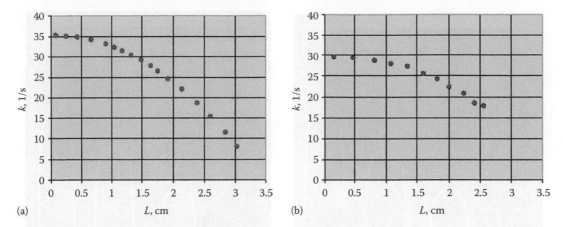

(a)   L, cm   (b)   L, cm

FIGURE 3.1.9
Stretch rate as a function of distance, from the top leading point along the flame front, (a) for a lean limit methane flame and (b) lean limit propane flame propagating upward in a standard cylindrical tube.

and propane flames in Figure 3.1.1), and the creation of a bubble-shaped stagnation zone. Measurements demonstrated that the temperature in this zone is lowered in comparison with that of the flame front (see Figure 3.1.8), because of heat loss by radiation. These conclusions were confirmed by numerical simulations. The computations were carried out for a flame propagating in a lean limit methane/air mixture, with and without heat transfer by radiation. The resulting temperature profiles are shown in Figure 3.1.10 and the reaction rate profiles in Figure 3.1.11. All experimental and numerical results presented here indicate that for Le < 1, flame extinction is brought about by radiative heat loss.

In the case of flame propagation in the lean limit methane/air mixture, the local laminar burning velocity at

the leading point of the flame is dramatically reduced, but it cannot fall below some critical value [6,19,27]. The flame goes to extinction owing to its configuration around the stagnation zone of reduced temperature (see Figures 3.1.2b, 3.1.4, and 3.1.8). Heat loss from the flame to this zone reduces the reaction temperature and the laminar burning velocity. Additionally, the location of the maximum heat release rate lies very close to the stagnation zone, which may contribute to flame extinction through incomplete reaction. The empirical flammability limit is slightly narrower ($\phi = 0.51$) than the present numerical simulations ($\phi = 0.50$) and the fundamental flammability limit ($\phi = 0.493$) indicated in Ref. [21].

In the experiments, limit flames usually propagate through the whole length of the tube and only in some

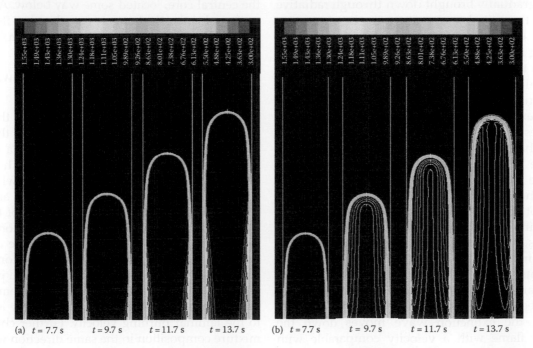

(a) $t = 7.7$ s $t = 9.7$ s $t = 11.7$ s $t = 13.7$ s   (b) $t = 7.7$ s $t = 9.7$ s $t = 11.7$ s $t = 13.7$ s

FIGURE 3.1.10
Temperature profiles for methane–air mixture with $\Phi = 0.50$ in 50 mm tube diameter (a) neglecting and (b) considering radiation heat transfer.

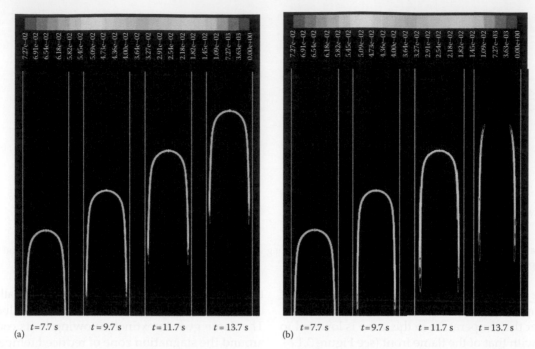

FIGURE 3.1.11
Reaction rate profiles for methane–air mixture with $\Phi = 0.50$ in 50 mm tube diameter (a) neglecting and (b) considering radiation heat transfer.

particular cases extinction is observed lower down the tube. It has been experimentally confirmed that the thermal conditions at the moment of ignition affect the location along the tube where flame extinction occurs. This has been demonstrated by using a wide range of ignition energies. The initial temperature in the stagnation zone is very important. During upward propagation, it is gradually brought down through radiative heat loss. If the initial temperature in the stagnation zone is low enough, its later reduction to a critical value during flame propagation in the end leads to the extinction of the flame.

Earlier schlieren observations of a flame propagating in a mixture at the flammability limit showed that after flame extinction, the rising bubble of hot combustion gases still travels some distance without any significant change in shape and at the same velocity as the flame [9,15]. PIV measurements were used to elucidate the cause of this phenomenon. Figure 3.1.12 illustrates the typical history of flame extinction. In some photographs, the tube was flashed by a laser sheet, illuminating a cloud of seeding particles. In these images, a dense cloud of particles can be seen in the fresh mixture, becoming thin in the hot region. This confirms earlier schlieren observations. Flame extinction starts at the flame tip, in Figure 3.1.12 somewhere between frames (shown as (c) and (d) in the figure), and the extinction wave moves down the sides of the flame with a velocity comparable with the bubble velocity. The gradually shrinking flame surface progressively leads to less production of hot

combustion gases at the top of the hot gas bubble. The evolution of the flow structure is illustrated in Figure 3.1.13. It can be seen that there is a heat deficit in this region of the bubble, initially supplied by the reaction gases from the reaction zone, located in the region of the flame skirts (recirculation). After complete flame extinction, heat comes from hot combustion gases of the central core, located some way below. As the flame surface shrinks, the hot gases accelerate. After complete flame extinction, their velocity exceeds 1 m/s.

3.1.7 Extinction Mechanism of a Downward Propagating Flame

Downward propagation is studied with the standard tube closed at the bottom and ignition at the open top end. A flame propagating downward in a mixture far from the flammability limits is convex. The convexity, expressed by the parameter $\varepsilon = w/S_L - 1$ (where w is the velocity of the flame at the leading point and S_L is the laminar burning velocity), is a function of the thermal expansion ratio $1/\alpha = T_b/T_a$, dependent on the equivalence ratio ϕ [28]. Gradual change in the equivalence ratio toward the flammability limit is accompanied by decreasing flame convexity. For downward propagating flames, for the limit mixture the flame becomes approximately flat ($\varepsilon \approx 0$). Its propagation velocity is then close to the laminar burning velocity. Further change in the mixture composition in the same direction would bring down the laminar burning velocity to below the adiabatic value. Extinction near the wall would follow.

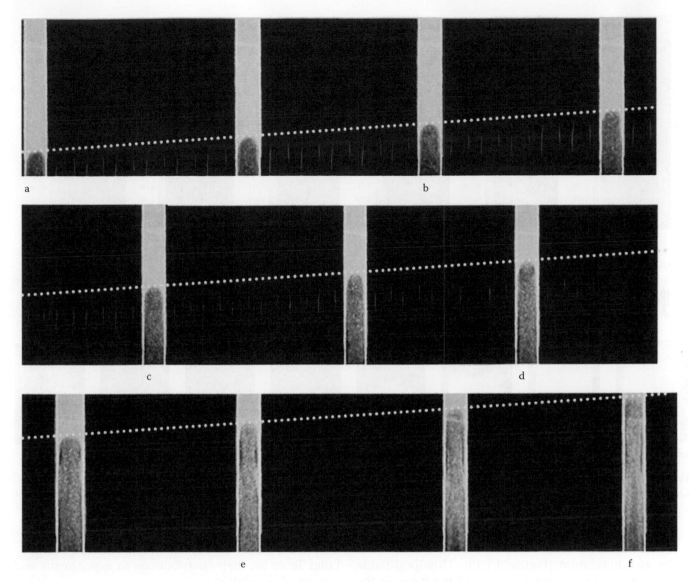

FIGURE 3.1.12
History of upward flame propagation and extinction in lean limit methane/air mixture. Square 5 cm × 5 cm vertical tube. Green color frames indicate PIV flow images. Red color represents direct photography of propagating flame. Extinction starts just after frame c. Framing rate 50 frames/s.

Near a wall, the conditions are more favorable for the propagation of a convex flame than that of a flat one. A convex flame has the following advantages: (1) its effective area of contact with the wall is reduced since the surface is inclined, (2) its burning velocity is higher than that of the limit flame, and (3) the edge of the flame near the wall is continuously supplied with chemically active species (radicals). A limit flat flame has no such advantages. The area of contact of such a flame (the zone of combustion products and adjacent to it) is large, the burning velocity is at a minimum and the flame stretch mechanism cannot help to supply its edges with reactive species. Under these conditions, the flame is locally effectively cooled by the walls. Temperature lowering near the wall effectively quenches chemical reactions at a distance of similar order of magnitude as the flame thickness. Hence, local flame extinction occurs. Thus, extinction of a propagating flat flame, perpendicular to the wall, is triggered by heat loss to the wall. After some time, the flame loses contact with the wall and floats freely in the interior of the tube. This residual flame, with hot gases behind, is finally driven to extinction by buoyancy, forcing the cooler product gases near the walls ahead of the flame.

Experiments confirm this mechanism. It was observed that before the extinction events set in, the speed of a limit flame propagating downward falls and the flame partially loses contact with the walls (Figure 3.1.14). In a square tube, local extinction starts in the corners, where heat loss to the walls is expected

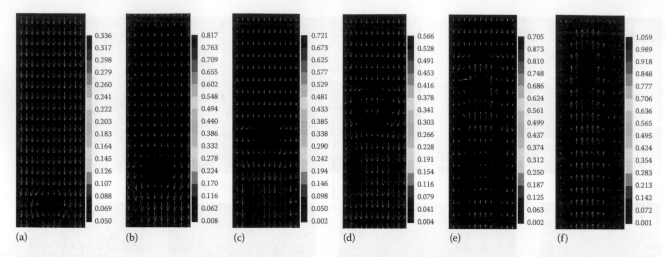

FIGURE 3.1.13

Evolution of flow velocity field in flame coordinates during extinction of upward propagating lean limit methane flame. Frames selected from Figure 3.1.12.

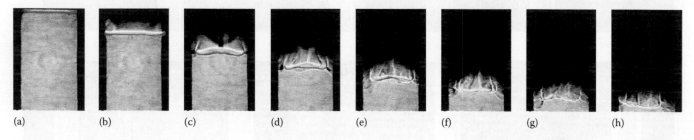

FIGURE 3.1.14

Quenching process of a flat limit flame, propagating downward from the open end of the tube in mixture with 2.20% C_3H_8 observed by schlieren system with superimposed direct photography. Square tube 125 mm × 125 mm × 500 mm. Time interval between frames 0.3 s.

to be at a maximum (Figure 3.1.15). This local flame extinction gradually spreads out and the flame in the central core of the tube begins to float. Small tongues of the combustion products, which are cooled by heat loss to the tube walls, move down into the fresh mixture (Figure 3.1.14). The flame occupies less and less area in the central, hot region of the tube and its propagation is balanced by buoyancy. Usually, a residual flame propagating down would rise just before final extinction.

Acknowledgments

This work was sponsored by the Marie Curie ToK project No MTKD-CT-2004-509847 and by the State Committee for Scientific Research project No 4T12D 035 27. The author thanks Yuriy Shoshin, Grzegorz Gorecki, and Luigi Tecce for their contributions to experiments and numerical simulation.

References

1. Humboldt A. and Gay Lussac J.F., Experiénce sur les moyens oediométriques et sur la proportion des principes constituants de l'atmosphére, *J. de Physique*, 60: 129–168, 1805.
2. Mallard E. and Le Chatelier H.L., On the propagation velocity of burning in gaseous explosive mixture, *Compt. Rend.*, 93: 145–148, 1881.
3. Jouguet E., Sur la propagation des deflagration dans les mélanges, *Compt. Rend.*, 156: 872–875, 1913.
4. Daniell P., The theory of flame motion, *Proc. R. Soc. A*, 126, 393–405, 1930.
5. Lewis B. and Elbe G., On the theory of flame propagation, *J. Chem. Phys.*, 2: 537–546, 1934.

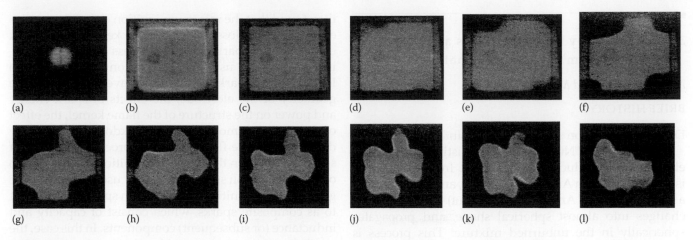

FIGURE 3.1.15
Quenching process of a flat limit flame, propagating downward from the open end of the tube in mixture with 2.25% C_3H_8, observed in a mirror located at the bottom closed end of it. Square tube 125 mm × 125 mm × 500 mm. Time interval between frames 0.2 s.

6. Zel'dovich Ya.B., Theory of combustion and gas detonation, *Moscow Akad. Nauk SSSR*, 1944.

7. Spalding D.B., A theory of inflammability limit and flame quenching, *Proc. R. Soc. London A*, 240, 83–100, 1957.

8. Coward H.F. and Jones G.W., Limits of flammability of gases and vapors, Bureau of Mines Report, #503, 1952.

9. Levy A., An optical study of flammability limits, *Proc. R. Soc. London A*, 283: 134–145, 1965.

10. Davies R.M. and Taylor F.R.S., The mechanism of large bubbles rising through extended liquids and through liquids in tubes, *Proc. R. Soc. A*, 200: 375–390, 1950.

11. Lovachev L.A., The theory of limits on flame propagation in gases, *Combust. Flame*, 17: 275–278, 1971.

12. Gerstein M. and Stine W.B., Analytical criteria for flammability limits, *Proc. Combust. Inst.*, 14: 1109–1118, 1973.

13. Buckmaster J., The quenching of deflagration waves, *Combust. Flame*, 26: 151–162, 1976.

14. Bregeon B., Gordon A.S., and Williams F.H., Near-limit downward propagation of hydrogen and methane flames in oxygen-nitrogen mixtures, *Combust. Flame*, 33: 33–45, 1978.

15. Jarosinski J., Strehlow R.A., and Azarbarzin A., The mechanisms of lean limit extinguishment of an upward and downward propagating flame in a standard flammability tube, *Proc. Combust. Inst.*, 19: 1549–1557, 1982.

16. von Lavante E. and Strehlow R.A., The mechanism of lean limit flame extinction, *Combust. Flame*, 49: 123–140, 1983.

17. Strehlow R.A., Noe K.A., and Wherley B.L., The effect of gravity on premixed flame propagation and extinction in a vertical standard flammability tube, *Proc. Combust. Inst.*, 21: 1899–1908, 1986.

18. Law C.K., Dynamics of stretched flames, *Proc. Combust. Inst.*, 22: 1381–1402, 1989.

19. Law C.K. and Egolfopoulos F.N., A kinetic criterion of flammability limits: The C-H-O-inert system, *Proc. Combust. Inst.*, 23: 413–421, 1990.

20. Law C.K. and Egolfopoulos F.N., A unified chain-thermal theory of fundamental flammability limits, *Proc. Combust. Inst.*, 24: 137–144, 1992.

21. Sung C.J. and Law C.K., Extinction mechanisms of near-limit premixed flames and extended limits of flammability, *Proc. Combust. Inst.*, 26: 865–873, 1996.

22. Dixon-Lewis G., Structure of laminar flames, *Proc. Combust. Inst.*, 23: 305–324, 1990.

23. Gorecki G., Analysis of laminar flames propagating in a vertical tube based on PIV measurements, PhD dissertation, Technical University of Łódź, Łódź, Poland, 2007.

24. Shoshin Y. and Jarosinski J., On extinction mechanism of lean limit methane-air flame in a standard flammability tube, paper accepted for publication in the 32nd Proceedings of the Combustion Institute, 2009.

25. Rimai L., Marko K.A., and Klick D., Optical study of a 2-dimensional laminar flame: Relation between temperature and flow-velocity fields, *Proc. Combust. Inst.*, 19: 259–265, 1982.

26. Law C.K. and Sung C.J., Structure, aerodynamics, and geometry of premixed flamelets, *Prog. Energy Combust. Sci.*, 26: 459–505, 2000.

27. Dixon-Lewis G., Aspects of laminar premixed flame extinction limits, *Proc. Combust. Inst.*, 25: 1325–1332, 1994.

28. Zel'dovich Ya.B., Istratov A.G., Kidin N.I., and Librovich V.B., Hydrodynamics and stability of curved flame front propagating in channels, Institute for Problems in Mechanics, The USSR Academy of Sciences, preprint nr 143, Moscow 1980.

3.2   Ignition by Electric Sparks and Its Mechanism of Flame Formation

Michikata Kono and Mitsuhiro Tsue

BRIEF HISTORY

Does spark ignition start from the point?

The answer is "NO." In the combustible mixture, an electric spark produces a flame kernel. Initially, its shape is elliptical (like an American football), and then becomes a torus (like an American doughnut). Afterwards, it changes into almost spherical shape, and propagates spherically in the unburned mixture. This process is formed by the existence of spark electrodes, which is necessary for spark discharge. Spark electrodes lead not only to heat loss from the flame kernel but also a change in the kernel shape. Both affect the minimum ignition energy.

3.2.1   Introduction

In the combustible mixture, an electric spark simultaneously produces a flame kernel. If the spark energy is sufficient, the flame kernel grows and propagates outward as a self-sustained flame; this is called success in ignition or simply ignition. If the spark energy is insufficient, the flame kernel disappears after a certain time period, known as misfire or extinction. The minimum value of spark energy that just produces the ignition is defined as the minimum ignition energy. There have been many investigations on the spark ignition since the beginning of twentieth century. Above all, experimental works by Lewis and von Elbe [1] introduced the very important concept of the relation between the minimum ignition energy and the quenching distance. Although they used electric sparks of very short duration (the capacity spark),

they found that the minimum ignition energy was influenced by a heat loss from the flame kernel to the spark electrodes at a spark gap distance less than the quenching distance. By using short-duration sparks and also long-duration sparks, the authors have investigated the geometry of the electrode, the effects of spark energy and power on the structure of the flame kernel, the effect of the discharge mode such as breakdown, arc, or glow discharges on the flame initiation process, the effect of spark duration on the minimum ignition energy, and so on. Long-duration sparks are widely used in automobile gasoline spark ignition engines. Such sparks are referred to as composite sparks, which consist of capacity and inductance (or subsequent) components. In this case, the spark duration must be considered as another important parameter.

Figure 3.2.1 shows flame kernels of the schlieren photograph taken by a high-speed camera. These photographs can be compared with the calculated temperature distribution in Figure 3.2.4. As can be seen, both of them bear a close resemblance. From this result, the authors firmly believe that the numerical simulation is a significant tool for grasping the mechanism of spark ignition. Of course, the experimental work should also be of importance to verify the results obtained by numerical simulations. In this work, the authors mainly introduce the results of numerical simulations that have been obtained until then in their laboratory.

3.2.2   Numerical Model Description

3.2.2.1   Numerical Method

A model that employs a two-dimensional cylindrical coordinate system and assumes axial symmetry with respect to r- and z-axes is developed. Figure 3.2.2 shows the coordinate system, computing region, and

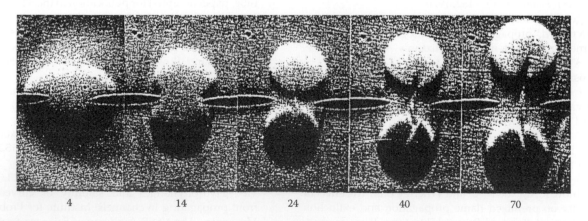

4	14	24	40	70

FIGURE 3.2.1
Schlieren photograph of flame kernels by a high-speed camera. Time is given from the onset of spark discharge in microseconds. Spark electrode diameter: 0.2 mm; spark gap width: 1 mm.

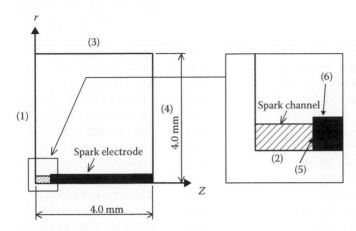

FIGURE 3.2.2
Computing region, notation of initial and boundary conditions.

location of the electrode. The size of the computing region is 4 mm × 4 mm. The diameter of spark electrodes is 0.5 mm, and the spark gap is 1.0 mm. The mass conservation, momentum conservation, energy conservation, and species conservation equations are found in Ref. [2], which are solved by a part of RICE code [3]. Briefly, at a given time step, the conservation equations for mass and momentum are solved in an iterative manner by a first-order implicit scheme, then the conservation equation for energy and species are solved by an explicit scheme, and afterwards the values of density and internal energy are updated. Final pressure is determined from the equation of state. Based on numerical stability, the time step is varied from 100 to 300 ns, and the cell size is taken to be 50 μm. It is confirmed that the numerical stability is fairy good. It should be noted that the time step for the explicit scheme calculation is reduced by a factor of 1000. It has been well known that the RICE code is not appropriate for the simulation of the rapid change behavior such as the sudden pressure rise in shock waves. Thus, another calculation method has been developed: The governing equations are solved with the second-order Harten–Yee upwind total variation diminishing (TVD) scheme [4] as the spatial differential method and the fourth-order Runge–Kutta method as the temporal differential method for convection terms. The second-order central difference scheme and the second-order two-step predictor–corrector method are employed for viscous terms.

The boundary conditions are as follows: In Figure 3.2.2, z-axis component and r-axis component velocities are zero for (1) and (2), respectively. The gradients of other variables are zero for both the boundaries. The gradients of all variables are zero for (3) and (4). No slip condition and heat transfer from the flame kernel to the spark electrode are assumed for (5) and (6), at the surface of spark electrode.

For simplicity of the model, it is assumed that the natural convection, radiation, and ionic wind effect are ignored. The ignorance of the radiation loss from the spark channel during the discharge may be reasonable, because the radiation heat loss is found to be negligibly small in the previous studies [5,6]. The amount of heat transfer from the flame kernel to the spark electrodes, whose temperature is 300 K, is estimated by Fourier's law between the electrode surface and an adjacent cell.

Specific heat of each species is assumed to be the function of temperature by using JANAF [7]. Transport coefficients for the mixture gas such as viscosity, thermal conductivity, and diffusion coefficient are calculated by using the approximation formula based on the kinetic theory of gas [8]. As for the initial condition, a mixture is quiescent and its temperature and pressure are 300 K and 0.1 MPa, respectively.

3.2.2.2 Chemical Reaction Scheme

The mixture used in the present simulation is stoichiometric methane–air. Table 3.2.1 shows the chemical reaction schemes for a methane–air mixture, which has 27 species, including 5 ion molecules such as CH^+, CHO^+, H_3O^+, CH_3^+, and $C_2H_3O^+$ and electron and 81 elementary reactions with ion–molecule reactions [9–11]. The reaction rate constants for elementary reaction with ion molecules have been reported in Refs. [10,11].

3.2.2.3 Spark Ignition

Spark discharge is simulated by the supply of energy in the spark channel, as shown in Figure 3.2.2. The spark energy is given at a certain rate in a spark channel, as shown in Figure 3.2.3, which simulated the energy deposition schedule of composite spark [12]. The composite spark, which consists of capacitance spark and inductance spark, has been mainly employed in the ignition system for automobile spark ignition engines. When the applied voltage between the electrodes increases and reaches the required voltage, the breakdown occurs and spark discharge begins. The discharge at the initial stage is caused by the release of the electric energy stored in the capacitance of the ignition circuit and the discharge duration is within 1 μs. This is known as the capacity spark. Then, the discharge with almost constant spark voltage continues for several milliseconds. This discharge originates from the electric energy stored in the inductance of the ignition coil, which is known as the inductance spark. As shown in Figure 3.2.3, the simulated capacity spark has high spark energy with the duration of 1 μs. The simulated inductance spark that has relatively low spark energy continues from 1 to 100 μs after the onset of discharge. The ratio of capacity spark energy is defined as the ratio of capacitance component

TABLE 3.2.1

Elementary Reactions for Numerical Simulations

(1)	$H + O_2 = OH + O$	(28)	$CHO + O2 = CO + HO_2$	(55)	$CH_2 + O_2 = CH_2O + O$
(2)	$O + H_2 = OH + H$	(29)	$CHO + OH = CO + H_2O$	(56)	$CH_2 + O_2 = CO_2 + H_2$
(3)	$H_2 + OH = H + H_2O$	(30)	$CHO + M = CO + H + M$	(57)	$CH_2 + H = CH + H_2$
(4)	$OH + OH = H_2O + O$	(31)	$CO + OH = CO_2 + H$	(58)	$CH + O = CO + H$
(5)	$H + H + H_2 = H_2 + H_2$	(32)	$CO + O + M = CO_2 + M$	(59)	$CH + O_2 = CO + OH$
(6)	$H + H + N_2 = H_2 + N_2$	(33)	$CH_3 + CH_3 = C_2H_6$	(60)	$C_2H + O = CO + CH$
(7)	$H + H + O_2 = H_2 + O_2$	(34)	$C_2H_6 + O = C_2H_5 + OH$	(61)	$CH^* + M = CH + M$
(8)	$H + H + H_2O = H_2 + H_2O$	(35)	$C_2H_6 + H = C_2H_5 + H_2$	(62)	$CH^* + O_2 = CH + O_2$
(9)	$O + O + M_1 = O_2 + M_1$	(36)	$C_2H_6 + OH = C_2H_5 + H_2O$	(63)	$CH^* = CH$
(10)	$OH + H + M_2 = H_2O + M_2$	(37)	$C_2H_5 + H = C_2H_6$	(64)	$C_2H + O_2 = CH^* + CO_2$
(11)	$H + O_2 + M_3 = HO_2 + M_3$	(38)	$C_2H_5 + H = CH_3 + CH_3$	(65)	$C_2H + O = CH^* + CO$
(12)	$H + HO_2 = OH + OH$	(39)	$C_2H_5 = C_2H_4 + H$	(66)	$C_2H_2 + H = C_2H + H_2$
(13)	$H + HO_2 = O_2 + H_2$	(40)	$C_2H_5 + O_2 = C_2H_4 + HO_2$	(67)	$C_2H_2 + OH = C_2H + H_2O$
(14)	$H + HO_2 = H_2O + O$	(41)	$C_2H_4 + O = CH_2 + CH_2O$	(68)	$C_2H + O_2 = CO + CHO$
(15)	$O + HO_2 = OH + O_2$	(42)	$C_2H_4 + OH = CH_2O + CH_3$	(69)	$CH + O = CHO^+ + e^-$
(16)	$OH + HO_2 = H_2O + O_2$	(43)	$C_2H_4 + O = C_2H_3 + OH$	(70)	$CH^* + O = CHO^+ + e^-$
(17)	$CH_4 + H = CH_3 + H_2$	(44)	$C_2H_4 + O_2 = C_2H_3 + HO_2$	(71)	$CHO^+ + H_2O = H_3O^+ + CO$
(18)	$CH_4 + OH = CH_3 + H_2O$	(45)	$C_2H_4 + H = C_2H_3 + H_2$	(72)	$H_3O^+ + C_2H_2 = C_2H_3O^+ + H_2$
(19)	$CH_4 + O = CH_3 + OH$	(46)	$C_2H_4 + OH = C_2H_3 + H_2O$	(73)	$CHO^+ + CH_2 = CH_3^+ + CO$
(20)	$CH_3 + O = CH_2O + H$	(47)	$C_2H_3 + M = C_2H_2 + H + M$	(74)	$H_3O^+ + CH_2 = CH_3^+ + H_2O$
(21)	$CH_3 + O_2 = CH_2O + OH$	(48)	$C_2H_3 + O_2 = C_2H_2 + HO_2$	(75)	$CH_3^+ + C_2H_2 = C_3H_3^+ + H_2$
(22)	$CH_3 + OH = CH_2O + H_2$	(49)	$C_2H_3 + H = C_2H_2 + H_2$	(76)	$CH_3^+ + H_2O = C_2H_3O + CH_2$
(23)	$CH_2O + H = CHO + H_2$	(50)	$C_2H_3 + OH = C_2H_2 + H_2O$	(77)	$CH_3^+ + CO_2 = C_2H_3O^+ + O$
(24)	$CH_2O + O = CHO + OH$	(51)	$C_2H_2 + OH = CH_3 + CO$	(78)	$H_3O^+ + e^- = H_2O + H$
(25)	$CH_2O + OH = CHO + H_2O$	(52)	$CH_3 + H = CH_2 + H_2$	(79)	$CH + e^- = products$
(26)	$CHO + O = CO + OH$	(53)	$CH_3 + OH = CH_2 + H_2O$	(80)	$CH + e^- = CH + H$
(27)	$CHO + H = CO + H_2$	(54)	$CH_2 + O_2 = CHO + OH$	(81)	$CHO + e^- = products$

Third body and factor of reaction rate

(9) $M_1 = N_2$

(10) $M_2 = H_2O + 0.25H_2 + 0.25O_2 + 0.2N_2$

(11) $M_3 = H_2 + 0.44N_2 + 0.35O_2 + 6.5H_2O$

(30), (32)   M=all species with factor of unity.

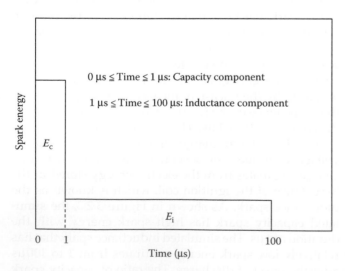

FIGURE 3.2.3

Energy distribution of composite spark. (E_c: Capacity spark energy; E_i: inductance spark energy.)

energy E_c to the total spark energy ($E_c + E_i$). Calculations are performed by varying the capacity spark energy and inductance spark energy independently. The durations of capacity spark and inductance spark are kept to be 1 and 99 μs, respectively. The energy density, i.e., the energy per unit time, keeps constant with time for both the components. Each energy component is varied by changing the energy density.

3.2.3 Calculated Results

3.2.3.1 Spark Ignition Process

Figure 3.2.4 shows the calculated results of the time series of temperature distribution with the total spark energy of 0.7 mJ and the ratio of capacity spark of 100%. The hot kernel is initially an ellipsoid and the maximum temperature region is located in the center of the spark gap. Afterwards, the hot kernel develops into a torus and the highest temperature region moves into the ring

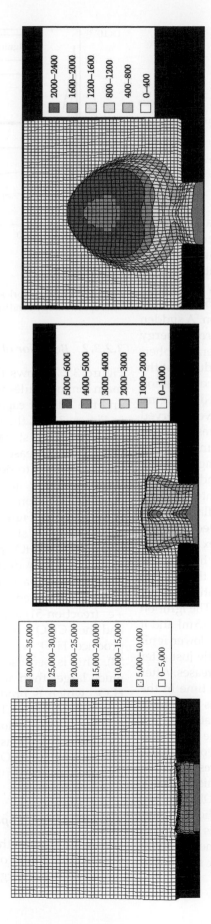

FIGURE 3.2.4

Calculated time series of temperature distribution. Spark energy: 0.70mJ; ratio of capacity spark energy: 100%. Left: time = 1 μs, central: time = 10 μs, right: time = 100 μs.

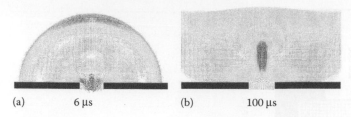

FIGURE 3.2.5
Calculated velocity distributions for total spark energy of 0.7 mJ. Ratio of capacity spark energy: 100%. (a) Time = 6 μs, (b) time = 100 μs.

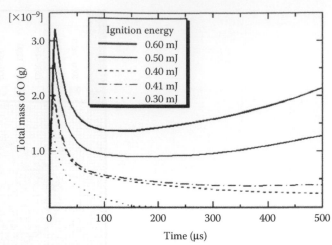

FIGURE 3.2.6
Calculated time histories of the total mass of O radical for various total spark energy conditions.

of torus. Then, the ring of the torus grows and the center of the ring extends outward. The schlieren photography by Ishii et al. [2] and Kono et al. [13] showed similar development of the flame kernel. This formation process of the flame kernel has been explained by gas movement near the electrodes [2,14]. Figure 3.2.5 shows the calculated time series of velocity distribution. A sudden pressure rise in the spark gap due to the strong energy release by the capacity spark discharge generates spherical shock wave and the outward flow is induced at the initial stage of the spark discharge. The compression wave with relatively thick front is indicated in Figure 3.2.5a, which corresponds to the shock wave observed in experiments. When the shock wave moves outward, the pressure in the spark gap gets low and inward flow is generated [2,14,15]. The inward flows along the spark electrode surfaces collide with each other and turn radially outward, which forms a pair of counter-rotating vortex structure outside the spark gap. The flame kernel that produces inside the spark gap is moved outward and is transformed into a torus by the vortex flows.

Figure 3.2.6 shows the calculated time histories of total mass of O radical, which is defined as the amount of O radical that exists in the whole computing region, for various ratios of capacity spark energy. The ignition is succeeded at the ignition energies above 0.5 mJ and it is failed (which is known as misfire) at the lower ignition energy. The total mass increases rapidly just after the onset of spark discharge and then it decreases. After 100 μs, the total mass increases again with time in the case of success of ignition, while it decreases monotonically in the case of misfire. This means that when the mass of O radical that is formed during the spark discharge is sufficient, it increases after the end of the spark discharge by the chemical reaction, which results in the formation of the self-propagating flame. This tendency for O radical is same as those for H and OH radicals, which suggests that the formation of radical species during the spark discharge largely affects the spark ignition process. It is also found that the time history of total mass of radical species is available for the simplified criterion of ignition in calculations.

3.2.3.2 Behavior of Ion Molecules

Figure 3.2.7 shows the time histories of total mass of main ion molecules with the total spark energy of 0.5 mJ and the ratio of capacity spark of 100% [16]. As shown in Figure 3.2.7a, the mass of CHO^+ and H_3O^+ decreases after the initial stage of the spark discharge. The mass of $C_2H_3O^+$ increases from the time when both CHO^+ and H_3O^+ start to decrease, which is much larger than those of CHO^+ and H_3O^+. The detailed behavior of mass of these ion molecules just after spark discharge is indicated in Figure 3.2.7b. First, CHO^+ starts to form and then H_3O^+ starts to increase after the time when the mass of CHO^+ reaches a maximum. As the H_3O^+ increases, CHO^+ decreases, which can be explained by the chemical reaction related with ion molecules. The CHO^+ is produced from CH and CH^* via $CH^+ O = CHO^+ + e^-$ (reaction 51) and $CH^* + O = CHO^+ + e^-$ (reaction 52), and then CHO^+ would be quickly consumed by H_2O to produce H_3O^+ via $CHO^+ + H_2O \rightarrow H_3O^+ + CO$ (reaction 53). This is because the reaction rate of reaction 53 is much larger than any other ion molecule reaction. The H_3O^+, which is produced by reaction 53, is consumed via $H_3O^+ + CH_2 = CH_3^+ + H_2O$ (reaction 56) and $H_3O^+ + e^- = H_2O + H$ (reaction 60). The H_3O^+ is gradually transformed by $C_2H_3O^+$ as the reaction proceeds, which is the reason why the H_3O^+ gradually decreases and $C_2H_3O^+$ increases gradually with time.

3.2.3.3 Effect of Spark Component on Ignition Process

Figure 3.2.8 shows the calculated result for the relationship between ratio of capacity spark energy and minimum ignition energy. The minimum ignition energy

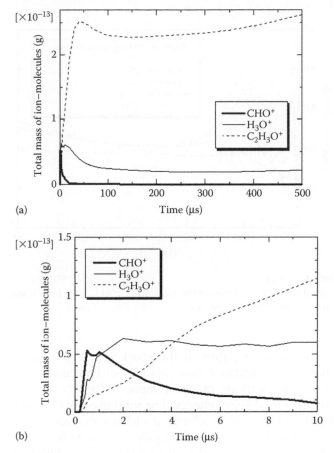

(a)

(b)

FIGURE 3.2.7
Calculated time histories of the total mass of ion molecules for various total spark energy conditions. Ignition energy: 0.70 mJ; ratio of capacity spark energy: 100%.

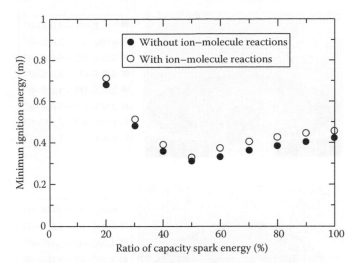

FIGURE 3.2.8
Relationship between the ratio of capacity spark energy and the minimum ignition energy obtained from calculations. Spark gap width: 1.0 mm; spark electrode diameter: 0.50 mm.

effect of ion–molecule reactions on the minimum ignition energy is not significant.

Figure 3.2.9 shows the relationship between the minimum ignition energy and the ratio of capacity spark energy obtained from experiments. The minimum ignition energy, which has been often evaluated by determining the spark energy for 50% ignition, is obtained conventionally from the line of minimum spark energies for which the ignition occurs for various ratios of capacity spark energy, because of the limitation of the number

is defined in the calculation as the lowest spark energy for which the flame kernel reaches the outer boundary of the computing region. As the ratio of capacity spark energy, the minimum ignition energy decreases and reaches a lowest value, and then it increases. The ratio of capacity spark energy at which the minimum ignition energy has a lowest value is about 50% in this case. The result obtained from the calculation without ion-molecule reactions is also indicated in Figure 3.2.8, which is qualitatively similar to that with ion–molecule reactions. The minimum ignition energy with ion–molecule reactions is slightly larger than that without ion-molecule reactions. The calculated results show that the temperature and heat release in the flame kernel is slightly larger with ion–molecule reactions than without them [16]. This may be due to the energy consumption for ionization because most of the ion–molecule reactions are endothermic. The difference between minimum ignition energy with and without ion–molecule reactions is found to be relatively small, which suggests that the

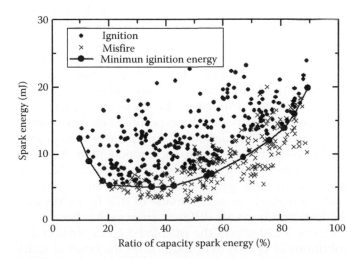

FIGURE 3.2.9
Relationship between the ratio of capacity spark energy and the minimum ignition energy obtained from experiments. Equivalence ratio of mixture: 0.62. Spark gap width: 0.5 mm; spark electrode diameter: 1.0 mm.

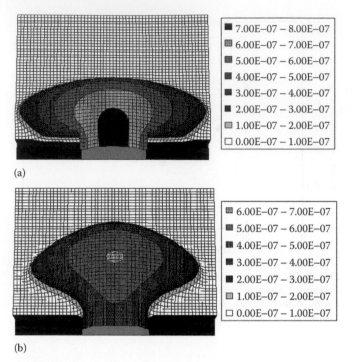

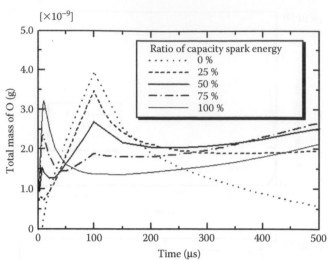

FIGURE 3.2.11
Calculated time histories of the total mass of O radical for various ratios of capacity spark energy.

FIGURE 3.2.10
Calculated O radical concentration distributions for total spark energy of 0.7 mJ. (a) Ratio of capacity spark energy: 20%. (b) Ratio of capacity spark energy: 80%.

of the test runs. The qualitative trend of calculated and experimental results in Figures 3.2.8 and 3.2.9 shows agreement, namely the optimum ratio of capacity spark energy, for which minimum ignition energy indicates the lowest value, exists for both cases. The discrepancy of the optimum ratio is mainly due to the difference in test conditions between the experiment and calculation. The optimum ratio is considered to be largely dependent on the test parameters, such as equivalence ratio of mixture, geometry of electrodes, and spark energy. As mentioned later, the existence of the optimum ratio is closely related to the formation of radicals and gas movement, which is largely influenced by these test parameters. The agreement between the experiment and calculation means that the calculation used in this work can predict qualitatively the effect of spark component on the minimum ignition energy. The comparison of absolute values of minimum ignition energy between experiment and calculation cannot be made in this figure because the test conditions are different from each other.

Figure 3.2.10 shows the calculated concentration distributions of O radical at 100 μs after the onset of spark discharge. The O radical concentration at lower ratio of capacity spark energy (high inductance component energy) is larger on the whole than that at higher ratio. This suggests that the inductance spark enhances the formation of radicals because high temperature region

keeps for a long time at the spark gap. Figure 3.2.11 shows the calculated time histories of total mass of O radical for various ratios of capacity spark energy at constant total spark energy. The total mass of O radical at the end of spark discharge ($t = 100$ μs) increases as the ratio of capacity spark energy decreases, which corresponds to the result shown in Figure 3.2.10. The total mass decreases monotonically with time after the end of discharge at the ratio of capacity spark energy of 0%. On the other hand, it increases with time again after it decreases when the ratio of capacity spark energy is relatively high. This tendency can be explained by the flow field near spark electrodes.

Figures 3.2.12 and 3.2.13 show calculated velocity and heat release distributions for different ratios of capacity spark energy, respectively. As the ratio of capacity spark energy increases, the strength of the shock wave increases. As a result, the generation of inward flows along the spark electrode surfaces is promoted and outward flow of the center of computing region becomes faster, as shown in Figure 3.2.12. Figure 3.2.13 shows that the constriction of the heat release distribution near tip end of electrodes is significant for high ratio of capacity spark energy, which corresponds to the promotion of inward flow along the electrodes. The heat release near tip end of electrodes for ratio for capacity spark of 80% is higher than that for 20%. In addition, it is confirmed that total heat release obtained by integration of the heat release in the whole computing region becomes larger as the ratio of capacity spark increases. These results suggest that the faster inward flow along the electrodes leads to the enhancement of supply of the unburned mixture in the flame kernel region, and

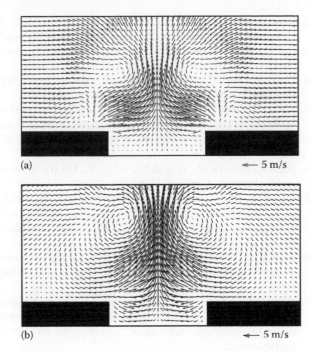

FIGURE 3.2.12
Calculated velocity distributions for total spark energy of 0.7 mJ at 100 μs. (a) Ratio of capacity spark energy: 20%. (b) Ratio of capacity spark energy: 80%.

the reaction occurs actively. This is because the capacity spark enhances the formation of O radical after the end of discharge, as shown in Figure 3.2.11.

The increase in the capacity spark energy leads to the decrease in the formation of radicals during the spark discharge. On the other hand, the increase in the capacity spark energy leads to the active reaction after spark discharge, which is due to the enhancement of supply of the unburned mixture by the inward flow along the electrodes. The optimum ratio of capacity spark energy where minimum ignition energy has a lowest value is determined by these two factors.

The present numerical simulation is insufficient for the quantitative prediction of the minimum ignition energy, which may be due to fact that the shock wave behavior just after the electric discharge cannot be simulated well in this calculation code. Since a large amount of spark energy is removed because of the shock wave [17], it is important to estimate precisely the strength of the shock wave for the estimation of minimum ignition energy. The energy removed by the shock wave is estimated roughly to be about 10% of the total spark energy in this calculation. In addition, the plasma channel formed between the electrodes during spark discharge is modeled as a cylinder of hot gases. This model may be somewhat oversimplified and further investigation will be needed.

3.2.3.4 Effect of Equivalence Ratio on Minimum Ignition Energy

It has been reported by Lewis et al. [1] that the equivalence ratio where the minimum ignition energy has a minimum is dependent on the fuel property for hydrocarbon fuel and air mixtures, and that it moves to the rich side as the molecular weight of the fuel increases. This equivalence ratio dependency has been explained by the preferential diffusion effect.

Figure 3.2.14 shows the calculated relationship between minimum ignition energy and equivalence ratio for hydrogen–air and methane–air mixtures. The minimum ignition energy has a minimum at the equivalence ratio below unity for the fuels with lower molecular weight. The simulation also predicts that the local equivalence ratio inside the spark gap becomes almost unity in the case that the overall equivalence ratio of the mixture is less than unity, which confirms that the preferential diffusion affects on the spark ignition process [18]. It is also shown that the minimum ignition energy increases rapidly as the equivalence ratio moves away from the

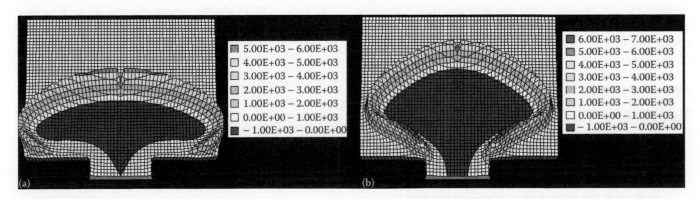

FIGURE 3.2.13
Calculated heat release distributions for total spark energy of 0.7 mJ at 500 μs. (a) Ratio of capacity spark energy: 20%. (b) Ratio of capacity spark energy: 80%.

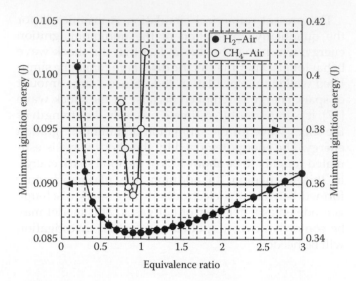

FIGURE 3.2.14
Relationship between minimum ignition energy and equivalence ratio for hydrogen–air and methane–air mixtures.

position where the minimum ignition energy has a minimum for both fuels. This tendency is significant for the methane–air mixture. This suggests that the improvement of the ignitability of mixtures is more effective for the success of ignition than the improvement in ability of the ignition system. The prediction for fuels with heavier molecular weight should be needed to confirm the preferential diffusion effect on the minimum ignition energy. It would be also interesting to understand whether the preferential diffusion effect is dominant or not for the ignition process in flowing conditions, which exists in the combustion chamber for practical engines.

3.2.4 Conclusion

In this work, as for the spark ignition of combustible mixtures, the authors mainly introduce numerical simulation results that have been obtained until then in their laboratory. As can be seen, the authors firmly believe that the numerical simulation has a significant tool to grasp the mechanism of spark ignition, though in the quiescent mixture.

References

1. Lewis, B. and von Elbe, B., *Combustion, Flames and Explosion of Gases*, 3rd ed., Academic Press, New York, p. 333, 1987.
2. Ishii, K., Tsukamoto, T., Ujiie, Y., and Kono, M., Analysis of ignition mechanism of combustible mixtures by composite spark, *Combust. Flame*, 91, 153, 1992.
3. Rivard, W. C., Farmar, O. A., and Butler, T. D., RICE: A computer program for multicomponent chemically reactive flows at all speeds, National Technical Information Service, LA-5812, 1975.
4. Yee, H., Upwind and symmetric shock capturing schemes, NASA TM 89464, NASA, 1987.
5. Maly, R. and Vogel, M., Initiation and propagation of flame fronts in lean CH₄-air mixtures by the three modes of the ignition spark, *Proc. Combust. Inst.*, 17, 821, 1979.
6. Sher, E., Ben-Ya'ish, J., and Kravchik, T., On the birth of spark channels, *Combust. Flame*, 89, 6, 1992.
7. Chase, Jr., M. W., Davies, C. A., Downey, Jr., J. R., Frulip, D. J., McDonald, R. A., and Syverud, A. N., JANAF thermochemical tables Third Edition, *J. Phys. Chem. Ref. Data*, 1985.
8. Reid, R. C., Prasnitz, J. M., and Poling, B. E., *The Properties of Gases and Liquids*, McGraw-Hill International Editions, 1988.
9. Takagi, T. and Xu, Z., Numerical analysis of hydrogen-methane diffusion flames (effects of preferential diffusion), *Trans. Jpn. Soc. Mech. Eng.*(B), 59, 607, 1993 (in Japanese).
10. Pedersen, T. and Brown, R. C., Simulation of electric field effects in premixed methane flame, *Combust. Flame*, 94, 433, 1994.
11. Eraslan, A. N., Chemiionization and ion-molecule reactions in fuel-rich acetylene flame, *Combust. Flame*, 74, 19, 1988.
12. Yuasa, T., Kadota, S., Tsue, M., Kono, M., Nomuta, H., and Ujiie, Y., Effects of energy deposition schedule on minimum ignition energy in spark ignition of methane-air mixtures, *Proc. Combust. Inst.*, 29, 743, 2002.
13. Kono, M., Kumagai, S., and Sakai, T., Ignition of gases by two successive sparks with reference to frequency effect of capacitance sparks, *Combust. Flame*, 27, 85 1976.
14. Kono, M., Niu, K., Tsukamoto, T., and Ujiie, Y., Mechanism of flame kernel formation produced by short duration sparks, *Proc. Combust. Inst.*, 22, 1643, 1988.
15. Kravchik, T. and Sher, E., Numerical modeling of spark ignition and flame initiation in a quiescent methane-air mixture, *Combust. Flame*, 99, 635, 1994.
16. Kadota, S., et al., Numerical analysis of spark ignition process in a quiescent methane-air mixture with ion-molecule reactions, *The 2nd Asia-Pacific Conference on Combustion*, p. 617, 1999.
17. Ballal, D. R. and Lefebvre, A. H., A general model of spark ignition for gaseous and liquid fuel-air mixtures, *Combust. Flame*, 24, 99, 1975.
18. Nakaya, S., et al., A numerical study on early stage of flame kernel development in spark ignition process for methane/air combustible mixtures, *Trans. Jpn. Soc. Mech. Eng.*(B), 73–732, 1745, 2007 (in Japanese).

4

Influence of Boundary Conditions on Flame Propagation

CONTENTS

4.1 Propagation of Counterflow Premixed Flames

Chih-Jen Sung

4.1.1 Introduction

The stagnation-point flow has been extensively used for stretched flame analysis, because the flow field is well defined and often can be represented by a single parameter, namely the stretch rate. Notable stagnation-flow configurations that have been widely used in premixed combustion experiments include flow impinging on a surface or single jet-wall (e.g., Refs. [1,2]), porous burner in uniform flow or Tsuji burner (e.g., Refs. [3,4]), opposing jets or counterflow (e.g., Refs. [5,6]), tubular flows (e.g., Refs. [7–9]), etc. In the single jet-wall geometry, a premixed reactant flow from a nozzle impinges on to a flat stagnation plate. If the premixed flame front is situated sufficiently far away from the stagnation plate, then the downstream heat loss from the flame to the plate has a minimal effect on the flame propagation. In the Tsuji burner, a cylindrical or spherical porous body is placed horizontally in a wind tunnel. A premixed reactant mixture is injected out uniformly from the upstream portion of the porous body. The stretch rate in the forward stagnation region is proportional to U/R, where U is the forced flow velocity in the tunnel and R is the radius of the cylinder or sphere. Unlike the stagnation-flow field produced between the two counterflowing coaxial jets, in a tubular burner, the inlet flow is directed radially, while the exhaust flow is axial. The tubular premixed flame formed in a tubular flow is subjected to the combined effects of flame stretch and flame curvature; whereas, for the planar flame established in the opposed-jet configuration, the effect of aerodynamic straining on the flame response can be studied without complications owing to flame curvature. Although the above-mentioned stagnation-point flow configurations in the laboratory practice are generally multidimensional in nature owing to the inevitable edge effects, the existence of stagnation-flow similarity for most part of the flamelet, when valid, would greatly facilitate the associated mathematical analysis and computation with detailed chemistry in that the flame scalars, such as temperature and species concentrations, can be assumed to depend on only one independent variable.

In this chapter, laminar flame propagation in an opposed-jet configuration is of particular interest. The counterflow is ideally suited for the study of the effects of aerodynamics on flames, because the flame in this flow field is planar and the stretch rate is well defined by the velocity gradient ahead of the flame. However, experimental observations [10,11] and theoretical studies [12,13] demonstrated several nonplanar flame configurations that are also possible in opposed-jet flow fields, including cellular flames, star-shaped flames, groove-shaped flames, open groove-shaped flames, rim-stabilized Bunsen flames, rim-stabilized uniform flames, and rim-stabilized polyhedral flames. While these nonplanar counterflow flame phenomena are interesting and challenging topics, they are beyond the scope of this chapter.

Both premixed and diffusion flames can be studied by using the same apparatus. For example, by having identical combustible mixtures flow from the two opposing circular nozzles, two symmetrical premixed flames can be established. The symmetry condition at the midplane assures reasonable downstream adiabaticity for each flame. The effect of downstream heat loss/gain on premixed flames can be studied by impinging a combustible jet against an inert jet or hot products of given temperature, which can control the extent of heat loss/gain. A diffusion flame can be established by impinging a fuel jet against an oxidizer jet. In addition, a triple flame consisting of a lean and a rich premixed flame, sandwiching a diffusion flame can also be established by impinging a rich mixture against a lean mixture, such that the excess fuel from the rich flame reacts with the excess oxidizer from the lean flame to form the diffusion flame. This allows the study of flame interaction effects (cf. Ref. [14]). In this chapter, we shall only be concerned with the symmetrical premixed twin flames.

Figure 4.1.1 shows a schematic representation of a counterflow burner system used in the author's laboratory. The stagnation-flow field is established by impinging two counterflowing uniform jets generated from two identical aerodynamically shaped high-contraction-ratio nozzles. Each stream is surrounded by a shroud of nitrogen flow to isolate the twin flames from the environment. For opposed-jet flames, the imposed stretch rate depends on the impinging jet velocities and the separation distance between two opposing jets. In general, the separation distance (L) is kept around the same

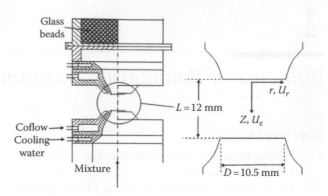

FIGURE 4.1.1
Schematic of the counterflow burner assembly.

order of the nozzle diameter (D) to ensure high-quality, quasi-one-dimensional, planar flamelets. If L is much lesser than D, the range of stretch rates to avoid conductive heat loss from the flame to the nozzle is limited. On the other hand, when L is much larger than D, the twin flames are more susceptible to the effect of entrainment. With a fixed separation distance, the variation of stretch rate can be achieved by varying the nozzle exit velocity (i.e., the mixture flow rate).

Figure 4.1.2 is a photograph of a counterflow burner assembly. The experimental particle paths in this cold, nonreacting, counterflow stagnation flow can be visualized by the illumination of a laser sheet. The flow is seeded by submicron droplets of a silicone fluid (polydimethylsiloxane) with a viscosity of 50 centistokes and density of $970 \, \text{kg/m}^3$, produced by a nebulizer. The well-defined stagnation-point flow is quite evident. A direct photograph of the counterflow, premixed, twin flames established in this burner system is shown in Figure 4.1.3. It can be observed that despite the edge effects,

FIGURE 4.1.2
Photograph of a counterflow burner system and the nonreacting flow visualization using a laser sheet.

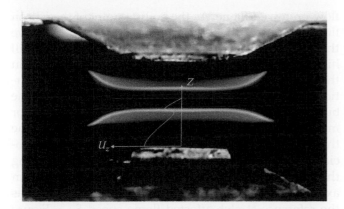

FIGURE 4.1.3
Close-up of premixed counterflow twin *iso*-octane/air flames. The representative profile of axial velocity (U_z) in the axial direction (Z) is also sketched and the red line indicates the location of stagnation surface.

the majority of the flamelet is quite flat, thereby demonstrating the similarity of the resulting reacting flow. The corresponding location of the stagnation surface and the typical axial velocity profile along the centerline are also sketched in Figure 4.1.3. It can be noted that the axial velocity first decreases along the axis when exiting the nozzle, and reaches a minimum as it approaches the upstream boundary of the luminous zone of the flame. As the flow enters the flame zone, the axial velocity increases as a consequence of thermal expansion, reaches a maximum, and eventually decreases as it approaches the stagnation surface.

Unlike the diffusion flames, a premixed flame is a wave phenomenon and hence it can freely adjust itself in response to stretch-rate variations so as to achieve a dynamic balance between the local flame propagation speed and the upstream flow velocity experienced by it (cf. Refs. [15,16]). As discussed earlier, for the present stagnation flow, the axial velocity upstream of the flame zone decreases from the nozzle exit. If we apply the dynamic equilibrium relation that results in the balancing of the local axial velocity by the upstream flame speed, then the flame would move closer to the stagnation surface (nozzle exit) with increasing (decreasing) flow straining/flow rate, as observed in Figure 4.1.4. Thus, the separation distance between the twin flamelets increases with decreasing stretch/flow rate. It can also be noted from Figure 4.1.4 that when the flame approaches the nozzle as the flow rate/stretch rate is reduced, its surface can develop some curvature near the center of the flamelet. This is because the exit flow tends to decelerate around the centerline and accelerate at the larger radii in the low-flow rate cases [17]. Therefore, the establishment of low-stretch, adiabatic, planar, counterflow, premixed twin flames is challenging because of the increasing conductive heat loss from

the flame to the nozzle, the stronger effect of pressure forces on the inertia forces, and other forms of flow/flame disturbances (e.g., buoyancy-induced flows).

Because of the planar nature of the counterflow flame and the relatively high Reynolds number associated with the flow, the flame/flow configuration can be considered to be "aerodynamically clean," where the quasi-one-dimensional and boundary-layer simplifications can be implemented in either analytical or computational studies. Useful insights into the thermochemical structure

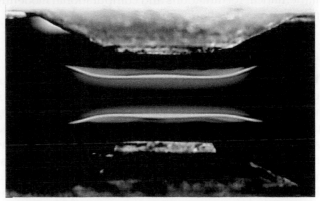

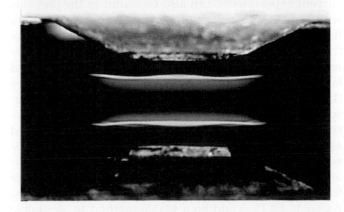

FIGURE 4.1.4
Direct images of the three counterflow twin flames with decreasing stretch rate through the reduction of the mixture flow rate (from top photo to bottom photo), while keeping the flow mixture composition constant.

of the laminar stretched flames can be obtained by comparing the experimental data with the numerically calculated ones, incorporating detailed chemistry and transport. In addition, global combustion properties of a fuel/oxidizer mixture, such as laminar flame speed and autoignition delay time, have been widely used as target responses for the development, validation, and optimization of a detailed reaction mechanism. In particular, laminar flame speed is a fundamental and practical parameter in combustion that depends on the reactivity, exothermicity, and transport properties of a given mixture.

Next, we shall sequentially present the experimental methodology of determining the laminar flame speed using counterflow twin-flame configuration, along with the results of the measured flame speeds for various hydrocarbon fuels. However, for the latter, we found a compilation of experimental flame-speed data of mixtures of hydrogen, methane, C_2-hydrocarbons, propane, and methanol documented by Law [18]. The present compilation emphasizes the measured flame speeds of liquid hydrocarbons with preheat. In addition, the extraction of overall activation energy will also be discussed. Furthermore, earlier studies, for example the study by Kee et al. [19], can be referred to for understanding the mathematical formulations and computational modeling of the counterflow flames.

4.1.2 Determination of Laminar Flame Speeds

The velocity field of the counterflow configuration can be obtained using a planar digital particle image velocimetry (DPIV) system. The mixture flow is seeded using submicron-sized particles of silicone fluid, as described earlier. The uniformly dispersed particles in the flow are illuminated by a light sheet of 0.2 mm thickness in the vertical plane using a dual-laser head, pulsed Nd:YAG laser. The details of DPIV, as applied in the current experiments, can also be found in the previous studies [20–23]. Applying DPIV for the measurement of flame speeds has two advantages. First, the mapping of the entire two-dimensional flow field can substantially reduce the test run-time than that of the point-based laser techniques, such as laser Doppler velocimetry (LDV). Second, it can reduce positioning error. In flame-speed measurements using the counterflow configuration, an accurate determination of the reference speed and stretch rate is crucial. The positioning error of DPIV is much smaller than that of LDV, because the fine-arrayed CCD camera is employed and translation of the probe volume to different points is not necessary. As such, DPIV could be a viable method to accurately measure the velocity at a particular point, along with its associated radial and axial velocity gradients.

Figure 4.1.5 shows an example of a PIV image and the resulting two-dimensional velocity map for a counterflow premixed flame. As the boiling point of the silicon fluid is about 570 K, the seeding droplets will not survive the post-flame region. However, this boiling point is still high enough to capture the minimum velocity point in the preheat zone. Unlike the use of solid seeding particles, the benefit of using silicone droplets is that the burner will not be "contaminated" or clogged. The vector map obtained by DPIV is further analyzed to determine the reference stretch-affected flame speed and the associated stretch rate. By plotting the axial velocity along the centerline, as shown in Figure 4.1.5, the minimum axial velocity upstream of the flame location is taken as the reference stretch-affected flame speed $S_{u,ref}$. Figure 4.1.5 also shows that the radial velocity profile at this reference location is linear. Hence, the radial velocity gradient (a) can be used to unambiguously characterize the flame stretch rate. The stretch rate (K) is conventionally defined using the axial velocity gradient and is equal to twice the radial velocity gradient, i.e., $K = 2a$. Based on the variation of $S_{u,ref}$ with K, the unstretched laminar flame speed (S_u^o) can be determined by the methodology of either linear or nonlinear extrapolation.

Figure 4.1.6 exhibits $S_{u,ref}$ as a function of Karlovitz numbers (Ka) for various *n*-heptane/air and *iso*-octane/air flames, with the unburned mixture temperature (T_u) of 360 K. The Karlovitz number is defined as $Ka = K\alpha_m/(S_u^o)^2$, where α_m is the thermal diffusivity of the unburned mixture. For each case, the linear extrapolation technique is compared with a nonlinear extrapolation, based on the theoretical analysis of Tien and Matalon [24], obtained using a potential flow field. The solid and dotted lines in Figure 4.1.6 represent the linear and nonlinear extrapolations, respectively. It can be observed from Figure 4.1.6 that for most of the experimental conditions, the value of Ka was taken as <0.1. Vagelopoulos et al. [25] and Chao et al. [26] demonstrated that when the Karlovitz numbers are retained to the order of $O(0.1)$, the accuracy of linear extrapolation is improved and the overprediction by linear extrapolation can be reduced to be within the experimental uncertainty. Figure 4.1.6 also shows that the linearly extrapolated laminar flame speed is not >3 cm/s higher than the value obtained by using a nonlinear extrapolation of Tien and Matalon [24]. Hence, in the subsequent sections, all the laminar flame speeds presented refer to the linearly extrapolated values.

It can also be noted that the slope of the $S_{u,ref}$–Ka plot reflects the combined effect of stretch rate and non-equidiffusion on the flame speed. Figure 4.1.6 clearly shows that the flame response with stretch rate variation differs for lean and rich mixtures. In particular, as Ka increases, the $S_{u,ref}$ for stoichiometric and rich mixtures increases, but decreases for the mixture of equivalence ratio $\phi = 0.7$. This is because the effective Lewis numbers of lean *n*-heptane/air and lean *iso*-octane/air flames are

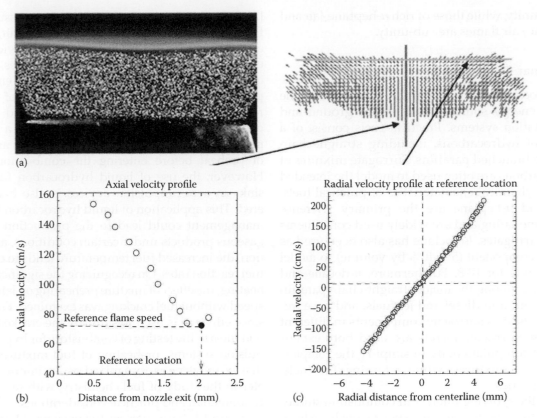

FIGURE 4.1.5
A sample PIV image and the corresponding two-dimensional velocity map. The axial velocity along with distance from nozzle exit is plotted accordingly. This minimum point is defined as the reference flame speed. At this reference point, the linearity of the radial velocity profile is illustrated.

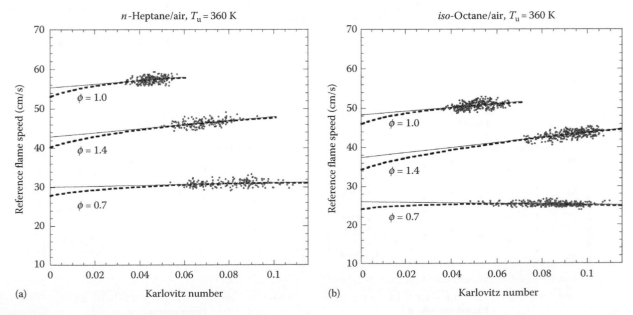

FIGURE 4.1.6
Reference stretch-affected flame speeds as a function of Karlovitz number for various (a) *n*-heptane/air and (b) *iso*-octane/air flames, showing how the reference stretch-affected flame speed is extrapolated to zero stretch to obtain the laminar flame speed. The unburned mixture temperature T_u is 360 K. Solid lines represent linear extrapolation, while dotted lines denote nonlinear extrapolation.

greater than unity, while those of rich *n*-heptane/air and rich *iso*-octane/air flames are sub-unity.

4.1.3 Laminar Flame Speeds with Preheat

Liquid hydrocarbon fuels constitute the bulk of the present-day energy sources available for ground and air transportation systems. Practical fuels consist of a wide array of hydrocarbons, including straight-chain paraffins and branched paraffins. Surrogate mixtures of pure hydrocarbons are often used to model the intended physical and chemical characteristics of practical fuels. *Iso*-octane and *n*-heptane are the primary reference fuels for octane rating and are widely used constituents of gasoline surrogates. *Iso*-octane has also been used as a significant component (5%–10% by volume) to model surrogate blends for JP-8. Furthermore, *n*-decane and *n*-dodecane are among the major straight-chain paraffin components found in diesel and jet fuels, and have frequently been used as surrogate components in different studies. These surrogate blends are used both experimentally and computationally to simplify the complex, naturally based mixtures, while retaining the fuels' essential properties.

In view of the growing interest in combustion studies of higher hydrocarbons, various experiments have been conducted with counterflow twin-flame configuration

[21–23] to determine the atmospheric-pressure laminar flame speeds of *n*-heptane/air, *iso*-octane/air, *n*-decane/air, and *n*-dodecane/air mixtures over a wide range of equivalence ratios and preheat temperatures. Preheating of air is one of the methods frequently employed in practical combustion devices as a mode of waste heat recovery. Fuel preheating is also employed for heavier oils to enable better atomization. Thus, in a majority of practical combustion devices, the reactants are in a state of preheat before entering the combustion chamber. However, the use of liquid hydrocarbon fuels as heat sinks in aero-propulsion devices is also being considered. This application of liquid hydrocarbon for thermal management could lead to the production of cracked gaseous products under certain conditions, and in addition, the increased fuel temperature tends to enhance the fuel reaction rates. On recognizing the significance of preheating, the effect of mixture preheating on laminar flame speed without fuel cracking was examined. Furthermore, since ethylene is used to simulate the cracked hydrocarbon fuels in the testing of combustors for hypersonic propulsion systems, preheating of fuel mixtures including that of ethylene is recognized as an important parameter in the study of fuel chemistry, with respect to both fundamental and practical considerations.

Figure 4.1.7 summarizes the measured laminar flame speeds of ethylene/air, *n*-heptane/air, *iso*-octane/air,

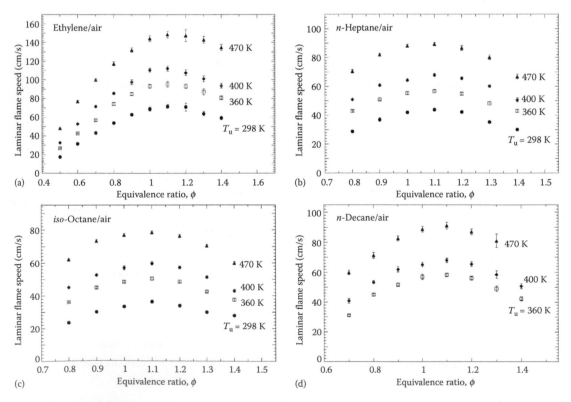

FIGURE 4.1.7
Measured laminar flame speeds of (a) ethylene/air, (b) *n*-heptane/air, (c) *iso*-octane/air, and (d) *n*-decane/air mixtures as a function of the equivalence ratio for various unburned mixture temperatures.

and *n*-decane/air mixtures as a function of the equivalence ratio for various preheat temperatures, obtained from earlier studies [21–23,27]. The error bars presented in Figure 4.1.7 indicate the 95% confidence interval estimate of the laminar flame speed obtained by using linear extrapolation [21]. It can be further noted that the flame speeds of *n*-heptane/air mixtures can be 5–10 cm/s higher than those of *iso*-octane/air mixtures. In accordance with the previous studies [21,28], the difference in the laminar flame speed was observed to be caused by the differences in the chemical kinetics, since the differences in the flame temperature and transport properties for both fuel/air mixtures were insignificant over the range of equivalence ratios investigated. In particular, the oxidation of *n*-heptane produced a large quantity of ethylene, while the main intermediates formed during the *iso*-octane oxidation were propene, *iso*-butene, and methyl radicals [21,28]. As a consequence, the flame speeds of *n*-heptane/air mixtures were higher.

When representing the dependence of laminar flame speed (S_u^o) on mixture preheat temperature (T_u) in the form of $S_u^o(T_u, \phi)/S_u^o(T_0, \phi) = (T_u/T_0)^n$, where T_0 is the lowest unburned mixture temperature investigated for a given fuel/air composition, the current experimental data can be correlated well with n in the range of 1.66–1.85. The exponent n is obtained by minimizing $\Sigma_\phi \Sigma_{T_u} \left[S_u^o(T_0, \phi) - S_u^o(T_u, \phi)/(T_u/T_0)^n \right]^2$, the sum of the squares of the errors for the mixture conditions investigated. Figure 4.1.8 further demonstrates that the correlated laminar flame speed, $S_u^o(T_u, \phi)/(T_u/T_0)^n$, reduces the experimental data to a single data set as a function of equivalence ratio.

It can be further noted that the mass burning flux, $m^o = \rho_u S_u^o$, is a fundamental parameter in the laminar flame propagation, where ρ_u is the unburned mixture density. Figure 4.1.9 demonstrates the effect of mixture preheat on the mass burning flux for various equivalence ratios. Furthermore, some experimental data of *n*-dodecane/air mixtures can also be observed. Significant enhancement of mass burning flux is observed with the increase in the preheat temperature, for all fuel/air mixtures studied. It can also be observed from Figure 4.1.9 that in the range of unburned mixture temperatures investigated, the mass burning flux increased linearly with T_u. As such, the increase in S_u^o with increasing T_u is not because of the reduction of ρ_u alone.

4.1.4 Determination of Overall Activation Energy

The effect of nitrogen concentration variation on laminar flame speed was also experimentally studied at

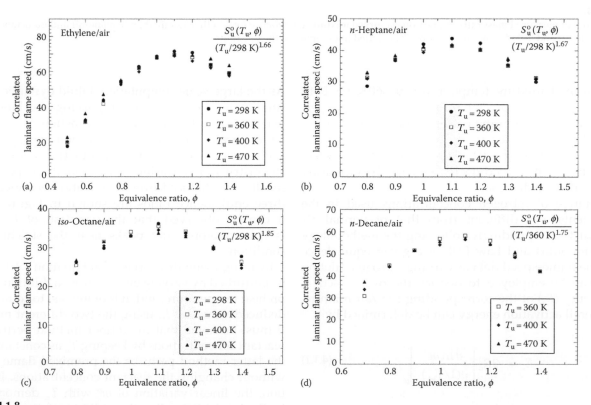

FIGURE 4.1.8
Correlated laminar flame speeds, $S_u^o(T_u, \phi)/(T_u/T_0)^n$, of (a) ethylene/air ($n= 1.66$), (b) *n*-heptane/air ($n= 1.67$), (c) *iso*-octane/air ($n= 1.85$), and (d) *n*-decane/air ($n= 1.75$) mixtures as a function of the equivalence ratio. T_0 is the lowest unburned mixture temperature tested for a fuel/air composition.

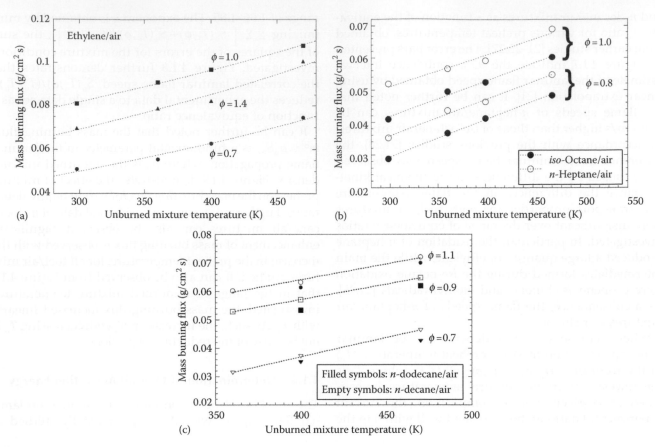

FIGURE 4.1.9

Dependence of measured mass burning flux on unburned mixture temperature for (a) ethylene/air, (b) *n*-heptane/air and *iso*-octane/air, and (c) *n*-decane/air and *n*-dodecane/air mixtures. The dotted lines represent the linear fits.

an unburned mixture temperature of 360 K for both *n*-heptane/air and *iso*-octane/air mixtures. Molar percentage of nitrogen in the oxidizer composed of nitrogen and oxygen varied from 78.5% to 80.5% for three equivalence ratios: ϕ = 0.8, 1.0, and 1.2. As expected, the laminar flame speed decreases with the increasing level of nitrogen dilution.

The range of nitrogen concentrations used in the experiments was different from that of normal air (79% N_2). Correspondingly, following the methodology of Egolfopoulos and Law [29], at a given equivalence ratio, the flame-speed data at varying nitrogen dilution levels can be employed to deduce the overall activation energy (E_a) of the corresponding fuel/air mixture. The overall activation energy can be determined by

$$E_a = -2R_u \left[\frac{\partial \ln m^\circ}{\partial (1/T_{ad})} \right]_p, \qquad (4.1.1)$$

where

T_{ad} is the adiabatic flame temperature
R_u is the universal gas constant

As the large-scale computational fluid dynamics (CFD) simulations often invoke simplifying the kinetics as one-step overall reaction, the extraction of such bulk flame parameter as overall activation energy is especially useful when the CFD calculation with detailed chemistry is not feasible. Based on the experimental results, the deduced overall activation energies of the three equivalence ratios are shown in Figure 4.1.10a. It can be observed that the variation of E_a with ϕ is nonmonotonic and peaks near the stoichiometric condition.

By recognizing this variation in Equation 4.1.1, T_{ad} can be perturbed by varying either nitrogen dilution level or preheat temperature; and, it is of interest to compare the deduced values of E_a using the two different methods. It must be noted that the former method perturbs the reactant concentrations by keeping T_u as constant, while the latter method perturbs the premixed flame system without changing the reactant concentrations. In addition, the linear variation of m° with T_u demonstrated in Figure 4.1.9 implies the validity of the extraction method through the changes in mixture preheat. Figure 4.1.10b shows the deduced E_a with varying T_u, based on

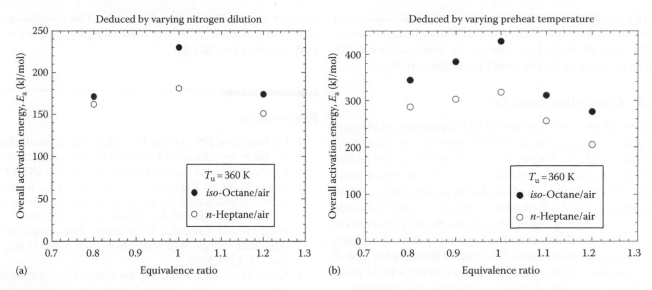

FIGURE 4.1.10
Experimentally deduced overall activation energies as a function of the equivalence ratio for *iso*-octane/air (filled symbols) and *n*-heptane/air (open symbols) mixtures obtained by varying the (a) N_2 concentration and (b) preheat temperature.

the experimentally determined m° of *iso*-octane/air and *n*-heptane/air mixtures and Equation 4.1.1.

Comparison of Figure 4.1.10a and b demonstrates that despite the quantitative differences in the deduced values, both the extraction methods yield a similar trend in the range of equivalence ratios investigated. The overall activation energy is observed to peak close to the stoichiometric condition and decrease on both the lean and rich sides. In addition, the overall activation energy values for *n*-heptane/air mixtures are observed to be lower when compared with *iso*-octane/air mixtures for all equivalence ratios under consideration. This similarity of trend and the differences in absolute values using two different extraction methods are also observed in the numerical computations with the available detailed

reaction mechanisms. Such a quantitative discrepancy possibly arises out of chemical interactions that occur as a consequence of varying N_2 concentration and consequently, the concentrations of fuel and O_2. However, the extraction method by varying T_u without involving the changes in the reactant composition (values shown in Figure 4.1.10b) is likely to be a better estimate of the overall activation energy.

By plotting the natural logarithm of the measured mass burning flux as a function of $1/T_{ad}$ for ethylene/air and *n*-decane/air mixtures of different equivalence ratios, the validity of the extraction method through the changes in mixture preheat can be demonstrated by the linear variation of $\ln m^\circ$ with $1/T_{ad}$. Figure 4.1.11 shows the experimentally deduced overall activation

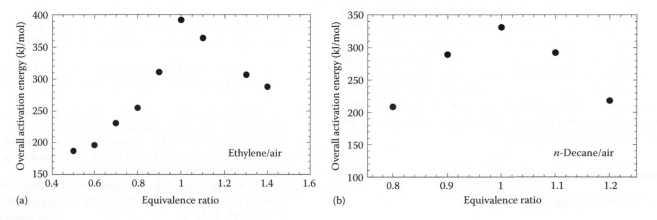

FIGURE 4.1.11
Experimentally deduced overall activation energies for the combustion of (a) ethylene/air and (b) *n*-decane/air as a function of the equivalence ratio, obtained by varying the preheat temperature.

energy as a function of the equivalence ratio. Again, the experimental results show a nonmonotonic dependence of the overall activation energy on equivalence ratio, with a peak close to the stoichiometric condition.

4.1.5 Concluding Remarks

Laminar flame speed is one of the fundamental properties characterizing the global combustion rate of a fuel/oxidizer mixture. Therefore, it frequently serves as the reference quantity in the study of the phenomena involving premixed flames, such as flammability limits, flame stabilization, blowoff, blowout, extinction, and turbulent combustion. Furthermore, it contains the information on the reaction mechanism in the high-temperature regime, in the presence of diffusive transport. Hence, at the global level, laminar flame-speed data have been widely used to validate a proposed chemical reaction mechanism.

For chemically reacting flows, perhaps one of the simplest as well as well-characterized flame configurations that are fundamentally affected by flow nonuniformity is the laminar counterflow flame. In particular, the counterflow twin-flame apparatus consists of two nitrogen-shrouded, convergent nozzle burners. In this nozzle-generated counterflow, two nearly planar and adiabatic flames can be established at varying flow straining. This type of flame and its characteristic propagation speed were demonstrated in this chapter. Furthermore, the methodologies for determining the laminar flame speeds and deducing the overall activation energy using the counterflow twin-flame technique were also discussed.

More research efforts are directed toward understanding the combustion kinetics of liquid hydrocarbon fuels to enable clean and efficient utilization of this depleting resource. It is therefore important to have a benchmark experimental database for the individual components, which is not only valuable by itself but also provides a data set for testing and validating the chemical kinetic mechanisms developed. Using the counterflow twin-flame experiments, recent results on the atmospheric pressure laminar flame speeds of ethylene/air, *n*-heptane/air, *iso*-octane/air, *n*-decane/air, and *n*-dodecane/air mixtures over a wide range of equivalence ratios and preheat temperatures were summarized. In addition, the experimentally determined overall reaction activation energies of varying equivalence ratios were also documented in this chapter.

Acknowledgments

This manuscript was prepared under the sponsorship of the National Aeronautics and Space Administration under Grant No. NNX07AB36Z, with the technical monitoring of Dr. Krishna P. Kundu. The assistance of Dr. Kamal Kumar with the manuscript preparation is very much appreciated.

References

1. Egolfopoulos, F.N., Zhang, H., and Zhang, Z., Wall effects on the propagation and extinction of steady, strained, laminar premixed flames, *Combust. Flame*, 109, 237, 1997.
2. Zhao, Z., Li, J., Kazakov, A., and Dryer, F.L., Burning velocities and a high temperature skeletal kinetic model for *n*-decane, *Combust. Sci. Technol.*, 177(1), 89, 2005.
3. Tsuji, H. and Yamaoka, I., Structure and extinction of near-limit flames in a stagnation flow, *Proc. Combust. Inst.*, 19, 1533, 1982.
4. Yamaoka, I. and Tsuji, H., Determination of burning velocity using counterflow flames, *Proc. Combust. Inst.*, 20, 1883, 1984.
5. Wu, C.K. and Law, C.K., On the determination of laminar flame speeds from stretched flames, *Proc. Combust. Inst.*, 20, 1941, 1984.
6. Law, C.K., Zhu, D.L., and Yu, G., Propagation and extinction of stretched premixed fames, *Proc. Combust. Inst.*, 21, 1419, 1988.
7. Ishizuka, S., On the behavior of premixed flames in a rotating flow field: Establishment of tubular flames, *Proc. Combust. Inst.*, 20, 287, 1984.
8. Kobayashi, H. and Kitano, M., Extinction characteristics of a stretched cylindrical premixed flame, *Combust. Flame*, 76, 285, 1989.
9. Mosbacher, D.M., Wehrmeyer, J.A., Pitz, R.W., Sung, C.J., and Byrd, J.L., Experimental and numerical investigation of premixed tubular flames, *Proc. Combust. Inst.*, 29, 1479, 2002.
10. Ishizuka, S., Miyasaka, K., and Law, C.K., Effects of heat loss, preferential diffusion, and flame stretch on flame-front instability and extinction of propane/air mixtures, *Combust. Flame*, 45, 293, 1982.
11. Ishizuka, S. and Law, C.K., An experimental study on extinction and stability of stretched premixed flames, *Proc. Combust. Inst.*, 19, 327, 1982.
12. Sheu, W.J. and Sivashinsky, G.I., Nonplanar flame configurations in stagnation point flow, *Combust. Flame*, 84, 221, 1991.
13. Sung, C.J., Trujillo, J.Y.D., and Law, C.K., On non-Huygens flame configuration in stagnation flow, *Combust. Flame*, 103, 247, 1995.
14. Sohrab, S.H., Ye, Z.Y., and Law, C.K., An experimental investigation on flame interaction and the existence of negative flame speeds, *Proc. Combust. Inst.*, 20, 1957, 1984.
15. Law, C.K., Dynamics of stretched flames, *Proc. Combust. Inst.*, 22, 1381, 1988.
16. Law, C.K. and Sung, C.J., Structure, aerodynamics, and geometry of premixed flamelets, *Prog. Energy Combust. Sci.*, 26, 459, 2000.
17. Vagelopoulos, C.M. and Egolfopoulos, F.N., Direct experimental determination of laminar flame speeds, *Proc. Combust. Inst.*, 27, 513, 1998.

18. Law, C.K., A compilation of experimental data on laminar burning velocities, *Reduced Kinetic Mechanisms for Application in Combustion*, Eds. N. Peters and B. Rogg, Springer-Verlag, Heidelberg, Germany, pp. 15–26, 1993.

19. Kee, R.J., Coltrin, M.E., and Glarborg, P., *Chemically Reacting Flow: Theory and Practice*, John Wiley & Sons, Hoboken, New Jersey Inc., 2003.

20. Hirasawa, T., Sung, C.J., Yang, Z., Joshi, A., Wang, H., and Law, C.K., Determination of laminar flame speeds of fuel blends using digital particle image velocimetry: Ethylene, *n*-butane, and toluene flames, *Proc. Combust. Inst.*, 29, 1427, 2002.

21. Huang, Y., Sung, C.J., and Eng, J.A., Laminar flame speeds of primary reference fuels and reformer gas mixtures, *Combust. Flame*, 139, 239, 2004.

22. Kumar, K., Freeh, J.E., Sung, C.J., and Huang, Y., Laminar flame speeds of preheated *iso*-octane/O_2/N_2 and *n*-heptane/O_2/N_2 mixtures, *J. Propulsion Power*, 23(2), 428, 2007.

23. Kumar, K. and Sung, C.J., Laminar flame speeds and extinction limits of preheated *n*-decane/O_2/N_2 and *n*-dodecane/O_2/N_2 mixtures, *Combust. Flame*, 151, 209, 2007.

24. Tien, J.H. and Matalon, M., On the burning velocity of stretched flames, *Combust. Flame*, 84, 238, 1991.

25. Vagelopoulos, C.M., Egolfopoulos, F.N., and Law, C.K., Further considerations on the determination of laminar flame speeds with the counterflow twin flame technique, *Proc. Combust. Inst.*, 25, 1341, 1994.

26. Chao, B.H., Egolfopoulos, F.N., and Law, C.K., Structure and propagation of premixed flame in nozzle-generated counterflow, *Combust. Flame*, 109, 620, 1997.

27. Kumar, K., Mittal, G., Sung, C.J., and Law, C.K., An experimental investigation on ethylene/O_2/diluent mixtures: Laminar flame speeds with preheat and ignition delays at high pressures, *Combust. Flame*, 153, 343, 2008.

28. Davis, S.G. and Law, C.K., Laminar flame speeds and oxidation kinetics of *iso*-octane-air and *n*-heptane-air flames, *Proc. Combust. Inst.*, 27, 521, 1998.

29. Egolfopoulos, F.N. and Law, C.K., Chain mechanisms in the overall reaction orders in laminar flame propagation, *Combust. Flame*, 80, 7, 1990.

4.2 Flame Propagation in Vortices: Propagation Velocity along a Vortex Core

Satoru Ishizuka

4.2.1 Introduction

The first literature concerning the flame propagation along a vortex core can be found in the journal *Fuel* in 1953. In Letters to the Editors, Moore and Martin [1] reported that when a combustible mixture was ejected with swirl from a closed end of a tube and ignited at the other open end, a tongue of flame projected within the tube mouth and extended eventually to the closed end. In this literature, a picture of a long flame, which was formed from the open mouth near to the closed end, was presented. They emphasized that such flame flashback occurred even when the flow rate exceeded the critical value for blow off, if the mixture was introduced in a straightforward manner and not tangentially. However, the flame speed was not reported.

The first determination of the flame speed in vortices was made in 1971 by McCormack [2] with the use of vortex ring. A combination of a pneumatically driven piston and a cylinder was used to make vortex rings. The determined flame speed was 300 cm/s for vortex rings of rich propane/air mixtures. Thereafter, McCormack collaborated with the Ohio State University and determined the flame speed as a function of the vortex strength [3]. The results are shown in Figure 4.2.1. The flame speed increases almost linearly, reaching about 1400 cm/s with an increase in the vortex strength. The mechanism for the high flame speed, however, was unknown. In McCormack's first paper [2], hydrodynamic instability, inherent to density gradient in a rotating flow was believed to be the underlying mechanisms; however, in his second paper [3], turbulence was considered as the potential mechanism.

A well-convincing mechanism for the rapid flame propagation was first proposed by Chomiak [4] in 1977.

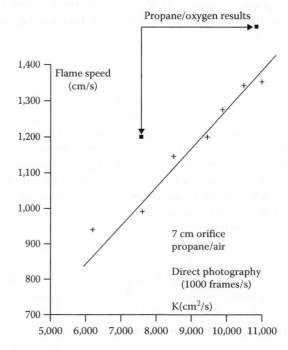

FIGURE 4.2.1
Flame speed versus vortex strength. (From McCormack, P.D., Scheller, K., Mueller, G., and Tisher, R., *Combust. Flame*, 19, 297, 1972.)

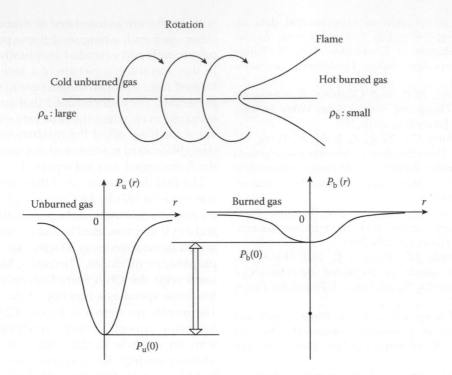

FIGURE 4.2.2
Vortex bursting mechanism proposed by Chomiak. (From Chomiak, J., *Proc. Combust. Inst.*, 16, 1665, 1977; Ishizuka, S., *Prog. Energy Combust. Sci.*, 28, 477, 2002.)

He considered that the rapid flame propagation could be achieved with the same mechanism as vortex breakdown. Figure 4.2.2 schematically shows his vortex bursting mechanism [4,5]. When a combustible mixture rotates, the pressure on the axis of rotation becomes lower than the ambient pressure. The amount of pressure decrease is equal to $\rho_u V_{\theta\,max}^2$ in Rankine's combined vortex, in which ρ_u denotes the unburned gas density and $V_{\theta\,max}$ denotes the maximum tangential velocity of the vortex. However, when combustion occurs, the pressure on the axis of rotation increases in the burned gas owing to the decrease in the density, and becomes close to the ambient pressure. Thus, there appears a pressure jump ΔP across the flame on the axis of rotation. This pressure jump may cause a rapid movement of the hot burned gas. By considering the momentum flux conservation across the flame, the following expression for the burned gas speed was derived:

$$V_f = V_{\theta\,max}\sqrt{\frac{\rho_u}{\rho_b}} \qquad (4.2.1)$$

in which ρ_b denotes the burned gas density [4].

Later, in 1987, Daneshyar and Hill [6] indicated that if the angular momentum conservation was assumed to be held across the flame front, then the pressure jump ΔP can be given as

$$\Delta P = \rho_u V_{\theta\,max}^2\left[1-\left(\frac{\rho_b}{\rho_u}\right)^2\right] \approx \rho_u V_{\theta\,max}^2 \qquad (4.2.2)$$

That is, once combustion is preceded, a pressure jump is invoked on the axis of rotation owing to angular momentum conservation. By further considering the pressure jump ΔP to be converted into the kinetic energy of the burned gas $\rho_b u_a^2/2$, an expression for the axial velocity of the hot gas can be obtained as:

$$u_a \approx V_{\theta\,max}\sqrt{\frac{2\rho_u}{\rho_b}} \qquad (4.2.3)$$

The predicted flame speeds, Equations 4.2.1 and 4.2.3, are in qualitative agreement with the flame speed determined by McCormack et al. [3] in the fact that the flame speed increases with an increase of vortex strength, and that the flame speed is further raised up if pure oxygen is used as an oxidizer, because the burned gas density is greatly reduced for combustion with pure oxygen.

Based on the rapid flame propagation along a vortex axis, turbulent combustion models have been constructed by Chomiak [4], Tabaczynski et al. [7], Klimov [8], Thomas [9], and Daneshyar and Hill [6]. In the model by Tabaczynski et al. [7], a flame was assumed to propagate instantaneously along a vortex of Kolmogorov scale, followed by combustion with a laminar burning velocity. In the hydrodynamic model by Klimov [8], the vortex scale was assumed to be much larger than the Kolmogorov scale. However, the validity of Equation 4.2.1 or Equation 4.2.3 has not been examined until recently.

Figure 4.2.3 shows the relation between the flame speed and the maximum tangential velocity obtained in a swirling flow in a tube [10]. In this experiment, a combustible mixture was injected tangentially at a closed end of a tube and exited from an open end. The tangential velocity decreased from the closed end toward the open end. Once ignited at the open end, a flame projected within the tube, as observed by Moore and Martin [1]. Local flame speeds were determined as a function of the local maximum tangential velocity at representative positions in the tube. It can be observed from Figure 4.2.3 that the flame speed increases with an increase in the

maximum tangential velocity, but they are much lower than the predictions, Equations 4.2.1 and 4.2.3. According to the predictions, the flame speed should be several times as high as the maximum tangential velocity, while the flame speeds measured were almost equal to or less than the maximum tangential velocity.

Further measurements on the flame speed have been obtained with the use of a rotating tube [11] and vortex ring combustion [12]. Figure 4.2.4 shows the flame speed in vortex rings [12]. The values of slope in the $V_f - V_{\theta \max}$ plane is nearly equal to unity for the near stoichiometric methane/air mixtures. Thus, this value is much lower than the predictions of $\sqrt{\rho_u/\rho_b}$ and $\sqrt{2\rho_u/\rho_b}$.

A theory, termed as the back-pressure drive flame propagation theory, has been proposed to account for the measured flame speeds [12]. This theory gives the momentum flux conservation on the axis of rotation in the form of

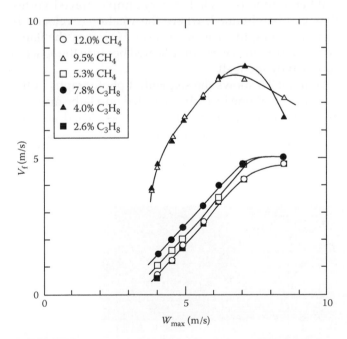

FIGURE 4.2.3
Relation between flame speed V_f and maximum tangential velocity $W_{\max}$ in an axially decaying vortex flow in a tube for various mixtures (tube diameter: 31 mm, the mean axial velocity: 3 m/s). (From Ishizuka, S., *Combust. Flame*, 82, 176, 1990.)

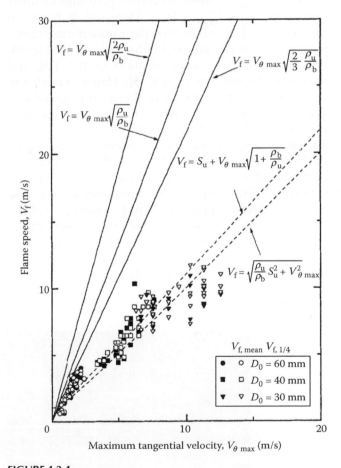

FIGURE 4.2.4
Relation between flame speed V_f and maximum tangential velocity $V_{\theta \max}$ in the vortex ring combustion. (From Ishizuka, S., Murakami, T., Hamasaki, T., Koumura, K., and Hasegawa, R., *Combust. Flame*, 113, 542, 1998.)

$$P_u(0) + \rho_u V_u^2 = P_b(0) + \rho_b V_b^2 \qquad (4.2.4)$$

in which

> $P(r)$ is the pressure as a function of radial distance r,
> whereas $r = 0$ is on the axis of rotation,
> V is the axial velocity, and
> subscripts u and b denote the unburned and
> burned gases, respectively.

The predicted flame speeds are given as

$$V_f = \sqrt{\frac{\rho_u}{\rho_b} S_u^2 + V_{\theta\max}^2} \quad \text{and} \quad S_u + V_{\theta\max}\sqrt{1 + \frac{\rho_b}{\rho_u}} \qquad (4.2.5)$$

for the cases when the burned gas is expanded in the axial and radial directions, respectively, in which S_u is the burning velocity.

Since the proportional factor ahead of $V_{\theta\,\max}$ is equal to unity or $\sqrt{1 + \rho_b/\rho_u}$, these formulas give almost unity slope in the $V_f - V_{\theta\max}$ plane. However, it has been pointed out by Lipatnikov that the momentum balance Equation 4.2.4 is not consistent for the Galilean transformation (personal communication in Ref. [13], and also see Lipatnikov note in Ref. [5]). Hence, a modified, steady-state back-pressure drive flame propagation theory was proposed, which however, again gave a near-unity slope in the $V_f - V_{\theta\max}$ plane [13].

To date, many theories and models have been proposed by various researchers, such as Atobiloye and Britter [14], Ashurust [15], Asato et al. [16], and Umemura [17,18]. Numerical simulations have also been conducted by Hasegawa and coworkers [19,20]. Recently, the phenomenon of rapid flame propagation has received keen interest from a practical viewpoint, to realize a new engine operated at increased compression ratios, far from the knock limit [21].

4.2.2 Appearance of Flame

Before discussing about the flame speed along a vortex core, it is first necessary to be familiar with the flames in various vortex flows. To date, four types of vortex flows have been used to study the flame behaviors. They are (1) a swirl flow in a tube [1,10], (2) vortex ring [2,3,12,13,16], (3) a forced vortex flow in a rotating tube [11], and (4) line vortex [22].

Figure 4.2.5 shows the appearance of the flame, which propagates in a vortex flow in a tube. In this case, a propane/air mixture was injected tangentially at a closed end of a tube. When the mixture was ignited at the other open end, a flame projected into the tube to reach the closed end. An interesting outlook is that the flame head appeared corrugated and highly disturbed. This was caused by the turbulence, which was very strong around the rotational axis. Another interesting feature is that the flame was more intensified in luminosity as the head was approached. Since the mixture was fuel rich (7.7% propane in volume), this intensification was caused by the so-called Lewis number effect [10]. However, in contrast to the rich mixtures, the heads of lean flames were weakened in luminosity and were dispersed (see Figure 4 of Ref. [10]). In this vortex flow, the rotational motion diminished axially owing to the viscosity, as the distance from the injector increased.

When a tube is rotated, a very simple, forced vortex flow can be obtained, where the rotational velocity is constant along the axis of rotation. However, the flame behavior becomes very complicated because the space is confined by the wall.

Figure 4.2.6 shows the sequential photographs of the propagating flame in a tube mounted on a lathe machine and rotated at 1210 rpm. One end was closed and the other end was open, where the mixture was ignited. Although a flame projected into the tube first, the flame

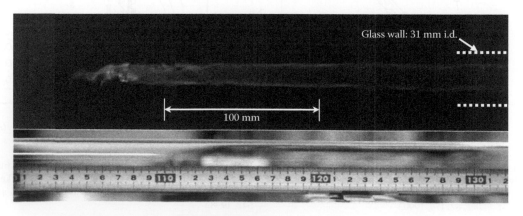

FIGURE 4.2.5
Appearance of the flame propagating in an axially decreasing vortex flow in a tube (fuel: propane, fuel concentration: 7.7%, tube: 31 mm in inner diameter and 1000 mm long, injector: 4 slits of 2 mm × 20 mm, mean axial velocity: 3 m/s).

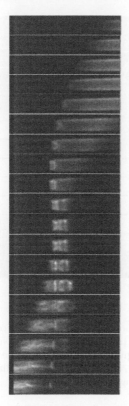

FIGURE 4.2.6
High-speed photographs of the propagating flame in a rotating tube (tube: 32 mm inner diameter and 2000 mm long, mixture: propane/air, equivalence ratio: 0.8, rotational speed: 1210 rpm, maximum tangential velocity at the wall: 2 m/s). (In collaboration with Prof. Yukio Sakai at Saitama Institute of Technology, October, 2003.)

speed was retarded and a typical tulip-shaped flame was formed around the middle of the rotating tube. After a while, reignition occurred, and consequently, the combustion zone projected again into the tube. Similar

tulip flame phenomenon also occurs in the vortex flow in the tube, as the space is surrounded by the solid wall (see Figure 12 of Ref. [5]).

In open space, however, such disturbances can be minimized. Figure 4.2.7 shows the Schlieren sequences of the vortex ring combustion, where two cases, weak and strong vortex rings, are illustrated. Through Schlieren photography, the boundary between the ambient air and the unburned combustible mixture of the vortex ring, and also the boundary between the burned gas and the unburned mixture can be visualized. The mixture was ignited by an electric spark at the bottom and the two flame fronts started out to propagate one another along the vortex axis in the opposite directions. As the vortex strength is increased, i.e., as the maximum tangential velocity of the vortex ring becomes larger, the flame becomes smaller in diameter and propagates faster along the core of the vortex ring.

The flame and the vortex core can be better visualized by other methods, as shown in Figure 4.2.8. The upper part of the vortex ring was illuminated by a laser sheet, while the mixture was ignited at the bottom and the propagating flame was photographed by a high-speed video camera with an image intensifier. To take these pictures, kerosene vapor was doped with methane/air mixtures and fine particles were obtained through condensation of the vapor. Owing to a centrifugal force of rotation, the number density of droplets was reduced in the core region; hence, the core was photographed as a dark zone in Figure 4.2.8. The diameter of the vortex ring (the distance between the centers of the core) is about 70 mm in this figure. In the case of lean mixture, whose equivalence ratio ϕ was 0.6, the flame diameter was observed to be small and the flame propagated within the dark, core region. However, in the stoichiometric mixture, the flame became larger

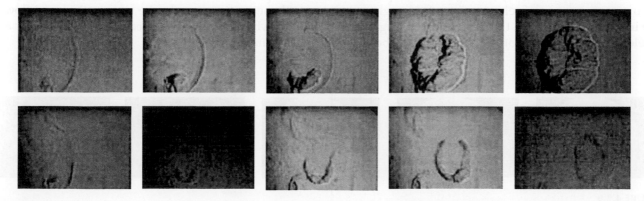

FIGURE 4.2.7
Schlieren sequences of vortex ring combustion of a stoichiometric propane/air mixture in an open air (Upper row: orifice diameter $D_0 = 60$ mm, driving pressure $P = 0.6$ MPa ($V_{\theta \max} = 11.4$ m/s), Lower row: orifice diameter $D_0 = 40$ mm, driving pressure $P = 1.0$ MPa ($V_{\theta \max} = 21.6$ m/s), diameter of vortex ring generator: 160 mm).

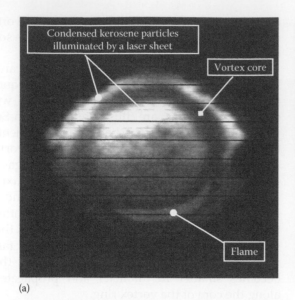
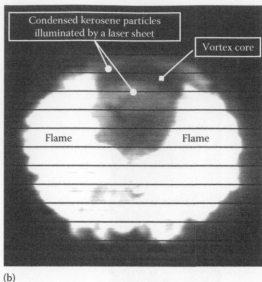

FIGURE 4.2.8
Photographs of the vortex ring combustion of propane/air mixture in air for equivalence ratio, (a) 0.6 and (b) 1.0. The combustible mixtures were doped with kerosene vapor and, through condensation, fine particles were formed in the rings. The particles were illuminated with a laser sheet and dark zones corresponded to the vortex core regions. The mixtures were ignited at the bottom of the vortex ring, and the propagating flames were recorded with a high-speed video camera with an image intensifier. A cylinder of 160 mm in diameter and an orifice of 60 mm in diameter were used to generate vortex rings. The mean diameter of the vortex rings were about 70 mm.

in diameter and the burning reached the free vortex region of the vortex ring. It is interesting to note that the vortex ring diameters are almost the same for the lean and stoichiometric mixtures. This signifies that the burnt gas expands mainly in the radial direction, i.e., in the direction normal to the axis of the vortex core. The gas expansion in the axial direction is restricted, resulting in the unchanged vortex ring diameter. This observation yields a useful insight on modeling the flame speed along a vortex core.

The existence of the vortex core can be observed in a very special case. Figure 4.2.9 exhibits a Schlieren sequence of vortex ring combustion of pure fuel. In this case, propane fuel was ejected through an orifice into an open air, and the vortex ring was ignited by an electric spark at the bottom. Although the boundary

between the fuel and the ambient air was disturbed, a yellow luminous flame appeared first at the bottom, and extended along the vortex core to become a circular string.

Although the vortex ring combustion in an open space is free from secondary effects, such as pressure waves or induced flows encountered in a confined space, the combustion suffers from other secondary effects that are inherent to vortex. For example, instability appears in vortex ring when the vortex motion becomes strong [23,24]. Figure 4.2.10 shows the time sequence of the flame propagation in such an unstable vortex ring. The flame speed is not constant along the core. Such a zigzag flame motion can also be seen in the swirl flow in a tube, which is caused by precession of the vortex axis (see Figure 18 of Ref. [5]).

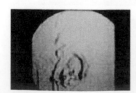
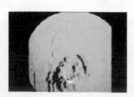

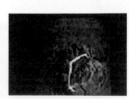

FIGURE 4.2.9
Schlieren sequence of vortex ring combustion of propane in an open air. After the mixture was ignited and the two flame fronts almost collided each other at the opposite side of the vortex ring, a yellow luminous zone appeared in the core region to extend along the core axis. (Cylinder diameter: 160 mm, orifice diameter: 60 mm, driving pressure: 1.0 MPa, piston stroke: 5 mm. The vortex ring was ignited at a position 500 mm apart from the orifice of the vortex ring generator.)

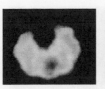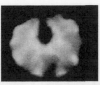

FIGURE 4.2.10

Time sequence of the intensified images of the vortex ring combustion (cylinder diameter: 160 mm, orifice diameter: 70 mm, driving pressure: 0.6 MPa, stoichiometric propane/air mixture, maximum tangential velocity: 7 m/s, Reynolds number of the vortex ring $Re \equiv UD/\nu$: 10^4 (U: translational speed of the vortex ring, D: vortex ring diameter, ν: kinematic viscosity).

Lastly, the recent PIV measurements on the flow field of vortex ring combustion are briefly presented in this chapter. Figure 4.2.11 shows the velocity vector profiles of a cold and a burning vortex ring. In the cold vortex ring, as illustrated by the right inset, the laser sheet was just on a plane that includes the axis of rotation of the vortex core. Thus, almost no velocity component can be observed along the vortex axis. However, in the burning vortex ring, the velocities are induced along the vortex axis. For this measurement, the position of the PIV laser sheet was simultaneously photographed by a CCD camera, which is shown in the right inset. It can be observed that the laser sheet remains exactly on the plane, which includes the core axis as well as the head of the flame propagating along the vortex core. It should be noted that axial flows are induced along the vortex axis.

Additional measurements have been made with close range arrangement. Figure 4.2.12 shows a velocity vector profile in the vicinity of the lean flame. The flame was photographed with another ICCD high-speed video camera from the same angle of the PIV camera, and its position, shown by a dotted line, was superimposed on the vector profile, as shown in Figure 4.2.12. The flame

position was slightly displaced from the dark zone, where the flame zone should be situated; however, it can be clearly observed that the velocities were induced along the vortex core. The magnitude of the induced velocities was around 11 m/s, which is slightly higher than the maximum tangential velocity of 9 m/s in this experiment; however, further measurements are being carried out.

4.2.3 Flame Speed

Although vortices of small scale, such as Kolmogorov scale or Taylor microscale, are significant in modeling turbulent combustion [4,6–9], vortices of large scale, in the order of millimeters, have been used in various experiments to determine the flame speed along a vortex axis.

With use of vortex ring combustion, the flame speed along a vortex axis has been determined by several researchers. McCormack et al. used a vortex ring generator of 220 mm in diameter and obtained the flame speed as the function of vortex strength [3]. Asato et al. [16] also used the same diameter of vortex ring generator to determine the flame speed. To obtain the value

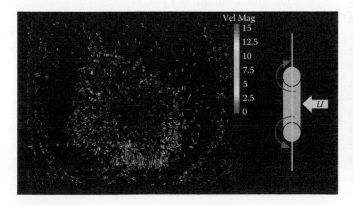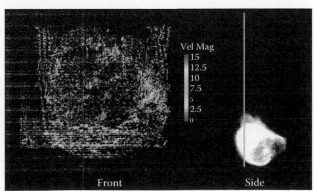

FIGURE 4.2.11

Vector profiles of (a) cold vortex ring and (b) burning vortex ring ($D_0 = 60$ mm, $P = 0.6$ MPa, stoichiometric mixture). Insets show the relative position of the PIV laser sheet to (a) the vortex ring and (b) the flame.

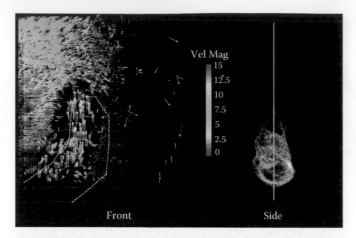

FIGURE 4.2.12
Vector profile of vortex ring combustion, showing induced velocities along the vortex core. (Lean propane/air mixture, equivalence ratio $\Phi = 0.8$, $D_0 = 60\,\text{mm}$, $P = 0.6\,\text{MPa}$, dotted lines show the flame front taken with the ICCD camera. The right inset shows the relative position of the PIV laser sheet relative to the flame.)

of $V_{\theta\max}$, they used Lamb's relation [25] and assumed a tangential velocity distribution of Rankine form (i.e., free/forced vortex) as

$$V_{\theta\max} = \frac{\Gamma}{\pi d} = 2U\frac{D/d}{\ln(8D/d) - 0.25} \qquad (4.2.6)$$

in which
  U is the translational velocity of the vortex ring
  Γ is the circulation
  D is the ring diameter
  d is the core diameter

However, this method overestimated the value of $V_{\theta\max}$ by a factor of 2 or 3, when the measured value of U and an assumed core/diameter ratio of 10% were used for the calculation. This is because most of the vortex rings available are turbulent vortex rings with a thick core [12,23,26]. Thus, measurements of the maximum tangential velocity are indispensable to obtain a quantitatively rigorous relationship between the flame speed and the maximum tangential velocity. So far, hot wire anemometry [12], laser Doppler velocimetry [13], and recently, PIV [22] have been used for the determination of the velocity. Furthermore, the vortex ring combustion have been conducted in an open air [2,3,12,16,27,28], in an inert gas atmosphere [13], and even in the same mixture atmosphere as the combustible mixture of the vortex ring [29]. However, diffusion burning occurs in the fuel-rich vortex ring combustion in an open air, and dilution by air and an inert gas occurs in the open air and the inert gas atmosphere, respectively.

Thus, the same mixture atmosphere may give pure results of vortex ring combustion.

Figure 4.2.13 shows the variation of the flame speed with the maximum tangential velocity obtained with vortex ring combustion in the same mixture atmosphere [29]. The cylinder diameter was 100 mm and various lean, stoichiometric, and rich methane/air and propane/air mixtures were examined. The diameter of the propagating flame was also determined and the ratio of the flame diameter to the core diameter was also plotted against the maximum tangential velocity.

In all the mixtures, the flame speed increased almost linearly with an increase in the maximum tangential velocity. The value of the slope in the $V_f - V_{\theta\max}$ plane was almost unity for the stoichiometric mixtures, however, the slope became smaller for the lean and rich mixtures. The flame to the core diameter ratio decreased with the increasing $V_{\theta\max}$. The ratio was around unity in the stoichiometric mixtures, while it was smaller than unity in the lean and rich mixtures.

At present, several theories have taken the finite flame diameter into consideration. Some of their predictions are presented in Figure 4.2.13 for comparison.

The curve 4 in Figure 4.2.13 is the prediction of a steady-state model by Umemura and Tomita [18]

$$4:\ V_f =$$

$$\sqrt{\left\{\left[2 + \frac{\rho_b}{\rho_u} + \frac{2\rho_u}{\rho_b}\ln\left[1 - \frac{\rho_b}{\rho_u}\left(1 - \frac{1}{k^2}\right)\right]\right]\right\}k^2 V_{\theta\max}^2 + \frac{\rho_u}{\rho_b}S_u^2} \quad (k\leq1)$$

$$(4.2.7)$$

in which
  d_u is the diameter of the unburned mixture, which is burned through the flame propagation
  d_c is the core diameter of the vortex
  k is the ratio of the unburned diameter to the core diameter, defined as $k \equiv d_u/d_c$

The line 5a in Figure 4.2.13 is the prediction by Asato et al. [16], who made a finite flame diameter approximation to obtain

$$5a:\ V_f = \frac{\alpha}{2a}V_{\theta\max}\sqrt{\left(\frac{\rho_u}{\rho_b} - 1\right)\left(1 - \frac{1}{4}\frac{\alpha^2}{a^2}\right)} \quad (\alpha \leq a)$$

$$(4.2.8)$$

in which
  α is the radius of the flame tip
  a is the core radius

The back-pressure drive flame propagation theory has been extended to a general case of variable burning

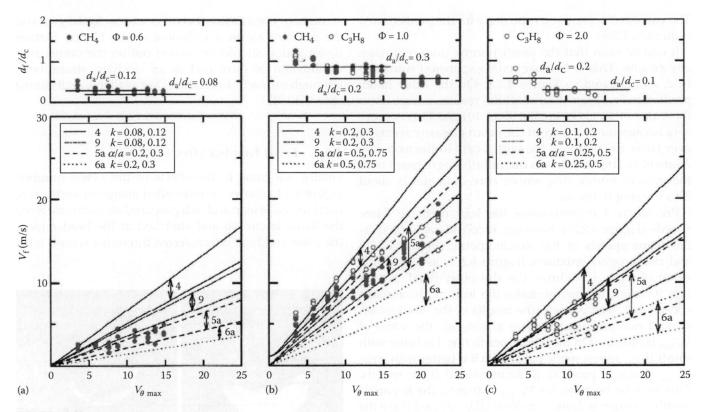

FIGURE 4.2.13
Measured flame speeds and theoretical predictions: (a) lean methane/air mixture ($\Phi = 0.6$, $S_u = 0.097\,\text{m/s}$, $\rho_u/\rho_b = 5.6$); (b) stoichiometric methane and propane mixtures ($\Phi = 1.0$, $S_u = 0.4\,\text{m/s}$, $\rho_u/\rho_b = 7.85(\text{mean})$); (c) rich propane/air mixture ($\Phi = 2.0$, $S_u = 0.18\,\text{m/s}$, $\rho_u/\rho_b = 7.7$). (From Ishizuka, S., Ikeda, M., and Kameda, K., *Proc. Combust. Inst.*, 29, 1705, 2002.)

region [27]. The curve 6a is one of its predictions for the axial expansion case,

$$6a:\ V_f = \sqrt{\rho_u (Y S_u)^2 / \rho_b + V_{\theta\max}^2\, f(k)}\ \text{(axial expansion)}$$

(4.2.9)

in which

Y is a ratio of the flame area to the cross-sectional area of the unburned mixture

the function $f(k)$ is given as $k^2/2$ for $k \le 1$ and $1 - 1/(2k^2)$ for $k \ge 1$

The line 9 is given by the steady-state, back-pressure drive flame propagation theory [29], which assumes the momentum flux balance between the upstream and downstream positions on the center streamline and the angular momentum conservation on each streamline,

$$9:\ V_f = k V_{\theta\max} \sqrt{1 + \frac{\rho_b}{2\rho_u} + \frac{\rho_u}{\rho_b} \ln\left[1 - \frac{\rho_b}{\rho_u}\left(1 - \frac{1}{k^2}\right)\right]}$$

(4.2.10)

This flame speed corresponds to the first term of Equation 4.2.7. Umemura and Tomita [18] used the

Bernoulli's equation on a center streamline ahead of and behind the flame and the momentum flux conservation across the flame front; however, the steady-state, back-pressure drive theory [29] used only the momentum flux balance across the flame front. These resulted in the $\sqrt{2}$ difference between Equation 4.2.10 and the first term of Equation 4.2.7.

When the burnt gas expands only in the radial direction, the parameter k can be estimated from a relation

$$k \equiv \frac{d_u}{d_c} = \frac{d_f}{d_c}\sqrt{\rho_b/\rho_u}$$

(4.2.11)

Using this relation, the values of k can be calculated. The representative values, $k = 0.08$ and 0.12, are shown in the upper figures in Figure 4.2.7a, while $k = 0.2$ and 0.3 and $k = 0.1$ and 0.2 are shown in Figure 4.2.7b and c, respectively. These values were used for the predictions 4 and 9.

In the model by Asato et al. (line 5a) and also in the case of the axial expansion (curve 6a), both the values of α/a and k are set to be equal to d_f/d_c. The values used are $k=0.2$ and 0.3 in Figure 4.2.7a, $k=0.5$ and 0.75 in Figure 4.2.7b, and $k=0.25$ and 0.5 in Figure 4.2.7c.

The value of Y is assumed to be unity for the prediction 6a (Equation 4.2.9).

It can be seen that the prediction 6a underestimates any results. This is because axial expansion is unrealistic, as indicated in Figure 4.2.8. On the other hand, prediction 5a covers almost all the results, except when the value of $V_{\theta\,max}$ is smaller than 10 m/s. This is probably because of the usage of the mean pressure averaged over twice the radius of the vortex core in the model by Asato et al. [16], which is in quantitative agreement with the present vortex ring whose core diameter is about 25% the ring diameter.

Prediction 4 overestimates the lean methane flame speeds (Figure 4.2.7a), however, it considerably predicts the flame speeds of the stoichiometric (Figure 4.2.7b) and rich propane mixtures (Figure 4.2.7c) as long as the value of $V_{\theta max}$ is <10 m/s. On the other hand, prediction 9 somewhat overestimates the lean methane flame speeds, but, well predicts the results of the stoichiometric and rich propane mixtures as long as the value of $V_{\theta\,max}$ is >10 m/s. On the whole, prediction 4 is better with small $V_{\theta\,max}$ range, while prediction 9 is better with large $V_{\theta\,max}$ range in predicting the measured flame speeds. This may be because, for $V_{\theta\,max} > 10$ m/s, the Reynolds number, which is defined as $\mathrm{Re} \equiv UD/\nu$ (U and D are the translational velocity and diameter of the vortex ring, respectively, and ν is the kinematic viscosity), develops into the order of 10^4 and the so-called turbulent vortex rings are formed [26,29].

Thus, it should be noted that the flame propagation in combustible vortex rings is not steady, but "quasi-steady" in the strict sense of the word. This may explain why prediction 9, based on the momentum flux conservation can better describe the flame speed for large values of $V_{\theta max}$ than prediction 4, which adopts the Bernoulli's equation on the axis of rotation.

It is also interesting to note that the angular momentum conservation is assumed in predictions 4 and 9; however, the viscosity increases owing to temperature rise in the burnt gas, and the vortex motion diminishes rapidly behind the flame. The pressure behind the flame is raised up and becomes nearly equal to the ambient pressure. This may explain why the hot, stagnant gas model by Asato et al., line 5a, can considerably predict the results.

From the above discussion, it is quite certain that the proposed theories can well describe the measured flame speeds, which are at the highest, nearly equal to the maximum tangential velocity of the vortex. However, it is interesting to note that in the case of vortex ring combustion of very rich hydrogen in an open air, the flame speed is much increased, up to the value predicted by the original vortex bursting theory, i.e., $V_{\theta\,max}\sqrt{\rho_u/\rho_b}$ [4,13]. At present, this enhancement of the flame speed has been attributed to pressure increase owing to the

secondary combustion between excess hydrogen and the ambient air in a turbulent mode. Hence, further investigation should be carried out on the vortex ring combustion of pure fuel in an oxidizer atmosphere, although only a few literatures are available on its flame speed [30,31].

4.2.4 Lewis Number Effects

Finally, we come to the effects of the Lewis number. Figure 4.2.14 shows the intensified images of vortex ring combustion of lean and rich propane/air mixtures. Since the flame is curved and stretched at the head region, the mass and heat is transferred through a stream tube.

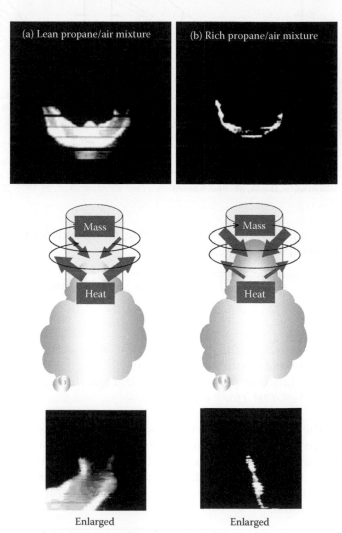

FIGURE 4.2.14
Intensified images of the vortex ring combustion, showing the Lewis number effect of a deficient species (Fuel: propane, vortex ring generator: 160 mm in diameter, condition: the same combustible mixture atmosphere), (a) equivalence ratio: 0.8, $D_0 = 60$ mm, $P = 0.8$ MPa ($V_{\theta max}\cong 5.8$ m/s, 4.9 ms after ignition), (b) equivalence ratio: 2.0, $D_0 = 50$ mm, $P = 0.8$ MPa ($V_{\theta max}\cong 8.8$ m/s, 9.1 ms after ignition).

In the lean mixtures, propane is deficient, hence, act as a controlling species. Since the mass diffusivity of propane is lesser than the thermal diffusivity of the mixture, burning is weakened at the head region. The flame is easily extinguished even if the vortex is not very strong. In this mixture, the flame diameter at extinction is large. On the other hand, in the rich mixtures, oxygen is a deficient species, whose mass diffusivity is larger than the thermal diffusivity of the mixture. Thus, burning is intensified at the head region, and the flame can propagate up to large values of the maximum tangential velocity. These Lewis number effects have also been observed for flames in the swirl flow in a tube [10], and also for flames in a rotating tube [11]. According to the theoretical study on a spherically expanding flame [32], unsteady terms that have heat or mass diffusivity coefficients in the governing equations, work as the external heat or mass source/sink in the steady-state case; thus, the Lewis number effect appears at an early stage of flame propagation. Since the flame propagation along a vortex axis suffers from unsteadiness inherent to vortex motion, the Lewis number effect cannot be ignored while discussing the flame behavior. Thus, it should be noted that although the flame speed along a vortex axis is mainly controlled by an aerodynamic factor, $V_{\theta \max}$, the limit flame behaviors, such as flame diameter at extinction and the maximum tangential velocity at extinction are also influenced by the physical and chemical properties of the mixture.

References

1. Moore, N. P. W. and Martin, D. G., Flame propagation in vortex flow, *Fuel*, 32, 393–394, 1953.
2. McCormack, P. D., Combustible vortex rings, *Proceedings of the Royal Irish Academy*, 71, Section A(6) 73–83, 1971.
3. McCormack, P. D., Scheller, K., Mueller, G., and Tisher, R., Flame propagation in a vortex core, *Combustion and Flame*, 19(2), 297–303, 1972.
4. Chomiak, J., Dissipation fluctuations and the structure and propagation of turbulent flames in premixed gases at high Reynolds numbers, *Proceedings of the Combustion Institute*, 16, 1665–1673, 1977.
5. Ishizuka, S., Flame propagation along a vortex axis, *Progress in Energy and Combustion Science*, 28, 477–542, 2002.
6. Daneshyar, H. D. and Hill, P. G., The structure of small-scale turbulence and its effect on combustion in spark ignition engines, *Progress in Energy and Combustion Science*, 13, 47–73, 1987.
7. Tabaczynski, R. J., Trinker, F. H., and Shannon, B. A. S., Further refinement and validation of a turbulent flame propagation model for spark-ignition engines, *Combustion and Flame*, 39, 111–121, 1980.
8. Klimov, A. M., Premixed turbulent flames–interplay of hydrodynamic and chemical phenomena, *Progress in Astronautics and Aeronautics*, Volume 88, Bowen, J. R., Manson, N., Oppenheim, A. K., and Soloukhin, R. I., eds., American Institute of Aeronautics and Astronautics, New York, pp.133–146, 1983.
9. Thomas, A., The development of wrinkled turbulent premixed flames, *Combustion and Flame*, 65, 291–312, 1986.
10. Ishizuka, S., On the flame propagation in a rotating flow field, *Combustion and Flame*, 82, 176–190, 1990.
11. Sakai, Y. and Ishizuka, S., The phenomena of flame propagation in a rotating tube, *Proceedings of the Combustion Institute*, 26, 847–853, 1996.
12. Ishizuka, S., Murakami, T., Hamasaki, T., Koumura, K., and Hasegawa, R., Flame speeds in combustible vortex rings, *Combustion and Flame*, 113, 542–553, 1998.
13. Ishizuka, S., Koumura, K., and Hasegawa, R., Enhancement of flame speed in vortex rings of rich hydrogen/air mixtures in the air, *Proceedings of the Combustion Institute*, 28, 1949–1956, 2000.
14. Atobiloye, R. Z. and Britter, R. E., On flame propagation along vortex tubes, *Combustion and Flame*, 98, 220–230, 1994.
15. Ashurst, Wm. T., Flame propagation along a vortex: The baroclinic push, *Combustion Science and Technology*, 112, 175–185, 1996.
16. Asato, K., Wada, H., Himura, T., and Takeuchi, Y., Characteristics of flame propagation in a vortex core: Validity of a model for flame propagation, *Combustion and Flame*, 110, 418–428, 1997.
17. Umemura, A. and Takamori, S., Wave nature in vortex bursting initiation, *Proceedings of the Combustion Institute*, 28, 1941–1948, 2000.
18. Umemura, A. and Tomita, K., Rapid flame propagation in a vortex tube in perspective of vortex breakdown phenomena, *Combustion and Flame*, 125, 820–838, 2001.
19. Hasegawa, T., Nishikado, K., and Chomiak, J., Flame propagation along a fine vortex tube, *Combustion Science and Technology*, 108, 67–80, 1995.
20. Hasegawa, T. and Nishikado, K., Effect of density ratio on flame propagation along a vortex tube, *Proceedings of the Combustion Institute*, 26, 291–297, 1996.
21. Gorczakowski, A., Zawadzki, A., and Jarosinski, J., Combustion mechanism of flame propagation and extinction in a rotating cylindrical vessel, *Combustion and Flame*, 120, 359–371, 2000.
22. Hasegawa, T., Michikami, S., Nomura, T., Gotoh, D., and Sato, T., Flame development along a straight vortex, *Combustion and Flame*, 129, 294–304, 2002.
23. Sullivan, J. P., Windnall, S. E., and Ezekiel, S., Study of vortex rings using a laser Doppler velocimeter, *AIAA Journal*, 11, 1384–1389, 1973.
24. Windnall, S. E., Bliss, D. B., and Tsai, C. Y., The instability of short waves on a vortex ring, *Journal of Fluid Mechanics*, 66, 35–47, 1974.
25. Lamb, H., *Hydrodynamics*, 6th ed., Dover, London, p. 241, 1945.
26. Maxworthy, T., Turbulent vortex rings, *Journal of Fluids Mechanics*, 64, 227–239, 1974.

27. Ishizuka, S., Hamasaki, T., Koumura, K., and Hasegawa, R., Measurements of flame speeds in combustible vortex rings: Validity of the back-pressure drive flame propagation mechanism, *Proceedings of the Combustion Institute*, 27, 727–734, 1998.

28. Morimoto, Y., Ikeda, M., Maekawa, T., Ishizuka, S., and Taki, S., Observations of the propagating flames in vortex rings, *Transactions of JSME*, 67(653), 219–225, 2001 (in Japanese).

29. Ishizuka, S., Ikeda, M., and Kameda, K., Vortex combustion in an atmosphere of the same mixture as the combustible, *Proceedings of the Combustion Institute*, 29, 1705–1712, 2002.

30. Choi, H. J., Ko, Y. S., and Chung, S. H., Flame propagation along a nonpremixed vortex ring combustion, *Combustion Science and Technology*, 139, 277–292, 1998.

31. Maekawa, T., Ikeda, M., Morimoto, Y., Ishizuka, S., and Taki, S., Flame speeds in vortex rings of combustible gases (4th Report), *Proceedings of the 37th Japanese Symposium on Combustion*, pp. 37–38, 1999 (in Japanese).

32. Frankel, M. L. and Sivashinsky, G. I., On effects due to thermal expansion and Lewis number in spherical flame propagation, *Combustion Science and Technology*, 31, 131–138, 1983.

4.3 Edge Flames

Suk Ho Chung

4.3.1 Introduction

A flame edge can be defined as the boundary between the burning and the nonburning states along the tangential direction on a flame surface, which could exist in both premixed and nonpremixed flames [1,2]. The base of a nozzle-attached flame, either premixed or nonpremixed systems, is a typical example.

Edge flames in premixed flames can be formed by various factors, including flow nonuniformity characterized by flame stretch [3], preferential diffusion effect characterized by the Lewis numbers of fuel and oxidizer, and heat loss. Some of the examples of premixed edge flames observed in the laboratory experiments are shown in Figure 4.3.1. The tip opening of a premixed Bunsen flame (Figure 4.3.1a) can occur by the combined effects of preferential diffusion and flame stretch (or curvature) [4]. The flame base is also an example of edge flames, owing to the quenching near the wall of a nozzle exit by heat and radical losses. Edge flames have also been observed in premixed flames propagating in a tube near the flammability limit (Figure 4.3.1b) when ignited at the bottom of the tube, owing to buoyancy and stretch [5]. A planar premixed flame in a slanted counterflow configuration demonstrates the edge flames (Figure 4.3.1c) by varying strain fields [6]. Spinning premixed flames in a sudden expansion tube (Figure 4.3.1d) is also an example of edge flames [7].

Nonpremixed edge flames are frequently encountered in fuel/oxidizer mixing layers, including two-dimensional (2D) mixing layers, jets, and boundary layers, as shown in Figure 4.3.2. The flame propagated in a 2D mixing layer (Figure 4.3.2a) [8] exhibited a tribrachial (or sometimes called triple) structure, having a lean premixed flame (LPF) and a rich premixed flame (RPF) wings together with a trailing diffusion flame (DF). This type of flame is important in terms of safety issues in a stagnant mixing layer [9] or flame stabilization in laminar jets (Figure 4.3.2b) [10,11]. Nonpremixed edge flames can also be relevant to flame spread front (Figure 4.3.2c) [12,13], nozzle-attached flame [14], and composite propellant combustion [15]. Nonpremixed edge flames could exist in inhomogeneously charged premixed conditions, for example, autoignition front in diesel engines (Figure 4.3.2d) [16] or flame fronts in direct injection gasoline engines. Nonpremixed edge flames have also been

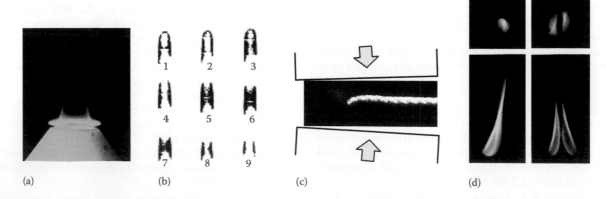

FIGURE 4.3.1
Observed premixed edge flames; (a) Bunsen flame-tip opening, (b) propagating premixed flame in tube (From Jarosinski, J., Strehlow, R.A., and Azarbarzin, A., *Proc. Combust. Inst.*, 19, 1549, 1982. With permission.), (c) slanted counterflow flame (From Liu, J.-B. and Ronney, P.D., *Combust. Sci. Tech.*, 144, 21, 1999. With permission.), and (d) spinning premixed flames in sudden expansion tube [7].

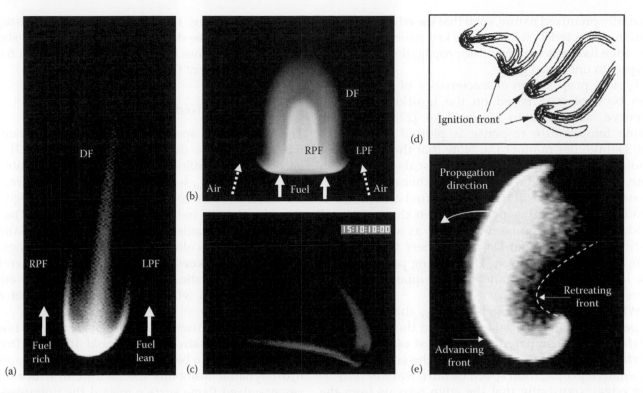

FIGURE 4.3.2
Nonpremixed edge flames; (a) 2D mixing layer (From Kīoni, P.N., Rogg, B., Bray, K.N.C., and Liñán, A., *Combust. Flame*, 95, 276, 1993. With permission.), (b) laminar jet (From Chung, S.H. and Lee, B.J., *Combust. Flame*, 86, 62, 1991.), (c) flame spread (From Miller, F.J., Easton, J.W., Marchese, A.J., and Ross, H.D., *Proc. Combust. Inst.*, 29, 2561, 2002. With permission.), (d) autoignition front (From Vervisch, L. and Poinsot, T., *Annu. Rev. Fluid Mech.*, 30, 655, 1998. With permission.), and (e) spiral flame in von Kármán swirling flow (From Nayagam, V. and Williams, F.A., *Combust. Sci. Tech.*, 176, 2125, 2004. With permission.). (LPF: lean premixed flame, RPF: rich premixed flame, DF: diffusion flame).

observed in rotating spiral flames in the von Kármán swirling flow with solid fuel rotation (Figure 4.3.2e) [17].

Edge flames in nonpremixed systems could exhibit diverse structures depending on the mixing length scale in front of an edge, which can be represented by the inverse of fuel concentration gradient or mixture fraction gradient. Numerical results of nonpremixed edge flames, represented in terms of reaction rate, are shown in Figure 4.3.3 for symmetric structures in a counterflow (Figure 4.3.3a through c) [18] and for asymmetric structures in a jet (Figure 4.3.3d and e) [19]. When the fuel concentration gradient is small, the nonpremixed edge flame exhibits the tribrachial structure, by having a rich and a lean premixed flame wing together with a trailing diffusion flame (Figure 4.3.3a and d). As the fuel concentration gradient increases, the radius of curvature of premixed flame wings becomes small and their intensity is weakened. Subsequently, either one or both of the wings begin to merge with the diffusion flame depending on the nature of symmetry, resulting in cotton-bud shape (Figure 4.3.3b) or bibrachial structure (Figure 4.3.3e). The asymmetry in edge-flame shape can be attributed to several factors, including the dependence of laminar burning velocity S_L° on fuel concentration, local velocity gradient, vortex structure [20], and Lewis numbers [21]. When the concentration gradient

is further increased, the edge flame has a monobrachial structure (Figure 4.3.3c).

When a premixed or a nonpremixed flame has an edge, the neighborhood of edge exhibits premixed or

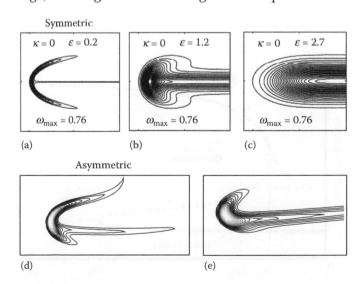

FIGURE 4.3.3
Various nonpremixed edge structures for symmetric cases (From Daou, R., Daou, J., and Dold, J., *Combust. Theory Model.*, 8, 683, 2004. With permission.); (a) tribrachial, (b) cotton-bud shape, and (c) monobrachial structures and for asymmetric cases [19]; (d) tribrachial and (e) bibrachial structures.

partially premixed nature, such that the edge has propagation characteristics. One of the key issues concerning the edge flames is their intrinsic propagation speed with respect to unburned gas.

Typical propagation characteristics of edge flames can be explained based on the ignition–extinction S-curve, as shown in Figure 4.3.4 [22], where the flame temperature response is plotted in terms of the Damköhler number Da. Note that the Damköhler could represent the inverse of the scalar dissipation rate, x^{-1}, which can be interpreted either to be proportional to the inverse of strain rate or fuel concentration gradient. The turning point between the upper and middle branches can be interpreted as a quasisteady extinction condition at $Da = Da_E$, where the subscript E indicates the extinction. The turning point between the lower and middle branches is the quasisteady ignition condition.

When a flame exists for $Da > Da_E$, the edge could experience an excess heat loss toward the nonreacting surface, when compared with the rest of the reacting surface. Thus, for Da close to Da_E, the edge can have a negative propagation speed (failure wave or retreating edge), signifying that the edge retreats from the nonreacting toward the reacting surface. For Da sufficiently larger than Da_E, the edge can overcome the extra heat loss, such that it can have a positive propagation speed (ignition wave or advancing edge). This implies the existence of a crossover Damköhler, Da_C. The existence of advancing and retreating edge speeds has been exhibited in spinning premixed flames (Figure 4.3.1d) and rotating spiral diffusion flames (Figure 4.3.2e).

In the following, the propagation characteristics of edge flames will be discussed together with the significance of edge flames in the stabilization of lifted flames in jets and turbulent flames.

4.3.2 Nonpremixed Edge Flames

The propagation speed of edge flame S_e in nonpremixed system, as depicted in Figure 4.3.5 [23], will be dominantly influenced by the fuel concentration gradient $dY_F/dy|_{st}$, where Y_F is the fuel mass fraction, y is the transverse coordinate in front of the edge, and the subscript "st" indicates the stoichiometry. The fuel concentration gradient influences the local laminar burning velocities along the premixed flame wings and the effective thickness of flammable region. Therefore, the fuel concentration gradient determines the curvature of premixed wings and thus, the shape of edge flame [24].

When the premixed flame wings are convex toward the upstream, the streamlines should diverge in front of the wing. This can be explained based on the Landau's hydrodynamic instability mechanism, in which a convex premixed flame surface toward the unburned gas induces the streamline divergence to satisfy the continuity of the tangential velocity component, and the jump in the normal velocity component due to gas expansion. Thereby, the local flow velocity in front of the edge decreases. Consequently, the propagation speed of tribrachial flame can be faster than the stoichiometric laminar burning velocity $S_L^o|_{st}$ owing to the flow redirection effect [23]. It has been shown that the propagation speed S_e decreases, relatively inversely with the fuel

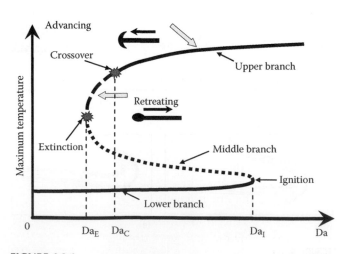

FIGURE 4.3.4
S-shaped flame temperature response with Damköhler number exhibiting edge propagation characteristics. (From Kim, J. and Kim, J.S., *Combust. Theory Model.*, 10, 21, 2006.)

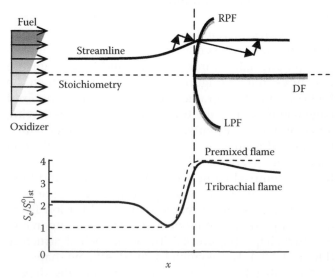

FIGURE 4.3.5
Tribrachial edge structure exhibiting flow redirection effect. (From Ruetsch, G.R., Vervisch, L., and Liñán, A., *Phys. Fluids*, 7, 1447, 1995.)

concentration gradient [25,26] owing to the decrease in the reaction intensity.

By combining this behavior together with the expected negative speed at high fuel concentration gradient, the overall behavior of the propagation speed of nonpremixed edge can be schematically represented as shown in Figure 4.3.6. First, the edge propagation speed of tribrachial flame S_e, which can be larger than the stoichiometric laminar burning velocity $S_L^o|_{st}$, leads to the existence of a transition regime in the propagation speed, when $dY_F/dy|_{st}$ becomes very small (regime A), since the propagation speed should be $S_L^o|_{st}$ for $dY_F/dy|_{st}= 0$, corresponding to a premixed flame propagation in the homogeneous stoichiometric premixture. The existence of the transition regime was confirmed in the counterflow [27] and 2D multi-slot burner experiments [28]. A tribrachial flame structure could exist in a limited range of $dY_F/dy|_{st}$ (regime B) with S_e reasonably inversely proportional to $dY_F/dy|_{st}$ [25,26].

As the concentration gradient further increases (regime C), the radius of curvature of premixed wings R_{cur} decreases. When it becomes comparable with, for example, the preheat zone thickness δ_T, typically O (1 mm), one or both of the premixed flame wings can be merged to the trailing diffusion flame by having a bibrachial or cotton-bud shaped structure.

As the concentration gradient further increases, both the premixed wings can be merged to trailing diffusion flame by having a monobrachial structure. In near-extinction conditions, the edge speed can be negative (retreating edge) in regime D, and can be expected to decrease rapidly. Based on the propagation characteristics in regimes B and D, it is expected that

the propagation speed could have a linear dependence on regime C.

If the strain rate exerted on an edge flame becomes very small, especially for weak intensity flames, the radiative heat loss could influence the edge behavior significantly. Consequently, the propagation speed could decrease appreciably in low-strained flames [29], as marked in the dash-dot line.

4.3.2.1 Edge Propagation Speed

The propagation speed of tribrachial flame was first analyzed in 2D mixing layers by Dold [25], by adopting activation energy asymptotics, assuming the concentration gradient to be small and neglecting the gas expansion. It has been demonstrated that S_e decreased with increasing concentration gradient, owing to the decrease in the reaction intensity. Note that a tribrachial edge could have a transition to a monobrachial edge at excessively high concentration gradients [30]. Owing to the limitation in neglecting the gas expansion, the maximum edge speed was bounded by $S_L^o|_{st}$.

As mentioned previously, S_e can be larger than $S_L^o|_{st}$ by the streamline divergence owing to the gas expansion, which has been confirmed experimentally [11,31,32]. This flow redirection effect was analyzed by Ruetsch et al. [23] by adopting (1) overall mass and momentum conservations, (2) Bernoulli equations in unburned and burnt regions, and (3) Rankine–Hugoniot relations across the premixed flame wings. The maximum limiting speed for small concentration gradient can be derived, demonstrating the dependence on the density ratio of unburned to burnt mixtures as follows:

$$\frac{S_{e,max}}{S_L^o\big|_{st}} \cong \left(\frac{\rho_u}{\rho_{b,st}}\right)^{1/2} \tag{4.3.1}$$

Later, an analytical closed-form solution for S_e was derived [26] by treating the density change as a small perturbation and assuming parabolic wing shape. Numerical studies with detailed reaction mechanisms [33,34] demonstrated that the enhancement of S_e can be primarily attributed to the flow redirection effect, and the contributions of the preferential diffusion and/or strain were ≤15%.

Ko and Chung [24] experimentally investigated the edge speed of tribrachial flames propagating in the laminar jets of methane fuel, by measuring the displacement speed S_d with respect to the laboratory coordinate and determining the local axial flow velocity u_e from the similarity solutions. The propagation speed S_e was determined from $(S_d + u_e)$. The results demonstrated that the edge speed S_e decreases with $dY_F/dy|_{st}$ by having a near inverse proportionality, in agreement with the

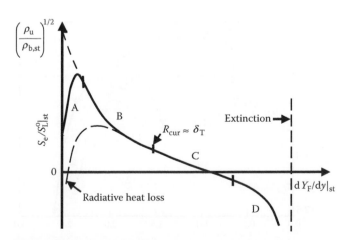

FIGURE 4.3.6
Behavior of edge propagation speed with concentration gradient (A: transition, B: tribrachial, C: bibrachial, D: monobrachial and near extinction regimes). (From Chung, S.H., *Proc. Combust. Inst.*, 31, 877, 2007.)

theoretical predictions [25,26]. Furthermore, a linear correlation between $dY_F/dy|_{st}$ and $1/R_{cur}$ was derived and substantiated in the experiment. Lee et al. [35] measured the propagation speed in propane jets under the microgravity condition, exhibiting similar behavior. In both the experiments, the extrapolated limiting propagation speeds for $dY_F/dr|_{st} \to 0$ were about 1 m/s, which are in good agreement with the predicted limiting speed in Equation 4.3.1, thus, substantiating the flow redirection effect.

The experimental results, when extrapolated for large $dY_F/dy|_{st}$, however, demonstrated that S_e can always be larger than $S_L^o|_{st}$, in contradiction with the existence of $S_e < S_L^o|_{st}$ as previously explained in Figure 4.3.6. To resolve this, the effect of velocity gradient at the upstream of the edge on S_e was investigated in coflow jets [36]. The velocity gradient could slant the premixed wing near the tribrachial point. When the edge is slanted with the angle θ to the transverse direction in jets (Figure 4.3.7), the propagation speed determined from $(S_d + u_e)$ must be corrected as $(S_d + u_e)\cos\theta$, considering the propagation direction of the edge. Also, the flow redirection effect needs to be modified, considering the effective heat conduction to the upstream stream tube, which is in the axial direction. The results are shown in Figure 4.3.7, where $(dY_F/dR|_{st})/(\cos\theta)$ is the correction, with respect to the effective heat conduction. The experimental results can be fitted to

$$\frac{S_e}{S_L^o|_{st}} = \frac{0.02018}{0.0077155 + dY_F/dR|_{st}/\cos\theta} + 0.155196$$

(4.3.2)

where $R = r/r_0$, r is the radial coordinate and r_0 is the nozzle radius of 0.172 mm. The result confirmed the limiting speed for small $dY_F/dr|_{st}$ corresponding to that in Equation 4.3.1. Also, it has been demonstrated that S_e can be smaller than $S_L^o|_{st}$, which is about 0.4 m/s, for large $dY_F/dr|_{st}$. The inset in Figure 4.3.7 shows the linear decreasing trend of S_e with $dY_F/dr|_{st}$ for large concentration gradients, corresponding to the cases with $R_{cur} < \delta_T$, where the flame edge exhibited bibrachial structures.

Experimental and numerical studies on negative edge speeds were conducted in opposed jet burners [37,38]. Recently, Cha and Ronney [29] observed edge propagation in a 2D opposed slot-jet burner, by varying strain rate and flame intensity with nitrogen dilution. The measured nondimensional edge speeds are shown in Figure 4.3.8 in terms of the normalized flame thickness, ε, which is proportional to the square root of strain rate divided by $S_L^o|_{st}$. The results exhibited negative edge speeds in the high strain regime. With excessive dilution (weak flames), only negative speeds existed. In the small strain regime, the edge speed decreases significantly with the decrease in the strain rate owing to the relative importance of radiative loss in this regime.

The maximum limiting speed of edge flames of about 1 m/s for methane was unable to enlighten the observed propagation speed of 1.8 m/s [9]. This contradiction may be resolved by considering a tribrachial flame propagation with respect to burnt gas, in the same way as the laminar burning velocity with respect to the burnt gas $S_{L,b}^o$. Ko et al. [27] carried out experiments in 2D counterflows by varying the equivalence ratios at the nozzle exits. The ignition was achieved by using a laser at the closed end (Figure 4.3.9 inset). In this situation,

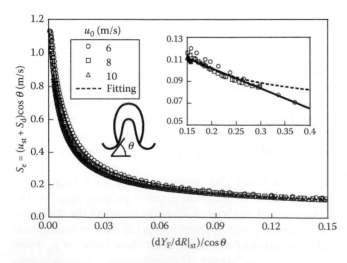

FIGURE 4.3.7
Propagation speed of nonpremixed edge flames of propane with concentration gradient. (From Kim, M.K., Won, S.H., and Chung, S.H., *Proc. Combust. Inst.*, 31, 901, 2007.)

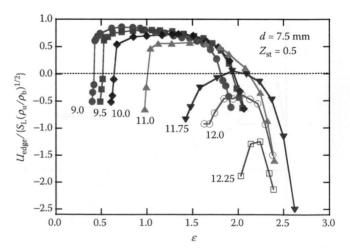

FIGURE 4.3.8
Effect of strain rate on edge propagation speed with N_2 dilution in $CH_4/O_2/N_2$ mixtures. (From Cha, M.S. and Ronney, P.D., *Combust. Flame*, 146, 312, 2006. With permission.)

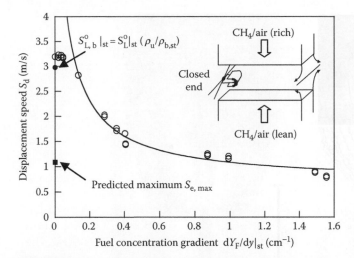

FIGURE 4.3.9
Displacement speed of tribrachial flames in 2D counterflow with fuel concentration gradient. (From Ko, Y.S., Chung, T.M., and Chung, S.H., *J. Mech. Sci. Technol.*, 16, 170, 2002.)

the burnt gas would accelerate the displacement of the edge flames. The displacement speeds S_d, shown in Figure 4.3.9, exhibited that the observed S_d was over 3 m/s for small $dY_F/dy|_{st}$, much higher than the $S_{e,max}$ of about 1 m/s. The transition regime from tribrachial flame to premixed flame was also exhibited, as $dY_F/dy|_{st}$ becomes sufficiently small.

From the analogy of Equation 4.3.1, phenomenological estimation for the maximum propagation speed of tribrachial flame with respect to burnt gas can be $S_{e,b,max}/S^o_{L,b}|_{st}=(\rho_u/\rho_{b,st})^{1/2}$. Considering $S^o_{L,b}|_{st}/S^o_L|_{st}=(\rho_u/\rho_{b,st})$, one can obtain $S_{e,b,max}/S^o_L|_{st}=(\rho_u/\rho_{b,st})^{3/2}$, which corresponds to $S_{e,b,max}=8.09$ m/s for methane. The best fit of experimental data with this value well-correlated, is marked as the solid line.

Various factors affecting the nonpremixed edge speed, such as flame stretch, preferential diffusion, and heat loss, have also been investigated, including cellular and oscillatory instabilities of edge flames [1,39–43].

4.3.2.2 Role of Edge Flames for Lifted Flame Stabilization

Laminar lifted flames were observed in free jets [10,44] and subsequently investigated extensively with respect to the characteristics of the tribrachial flames and their role in stabilization mechanism [11,45–51]. Figure 4.3.10 shows a typical lifted flame in a laminar jet [47], exhibiting Rayleigh scattering image representing fuel concentration in the cold region and reasonable temperature in the burnt region (Figure 4.3.10a) and that superimposed by CH* chemiluminescence representing the premixed flame wings and the OH LIF image representing the diffusion flame (Figure 4.3.10b), demonstrating

the tribrachial structure. The region between the nozzle and lifted flame edge is not influenced much by the existence of lifted flame, since the preheat zone thickness of premixed flame wings is $\delta_T=O$ (1 mm), while the liftoff height is typically O (1–10 cm). Thus, the behavior of laminar lifted flames can be described based on the similarity solutions of cold jets for axial velocity and fuel concentration.

The typical behavior of liftoff height with jet velocity is shown in Figure 4.3.11 for propane with the nozzle diameter $d=0.195$ mm [10]. When the jet velocity u_0 was small, the flame was attached to the nozzle with its flame length H_d increasing linearly with u_0, by having the quenching zones of H_q near the nozzle. At $u_0= 10.3$ m/s, the flame lifted off and the liftoff height H_L increased highly nonlinearly with u_0.

The existence of tribrachial structure at the base of lifted flame implies that the stabilization of liftoff flames is controlled by the characteristics of tribrachial edge flames. The coexistence of three different types of flames dictates that the edge is located along the stoichiometric contour [52] and the premixed wings have the propagation characteristics, whose speed should balance with the local flow velocity for the edge to be stationary. In the first approximation, the propagation speed was assumed to be constant [10].

The cold jet theory was applied in the region between the nozzle and the edge of lifted flame, which, for the axisymmetric jets, has the following similarity solutions [10,11]:

$$u = \frac{3}{8\pi\nu x}\frac{J}{\rho}\frac{1}{(1+\eta^2/8)^2},$$

$$Y_F = \frac{(2Sc+1)}{8\pi\nu x}\frac{I_F}{\rho}\frac{1}{(1+\eta^2/8)^{2Sc}} \qquad (4.3.3)$$

where η is the similarity variable defined as $\sqrt{u_{CL}x/\nu}\,(r/x)$, u_{CL} is the centerline velocity $(3/8\pi)(J/\rho\nu)/x$, x is the axial coordinate, ν is the kinematic viscosity, J is the momentum flux $(\pi\rho u_0^2 d^2)/(3+j)$, I_F is the fuel-mass flow rate $(\pi\rho u_0 d^2 Y_{F,0})/4$, Sc is the Schmidt number of fuel (ν/D_F), D_F is the mass diffusivity of fuel, and the subscript 0 indicates the condition at the nozzle exit. Here, j is 0 for the Poiseuille flow and 1 for uniform flow conditions at the nozzle exit.

A stationary lifted flame will be stabilized at (x^*,r^*), where $Y_F=Y_{F,st}$ and $u=S_e$. The liftoff height $H_L=x^*$ and the limitation on jet velocity can be derived from Equation 4.3.3 as follows:

$$\frac{H_L}{d^2}\frac{\nu}{S_e}=\left(\frac{3}{32}\right)\left(\frac{3}{2Sc+1}\frac{Y_{F,st}}{Y_{F,0}}\right)^{1/(Sc-1)}\left(\frac{u_0}{S_e}\right)^{(2Sc-1)/(Sc-1)} \qquad (4.3.4)$$

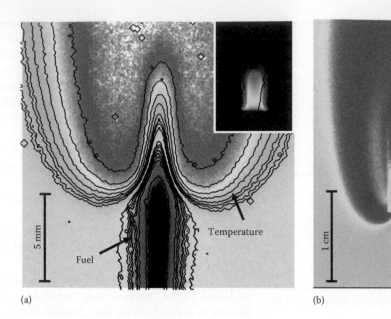

(a)   (b)

FIGURE 4.3.10

Lifted flame structure: Rayleigh image (a) and that superimposed with CH* and OH LIF (b). (From Lee, J., Won, S.H., Jin, S.H., and Chung, S.H., *Combust. Flame*, 135, 449, 2003.)

$$\eta^{*2}/8 = \left(\frac{3}{2\mathrm{Sc}+1} \frac{Y_{\mathrm{F,st}}}{Y_{\mathrm{F,0}}} \frac{u_0}{S_e} \right)^{-1/2(\mathrm{Sc}-1)} - 1 \qquad (4.3.5)$$

Note that the RHS of Equation 4.3.5 should be positive, which limits the jet velocity u_0.

For constant S_e and $Y_{\mathrm{F,st}}$, the liftoff height relation becomes

$$\left(H_{\mathrm{L}}/d^2 \right) Y_{\mathrm{F,0}}^{1/(\mathrm{Sc}-1)} = \mathrm{const} \times u_0^{(2\mathrm{Sc}-1)/(\mathrm{Sc}-1)} \qquad (4.3.6)$$

indicating that H_{L} is proportional to d^2, and the Schmidt number plays a crucial role for the dependence on jet velocity. For example, H_{L} increases with u_0 for $\mathrm{Sc} > 1$ and $\mathrm{Sc} < 0.5$, and decreases when $0.5 < \mathrm{Sc} < 1$. It has been shown that lifted flames for $\mathrm{Sc} < 1$ are unstable in free jets [11].

The best fit of velocity exponent n in $H_{\mathrm{L}} \propto u_0^n$ (Figure 4.3.11) for pure propane (n-butane) is $n = 4.733$ (3.638), corresponding to $\mathrm{Sc} = 1.37$ (1.61) from $n = (2\mathrm{Sc}-1)/(\mathrm{Sc}-1)$, which agreed well with the suggested value of $\mathrm{Sc} = 1.376$ (1.524). The experimental liftoff height data are shown in Figure 4.3.12 for various nozzle diameters and partial air dilutions to fuel [53]. It can be observed that the air dilution to fuel does not alter $Y_{\mathrm{F,st}}$ and $S_{\mathrm{L}}^{\mathrm{o}}|_{\mathrm{st}}$. The results substantiated the role of tribrachial flames on flame stabilization in laminar jets. As mentioned previously, Equation 4.3.5 limits the maximum velocity u_0 for $\mathrm{Sc} > 1$, which corresponds to blowout condition.

4.3.2.3 Relevance of Nonpremixed Edge Flames to Turbulent Flames

The relevance of nonpremixed edge flames to turbulent nonpremixed flames can be described in two aspects. One is the mechanism of turbulent nonpremixed lifted flames and the other, the flame-hole dynamics. For turbulent lifted flames in nonpremixed jets, the liftoff height is linearly dependent on jet velocity. There have

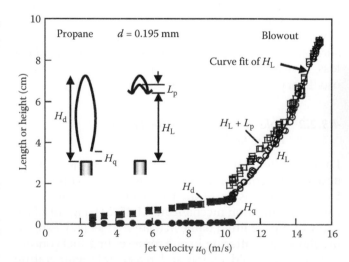

FIGURE 4.3.11

Liftoff height with jet velocity in free jet [10] (H_{d}: attached flame length, H_{q}: quenching distance, H_{L}: liftoff height, L_{p}: premixed flame length).

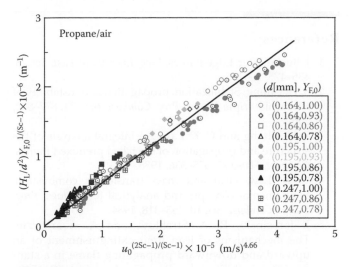

FIGURE 4.3.12
Correlation between liftoff height and jet velocity with partial premixing of air to fuel stream. (From Lee, B.J., Cha, M.S., and Chung, S.H., *Combust. Sci. Technol.*, 127, 55, 1997.)

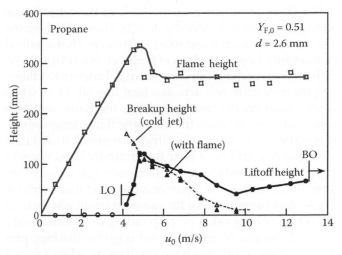

FIGURE 4.3.13
Jet-flame behavior demonstrating continuous transition from laminar-lifted to turbulent-lifted flames. (From Lee, B.J., Kim, J.S., and Chung, S.H., *Proc. Combust. Inst.*, 25, 1175, 1994.)

been several competing theories to explain this behavior [54], including the turbulent premixed flame model and the large-scale mixing model. Based on the observation of laminar lifted flames [10,44] having the tribrachial structure, the partially premixed flamelet model [55] was proposed.

A link between laminar and turbulent lifted flames has been demonstrated based on the observation of a continuous transition from laminar to turbulent lifted flames, as shown in Figure 4.3.13 [56]. The flame attached to the nozzle lifted off in the laminar regime, experienced the transition by the jet breakup characteristics, and became turbulent lifted flames as the nozzle flow became turbulent. Subsequently, the liftoff height increased linearly and finally blowout (BO) occurred. This continuous transition suggested that tribrachial flames observed in laminar lifted flames could play an important role in the stabilization of turbulent lifted flames. Recent measurements supported the existence of tribrachial structure at turbulent lifted edges [57], with the OH zone indicating that the diffusion reaction zone is surrounded by the rich and lean reaction zones.

Turbulent flames in practical combustors are frequent in the laminar flamelet regime, such that they can be viewed as an ensemble of laminar flamelets [58]. In general, a turbulent flame is subjected to random fluctuations of flow stretch and enthalpy gradient with a wide dynamic range. Therefore, each flamelet in the turbulent flame could undergo a distinct random-walk process between the reacting and nonreacting states, resulting in many locally quenched flame holes. In such cases, since quenching holes have edges, propagating and retreating edge flames could play an important role for the

re-ignition and extension of those quenching holes, in such a way that the overall characteristics of turbulent flames depend on the dynamics of the edge flames.

Based on the flame-hole dynamics [59], dynamic evolutions of flame holes were simulated to yield the statistical chance to determine the reacting or quenched flame surface under the randomly fluctuating 2D strain-rate field. The flame-hole dynamics have also been applied to turbulent flame stabilization by considering the realistic turbulence effects by introducing fluctuating 2D strain-rate field [22] and adopting the level-set method [60].

4.3.3 Premixed Edge Flames

When compared with the nonpremixed edge flames, studies on the propagation of premixed edge flames are rather limited [1]. This is because the propagation of an edge is in both the longitudinal and transverse directions on a flame surface. For example, for the edges with tip opening or counterflow with varying strain (Figure 4.3.1a and c), the remaining premixed flame has propagation characteristics normal to the flame surface and the edge has the propagation characteristics along the edge direction. The overall characteristics of premixed edge flames can also be explained based on Figure 4.3.4 of the characteristic S-curve, signifying that both advancing and retreating edges could be possible.

Furthermore, analytical solutions of premixed edges were studied and the dependence of edge speed with the Damköhler number was reported [61], in which the edge speed ranged from positive to negative values, in a similar fashion as nonpremixed edge flames.

The thermal-diffusive instabilities of the edge flames were also extensively investigated [1]. They arise from the imbalance of thermal and mass diffusions, characterized by nonunity Lewis numbers of fuel and/or oxidizer. Mass diffusion of reactants supplies chemical energy to a flame and thermal diffusion acts as a heat loss [4]. Thus, when Lewis numbers deviate from unity, the flame temperature could deviate from the adiabatic flame temperature. When the Lewis number is sufficiently small, three kinds of solutions could exist near the strain-induced quenching point: a periodic array of flame-strings, a single isolated flame-string, and a pair of interacting flame-strings. These structures can exist for values of strain greater than the 1D quenching value, corresponding to sublimit conditions [1]. Similar to nonpremixed edge oscillations, premixed edges could also have oscillations when Le $\gg$ 1 for the Damköhler number close to the extinction limit. This type of oscillation has been reported experimentally, in which the premixed edge flame occurred at the cracks of pyrolyzing HMX propellants and the oscillation frequency was in the range of 10^2–10^3 Hz [62].

The relevance of premixed edge flames to turbulent premixed flames can also be understood in parallel to the nonpremixed cases. In the laminar flamelet regime, turbulent premixed flames can be viewed as an ensemble of premixed flamelets, in which the premixed edge flames can have quenching holes by local high strain-rate or preferential diffusion, corresponding to the broken sheet regime [58].

4.3.4 Concluding Remarks

Edge flames are relevant to many flow situations, including stagnant mixing layers, 2D mixing layers, jets, and boundary layers, which are relevant to fire safety, flame spread, flame stabilization in jets, and propellant combustion. Although extensive studies have been conducted recently, many of the characteristics have not yet been clearly understood, such as the behavior of bibrachial and monobrachial nonpremixed edge flames and the propagation behavior of premixed edge flames. Particularly, the edge flame behavior with heat loss has not been identified clearly, which could lead to the understanding of nozzle-attached flames and their liftoff phenomena as well as the propagation of flame-spread front.

Acknowledgment

This work was supported by the CDRS Research Center.

References

1. J. Buckmaster, Edge-flames, *Prog. Energy Combust. Sci.* 28: 435–475, 2002.
2. S. H. Chung, Stabilization, propagation and instability of tribrachial triple flames, *Proc. Combust. Inst.* 31: 877–892, 2007.
3. S. H. Chung and C. K. Law, An integral analysis of the structure and propagation of stretched premixed flames, *Combust. Flame* 72: 325–336, 1988.
4. C. K. Law, Heat and mass transfer in combustion: Fundamental concepts and analytical techniques, *Prog. Energy Combust. Sci.* 10: 255–318, 1984.
5. J. Jarosinski, R. A. Strehlow, and A. Azarbarzin, The mechanisms of lean limit extinguishment of an upward and downward propagating flame in a standard flammability tube, *Proc. Combust. Inst.* 19: 1549–1557, 1982.
6. J. -B. Liu and P. D. Ronney, Premixed edge-flames in spatially-varying straining flows, *Combust. Sci. Tech.* 144: 21–45, 1999.
7. M. J. Kwon, B. J. Lee, and S. H. Chung, An observation of near-planar spinning premixed flames in a sudden expansion tube, *Combust. Flame* 105: 180–199, 1996.
8. P. N. Kioni, B. Rogg, K. N. C. Bray, and A. Liñán, Flame spread in laminar mixing layers: The triple flame, *Combust. Flame* 95: 276–290, 1993.
9. H. Phillips, Flame in a buoyant methane layer, *Proc. Combust. Inst.* 10: 1277–1283, 1965.
10. S. H. Chung and B. J. Lee, On the characteristics of laminar lifted flames in a nonpremixed jet, *Combust. Flame* 86: 62–72, 1991.
11. B. J. Lee and S. H. Chung, Stabilization of lifted tribrachial flames in a laminar nonpremixed jet, *Combust. Flame* 109: 163–172, 1997.
12. F. J. Miller, J. W. Easton, A. J. Marchese, and H. D. Ross, Gravitational effects on flame spread through non-homogeneous gas layers, *Proc. Combust. Inst.* 29(2): 2561–2567, 2002.
13. I. Wichman, Theory of opposed-flow flame spread, *Prog. Energy Combust. Sci.* 18: 553–593, 1992.
14. F. Takahashi and V. R. Katta, Further studies of the reaction kernel structure and stabilization of jet diffusion flames, *Proc. Combust. Inst.* 30: 383–390, 2005.
15. E. W. Price, Effect of multidimensional flamelets in composite propellant combustion, *J. Propulsion Power* 11: 717–728, 1995.
16. L. Vervisch and T. Poinsot, Direct numerical simulation of non-premixed turbulent flames, *Annu. Rev. Fluid Mech.* 30: 655–691, 1998.
17. V. Nayagam and F. A. Williams, Curvature effects on edge-flame propagation in the premixed-flame regime, *Combust. Sci. Tech.* 176: 2125–2142, 2004.
18. R. Daou, J. Daou, and J. Dold, The effect of heat loss on flame edges in a no-premixed counterflow within a thermo-diffusive model, *Combust. Theory Model.* 8(4): 683–699, 2004.

19. S. H. Won, J. Kim, K. J. Hong, M. S. Cha, and S. H. Chung, Stabilization mechanism of lifted flame edge in the near field of coflow jets for diluted methane, *Proc. Combust. Inst.* 30: 339–347, 2005.

20. D. Veynante, L. Vervisch, T. Poinsot, A. Liñán, and G. R. Ruetsch, Triple flame structure and diffusion flame stabilization, *Proceedings of the Summer Program, Center for Turbulent Research*: 55–73, 1994.

21. J. Buckmaster and M. Matalon, Anomalous Lewis number effects in tribrachial flames, *Proc. Combust. Inst.* 22: 1527–1535, 1988.

22. J. Kim and J. S. Kim, Modelling of lifted turbulent diffusion flames in a channel mixing layer by the flame hole dynamics, *Combust. Theory Model.* 10: 21–37, 2006.

23. G. R. Ruetsch, L. Vervisch, and A. Liñán, Effects of heat release on triple flames, *Phys. Fluids* 7: 1447–1454, 1995.

24. Y. S. Ko and S. H. Chung, Propagation of unsteady tribrachial flames in laminar nonpremixed jets, *Combust. Flame* 118: 151–163, 1999.

25. J. W. Dold, Flame propagation in a nonuniform mixture: Analysis of a slowly varying triple flame, *Combust. Flame* 76: 71–88, 1989.

26. S. Ghosal and L. Vervisch, Theoretical and numerical study of a symmetrical triple flame using the parabolic flame path approximation, *J. Fluid Mech.* 415: 227–260, 2000.

27. Y. S. Ko, T. M. Chung, and S. H. Chung, Characteristics of propagating tribrachial flames in counterflow, *J. Mech. Sci. Technol.* 16(12): 1710–1718, 2002.

28. N. I. Kim, J. I. Seo, K. C. Oh, and H. D. Shin, Lift-off characteristics of triple flame with concentration gradient, *Proc. Combust. Inst.* 30: 367–374, 2005.

29. M. S. Cha and P. D. Ronney, Propagation rates of nonpremixed edge flames, *Combust. Flame* 146(1–2): 312–328, 2006.

30. L. J. Hartley and J. W. Dold, Flame propagation in a nonuniform mixture: Analysis of a propagating triple-flame, *Combust. Sci. Technol.* 80: 23–46, 1991.

31. T. Plessing, P. Terhoven, N. Peters, and M. S. Mansour, An experimental and numerical study of a laminar triple flame, *Combust. Flame* 115: 335–353, 1998.

32. P. N. Kĩoni, K. N. C. Bray, D. A. Greenhalgh, and B. Rogg, Experimental and numerical studies of a triple flame, *Combust. Flame* 116: 192–206, 1999.

33. T. Echekki and J. H. Chen, Structure and propagation of methanol-air triple flame, *Combust. Flame* 114: 231–245, 1998.

34. H. G. Im and J. H. Chen, Structure and propagation of triple flames in partially premixed hydrogen-air mixtures, *Combust. Flame* 119: 436–454, 1999.

35. J. Lee, S. H. Won, S. H. Jin, S. H. Chung, O. Fujita, and K. Ito, Propagation speed of tribrachial (triple) flame of propane in laminar jets under normal and micro gravity conditions, *Combust. Flame* 134: 411–420, 2003.

36. M. K. Kim, S. H. Won, and S. H. Chung, Effect of velocity gradient on propagation speed of tribrachial flames in laminar coflow jets, *Proc. Combust. Inst.* 31: 901–908, 2007.

37. M. L. Shay and P. D. Ronney, Nonpremixed edge flames in spatially varying straining flows, *Combust. Flame* 112: 171–180, 1998.

38. V. S. Santoro, A. Liñán, and A. Gomez, Progapation of edge flames in counterflow mixing layers: Experiments and theory, *Proc. Combust. Inst.* 28: 2039–2046, 2000.

39. J. Daou and A. Liñán, The role of unequal diffusivities in ignition and extinction fronts in strained mixing layers, *Combust. Theory Model.* 2: 449–477, 1998.

40. R. W. Thatcher and J. W. Dold, Edges of flames that do not exit: Flame-edge dynamics in a non-premixed counterflow, *Combust. Theory Model.* 4: 435–457, 2000.

41. R. W. Thatcher, A. A. Omon-Arancibia, and J. W. Dold, Oscillatory flame edge propagation, isolated flame tubes and stability in a non-premixed counterflow, *Combust. Theory Model.* 6: 487–502, 2002.

42. S. R. Lee and J. S. Kim, On the sublimit solution branches of the stripe patterns formed in counterflow diffusion flames by diffusional–thermal instability, *Combust. Theory Model.* 6(2): 263–278, 2002.

43. V. N. Kurdyumov and M. Matalon, Radiation losses as a driving mechanism for flame oscillations, *Proc. Combust. Inst.* 29(1): 45–52, 2002.

44. Ö. Savas and S. R. Gollahalli, Flow structure in near-nozzle region of gas jet flames, *AIAA J.* 24: 1137–1140, 1986.

45. J. Lee and S. H. Chung, Characteristics of reattachment and blowout of laminar lifted flames in partially premixed jets, *Combust. Flame* 127: 2194–2204, 2001.

46. S. H. Won, J. Kim, M. K. Shin, S. H. Chung, O. Fujita, T. Mori, J. H. Choi, and K. Ito, Normal and micro gravity experiment of oscillating lifted flames in coflow, *Proc. Combust. Inst.* 29(1): 37–44, 2002.

47. J. Lee, S. H. Won, S. H. Jin, and S. H. Chung, Lifted flames in laminar jets of propane in coflow air, *Combust. Flame* 135: 449–462, 2003.

48. Y. -C. Chen and R. W. Bilger, Stabilization mechanism of lifted laminar flames in axisymmetric jet flows, *Combust. Flame* 122: 377–399, 2000.

49. S. Ghosal and L. Vervisch, Stability diagram for lift-off and blowout of a round jet laminar diffusion flame, *Combust. Flame* 124: 646–655, 2001.

50. A. Liñán, E. Frenández-Tarrrazo, M. Vera, and A. L. Sanchez, Lifted laminar jet diffusion flames, *Combust. Sci. Technol.* 177(5–6): 933–953, 2005.

51. T. Echekki, J. -Y. Chen, and U. Hedge, Numerical investigation of buoyancy effects on triple flame stability, *Combust. Sci. Technol.* 176(3): 381–407, 2004.

52. Y. S. Ko, S. H. Chung, G. S. Kim, and S. W. Kim, Stoichiometry at the leading edge of a tribrachial flame in laminar jets from Raman scattering technique, *Combust. Flame* 123: 430–433, 2000.

53. B. J. Lee, M. S. Cha, and S. H. Chung, Characteristics of laminar lifted flames in a partially premixed jet, *Combust. Sci. Technol.* 127: 55–70, 1997.

54. W. M. Pitts, Assessment of theories for the behavior and blowout of lifted turbulent jet diffusion flames, *Proc. Combust. Inst.* 22: 809–816, 1988.

55. C. M. Müller, H. Breitbach, and N. Peters, Partially pre-mixed turbulent flame propagation in jet flames, *Proc. Combust. Inst.* 25: 1099–1106, 1994.

56. B. J. Lee, J. S. Kim, and S. H. Chung, Effect of dilution of the liftoff of nonpremixed jet flames, *Proc. Combust. Inst.* 25: 1175–1181, 1994.

57. M. S. Mansour, Stability characteristics of lifted turbulent partially premixed jets, *Combust. Flame* 133: 411–420, 2003.

58. N. Peters, *Turbulent Combustion*, Cambridge: Cambridge University Press, 2000.

59. L. J. Hartley, Structure of laminar triple-flames: Implications for turbulent non-premixed combustion, PhD thesis, University of Bristol, 1991.

60. J. Kim, S. H. Chung, K. Y. Ahn, and J. S. Kim, Simulation of a diffusion flame in turbulent mixing layer by the flame hole dynamics model with level-set method, *Combust. Theory Model.* 10(2): 219–240, 2006.

61. T. Vedarajan and J. Buckmaster, Edge-flames in homogeneous mixtures, *Combust. Flame* 114: 267–273, 1998.

62. H. L. Berghout, S. F. Son, and B. W. Asay, Convective burning in gaps of PBX9501, *Proc. Combust. Inst.* 28: 911–917, 2000.

5

Instability Phenomena during Flame Propagation

CONTENTS

5.1 Instabilities of Flame Propagation

Geoff Searby

5.1.1 Introduction

Combustion features a wide range of instabilities, which have received considerable attention in recent years. The subject is of fundamental interest and it also has many practical implications. Combustion instability is often detrimental to the operation of the system and its dynamical effects can have serious consequences. It has become standard to distinguish three general classes of instabilities. In the first group, the instability is intrinsic to the combustion process. This is exemplified by the Darrieus–Landau hydrodynamic instability of premixed fronts or by the thermo-diffusive instabilities arising when the Lewis number departs from unity. The second group involves a coupling between combustion and the acoustics of the system. The resonant modes that assure the feedback are usually plane and their wavelength is commensurate with the total system longitudinal dimension. These "system" instabilities are generally characterized by low-frequency oscillations. In the third group, combustion is also coupled with acoustic modes, but these modes correspond to chamber resonances and the oscillation is often affected in the transverse or azimuthal direction. The wavelength corresponding to the third group is set by the chamber diameter, and in that case, the frequency of oscillation belongs to the high-frequency range.

This chapter considers the first group of instabilities and introduces the analysis of processes implying an interaction with external flow-field perturbations. This is exemplified by investigations of coupling between pressure waves and plane flames and also between an external acceleration field and flame fronts. The coupling between flow perturbations and flames giving rise to heat release unsteadiness and coupling with acoustic modes is considered in Chapter 5.2, which deals with the relationship between perturbed flame dynamics and radiated acoustic field, a fundamental process of thermo-acoustic instabilities.

5.1.2 Stability and Instability of Flame Fronts

Real-life premixed flame fronts are rarely planar. Of course, if the flow is turbulent, gas motion will continuously deform and modify the geometry of the flame front, see Chapter 7. However, even when a flame propagates in a quiescent mixture, the front rapidly becomes structured. In this chapter, we will discuss hydrodynamic flame instability, thermo-diffusive instability, and thermo-acoustic instability.

5.1.2.1 *The Darrieus–Landau Hydrodynamic Instability*

All premixed flames are unconditionally unstable to a hydrodynamic instability that has its origin in the expansion of the gas through the flame (but the flame may remain planar for other reasons). This phenomenon was first recognized by George Darrieus [1] and independently by Lev Landau [2], and is usually referred to as the Darrieus–Landau instability. The full derivation of the instability is arduous; here we will give a simple heuristic explanation.

Consider a planar premixed flame front, such as that sketched in Figure 5.1.1. For the moment, we will be interested only in long length scales and we will treat the flame as an infinitely thin interface that transforms cold reactive gas, at temperature and density T_o, ρ_o, into hot burnt gas at temperature and density T_b, ρ_b. The flame front propagates at speed S_L into the unburnt gas. We place ourselves in the reference frame of the front, so cold gas enters the front at speed $U_o = S_L$, and because of thermal expansion, the hot gases leave the front at velocity $U_b = S_L(\rho_o/\rho_b)$. The density ratio, ρ_o/ρ_b, is roughly equal to the

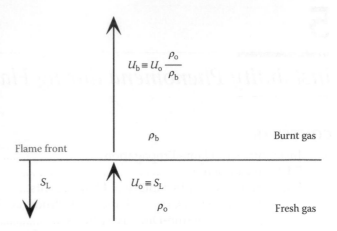

FIGURE 5.1.1
Gas expansion through a planar flame.

temperature ratio, and is typically of the order of 6–7 for standard hydrocarbon-air flames. To conserve momentum, this velocity jump must be accompanied by a small pressure jump, $\delta p = 1/2(\rho_b U_b^2 - \rho_o U_o^2) \equiv 1/2(\rho_o S_L^2)(\rho_o/\rho_b - 1)$, typically of the order of 1 Pa.

Let us now incline the flame front as shown in Figure 5.1.2a. The incoming gas flow can be decomposed into a vector component that is parallel to the front, with some speed $U_{//}$, and a component, U_n, that is normal to the front. If the front is stationary in our reference frame, then the speed of the normal component must be equal to the flame velocity, $U_n = S_L$. The burnt gas leaving the flame will have a normal component equal to $U_n(\rho_o/\rho_b)$ (gas expansion). The parallel component, $U_{//}$, is unaltered, since it is fairly obvious that there is no physical mechanism to sustain the parallel pressure jump that would be necessary to accelerate the parallel component of the flow. The inclined flame thus deviates the incoming gas flow toward the outgoing normal. If the flame is planar, this picture is invariant by translation. Figure 5.1.2b shows a visualization of the streamlines through a Bunsen flame front. The curvature of the streamlines in the hot burnt gas arises from buoyancy effects caused by gravity.

Consider now the situation shown in Figure 5.1.3, where the flame is no longer planar, but wrinkled at some wavelength λ. At places where the streamlines are normal to the front, they will be accelerated, but not deviated, as they cross the flame. At places where the front is inclined with respect to the incoming streamlines, they will be deviated toward the rear normal, as in Figure 5.1.2a. However, although the picture in Figure 5.1.3 is locally correct, it is globally wrong. The streamlines behind the front cannot cross; they must curve to become parallel again far downstream, as sketched in Figure 5.1.4.

Now if the streamlines are curved, there are pressure gradients in the flow: the wrinkled flame has introduced perturbations that are not local. It is this nonlocality that

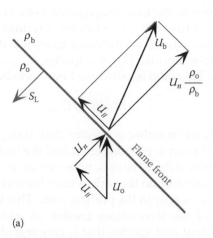

(a)

(b)

FIGURE 5.1.2
(a) Deviation of streamlines through an inclined flame. (b) Visualization of streamlines through an inclined Bunsen flame.

makes the mathematical solution difficult, for details, see for example [3] or [4]. For our purposes, we will simply admit that the presence of flame-induced pressure gradients will not only affect the downstream flow, but also the upstream flow. If the pressure gradient deviates a given downstream streamline to the right, then the upstream deviation will be in the same direction. The resulting streamlines are sketched in Figure 5.1.4. The effect of gas expansion through a curved flame causes the flow to converge at places where the front is concave to the unburnt gas, and to diverge where the front is convex. Mass conservation implies that the upstream flow is accelerated (decelerated) at places where the flame front lags behind (is ahead of) the mean position. Since we have supposed that the propagation velocity of the front S_L is constant, the situation is unconditionally unstable. The wrinkling will grow in time. This is the Darrieus–Landau instability.

In the analysis, there are only four parameters: the flame speed S_L, the wavelength λ (or wave number $k \equiv 2\pi/\lambda$), and the gas densities, ρ_o, ρ_b. Dimensional analysis tells us that there is only one way to construct a growth rate, σ (dimensions s^{-1}):

$$\sigma \propto kS_L f\left(\frac{\rho_o}{\rho_b}\right) \tag{5.1.1}$$

The exact expression is given by Landau [2,5]:

$$\sigma = kS_L \frac{E}{E+1}\left(\sqrt{\frac{E^2+E-1}{E}}-1\right) \tag{5.1.2}$$

where E is the gas expansion ratio $E = \rho_o/\rho_b$. This expression is valid in the linear limit when the amplitude of wrinkling is small when compared with the wavelength. The growth rate of the Darrieus–Landau instability

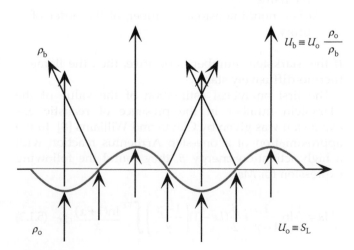

FIGURE 5.1.3
Local deviation of streamlines through a wrinkled flame.

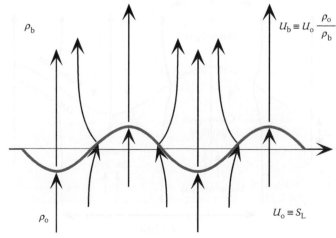

FIGURE 5.1.4
Curvature of streamlines through a wrinkled flame.

increases with the flame speed and with the wave number of wrinkling. For a typical stoichiometric hydrocarbon air flame, $S_L \approx 0.4\,\text{m/s}$, $E \approx 7$, the growth rate of 1 cm wrinkling is $\sigma \approx 400\,\text{s}^{-1}$. However, the growth rate cannot increase indefinitely for small wavelengths. Thermo-diffusive phenomena come into play when the wavelength of wrinkling is comparable with the flame thickness.

5.1.2.2 Thermo-Diffusive Effects

In the standard Zel'dovich–Frank-Kamenetskii (ZFK) model [6] of premixed flame propagation with a high activation energy, the chemical reactions are confined to a thin layer on the high-temperature side of the flame front, and the basic mechanism of propagation is controlled by the diffusion of the heat of combustion and species within the flame thickness, δ. If the flame is curved or wrinkled, the gradients of temperature and species concentration are no longer parallel to the average direction of propagation, see Figure 5.1.5, and the local flame velocity can change [7].

At places where the front is concave toward the unburnt gas, the heat flux is locally convergent. The local flame temperature increases and the local propagation velocity also increases, see the red arrows in Figure 5.1.5. The converse holds for portions of the front that are convex. The effect of thermal diffusion is to stabilize a wrinkled flame.

The gradient of species concentration is in a direction opposite to the thermal gradient, see the green arrows. At places where the front is concave toward the unburnt gas, the species flux is locally divergent. The flux of reactive species into the reactive zone decreases, leading in turn to

a decrease in the local propagation velocity. The effect of species diffusion is to destabilize a wrinkled flame. The net result of these two diffusive fluxes will depend on the ratio of the thermal, D_{th}, and species, D_{mol}, diffusion coefficients. This ratio is called the Lewis number,

$$Le = D_{th}/D_{mol}$$

If the Lewis number is greater than unity, the effect of heat diffusion is preponderant and the flame is thermo-diffusively stable. There is, however, an additional stabilizing contribution that arises from the inclination of the streamlines within the preheat zone. This internal inclination of the streamlines creates an additional transport of heat and species that is convergent or divergent with respect to the average direction of propagation. It has the effect of contributing an additional term in the expression for the local flame speed. This term is stabilizing, independent of the Lewis number, and increases with the gas expansion ratio. This effect of flame curvature on local flame velocity, S_n, was first recognized by Markstein [7] who wrote empirically:

$$\frac{S_n - S_L}{S_L} = \frac{\mathcal{L}}{R}$$

where
R is the radius of curvature of the flame
$\mathcal{L}$ is a characteristic length of the order of the flame thickness

This expression is often written in an equivalent form

$$\frac{S_n - S_L}{S_L} = Ma\frac{\delta}{R}$$

where
$\delta = D_{th}/S_L$ is a measure of the thermal thickness of the flame
Ma is a nondimensional number, of the order of unity

If the Markstein number is positive, then the flame is thermo-diffusively stable.

The first analytical estimation of the value of the Markstein number in the presence of realistic gas expansion was given by Clavin and Williams [8]. In the approximation of a one-step Arrhenius reaction with a high activation energy β, they found the following expression for Ma:

$$Ma = \frac{1}{\gamma}\ln\frac{1}{1-\gamma} + \frac{\beta}{2}(Le-1)\left(\frac{1-\gamma}{\gamma}\right)\int_0^{\frac{\gamma}{1-\gamma}}\frac{\ln(1+x)}{x}dx \quad (5.1.3)$$

where γ is the normalized nondimensional expansion ratio:

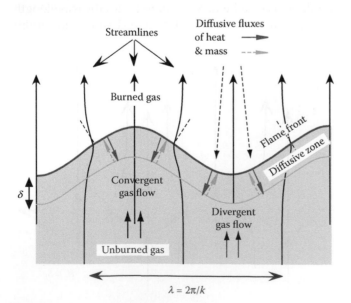

FIGURE 5.1.5
Internal structure of a wrinkled premixed flame.

$$\gamma = \frac{\rho_o - \rho_b}{\rho_o}$$

x is a dummy variable of integration.

The Lewis number, Le, is that of the deficient species (fuel or oxidant) in the mixture. In their analysis, Clavin and Williams used the simplifying approximation that the shear viscosity, the Lewis number, and the Prandtl numbers are all temperature-independent. They also showed that, at least for weak flame stretch and curvature, the change in local flame speed due to stretch and curvature is described by the same Markstein number:

$$\frac{S_n - S_L}{S_L} = Ma\left(\frac{\delta}{R} - \frac{\delta}{S_L}\frac{dU}{dx}\right)$$

where dU/dx is the longitudinal gas velocity gradient just upstream of the flame front, and is a measure of flame stretch. The first term on the r.h.s. of Equation 5.1.3 is strictly positive and for most hydrocarbon-air flames, the Markstein number is positive, even though the Lewis number may be smaller than unity. In fact, the Markstein number is generally positive for common hydrocarbon flames [9]. Dimensional analysis shows that the effect of diffusion on the growth rate of the wrinking must scale as some effective diffusion coefficient, D, times the square of the wave number [7,10]:

$$\sigma_{\text{Thermo-diffusive}} \propto -Dk^2$$

Thus, it is globally expected that long wavelength wrinkling is unstable, because of the Darrieus–Landau instability, with a growth rate that increases linearly with wave number. But for wavelengths comparable with the flame thickness, thermo-diffusive effects become predominant and restabilize the flame through a term proportional to $-k^2$.

5.1.2.3 Global Flame Stability

When both hydrodynamic and thermo-diffusive effects are simultaneously taken into account, it is found that the growth rate σ of wrinkling is given by the roots of the dispersion relation [11,12]:

$$(2-\gamma)(\sigma\tau_t)^2 + (2k\delta + (2-\gamma)(k\delta)^2 Ma\sigma\tau_t + \frac{\gamma}{Fr}k\delta - \frac{\gamma}{1-\gamma}(k\delta)^2$$

$$+ 2(k\delta)^3 Ma = 0 \tag{5.1.4}$$

The equation is written here in a nondimensional form, where $k\delta$ is the wave number nondimensionalized by the flame thickness, $\delta = D_{\text{th}}/S_L$, and the growth rate

$\sigma\tau_t$ is nondimensionalized by the flame transit time $\tau_t = \delta/S_L$. Here, γ is the normalized gas expansion ratio. Because of the difference in density between the fresh and burnt gas, gravity will also influence the dynamics of the flame front. The effect of gravity has been included through a Froude number, $Fr = S_L^2/(\delta g)$, positive for a flame propagating downward. Similar expressions have been derived by Matalon and Matkowski [13], and by Frankel and Sivashinsky [14].

Clavin and Garcia [15] have obtained a more general dispersion relation for an arbitrary temperature dependence of the diffusion coefficients. Their nondimensional result is qualitatively the same as Equation 5.1.4, but the coefficients contain information on the temperature dependence of the diffusivities:

$$A(k\delta)(\sigma\tau_t)^2 + B(k\delta)\sigma\tau_t + C(k\delta) = 0 \tag{5.1.5}$$

$$A(k\delta) = (2-\gamma) + \gamma k\delta\left(Ma - \frac{J}{\gamma}\right)$$

$$B(k\delta) = 2k\delta + \frac{2}{1-\gamma}(k\delta)^2(Ma - J)$$

$$C(k\delta) = \frac{\gamma}{Fr}k\delta - \frac{\gamma}{1-\gamma}(k\delta)^2\left\{1 + \frac{1-\gamma}{Fr}\left(Ma - \frac{J}{\gamma}\right)\right\}$$

$$+ \frac{\gamma}{1-\gamma}(k\delta)^3\left\{h_b + (2+\gamma)\frac{Ma}{\gamma} - \frac{2J}{\gamma} + (2Pr-1)H\right\}$$

where the quantities H and J are given by the integrals

$$H = \int_0^1 (h_b - h(\theta))d\theta \tag{5.1.6}$$

$$J = \frac{\gamma}{1-\gamma}\int_0^1 \frac{h(\theta)}{1+\theta\gamma/(1-\gamma)}d\theta \tag{5.1.7}$$

Ma is the Markstein number, of order unity, which for temperature-dependent diffusivities is now given by

$$Ma = \frac{J}{\gamma} - \frac{\beta}{2}(Le-1)\int_0^1 \frac{h(\theta)\ln(\theta)}{1+\theta\gamma/(1-\gamma)}d\theta \tag{5.1.8}$$

and

$h(\theta) = (\rho_\theta D_{\text{tho}})/(\rho_o D_{\text{tho}})$ is the ratio of thermal diffusivity times density at temperature θ to its value in the unburned gases

$\theta = (T-T_o)/(T_b-T_o)$ is the normalized temperature

h_b is the value of $h(\theta)$ in the burned gases

Pr is the Prandtl number, assumed to be independent of the temperature

Fr is the Froude number

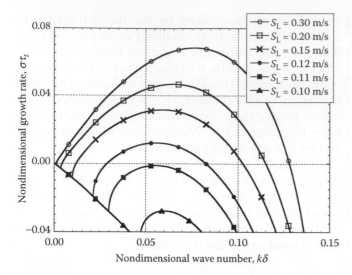

FIGURE 5.1.6
Example of dispersion relations calculated from Equation 5.1.5 for six lean propane-air flames. The wave number, k, is nondimensionalized by the flame thickness, δ, and the growth rate, σ, is nondimensionalized by the transit time through the flame, $\tau_t = \delta/S_L$.

In Figure 5.1.6, we have plotted typical curves for the dispersion relation calculated for propane-air flames using Equation 5.1.5. The temperature dependence of the thermal diffusivity was obtained using the JANAF tables. Six curves are plotted for laminar flame speeds ranging from 0.1 to 0.3 m/s (corresponding to equivalence ratios ranging from roughly 0.60 to 0.85). The curves are plotted for downward propagating flames, with the light burnt gas above the heavy unburnt gas, so that the effect of gravity is stabilizing. For the fastest flame, $S_L = 0.30$ m/s, the growth rate at first increases linearly with wave number, as expected, and then decreases quadratically as the thermo-diffusive effects become preponderant. The most unstable wave number occurs for $k \approx 0.075$ (corresponding to a dimensional wavelength of ≈ 6 mm) where the nondimensional growth rate is 0.07 (corresponding to a dimensional growth rate of $\approx 300\,\mathrm{s}^{-1}$). For slower flames,

not only does the Darrieus–Landau instability become weaker, but for downward propagating flames, the effect of gravity is to push the whole dispersion curve downward, with the result that sufficiently slow flames ($S_L < 0.11$ m/s for methane) are stable at all wavelengths. A similar behavior is observed for all hydrocarbon-air flames [16,17]. The seemingly curious linear behavior for very slow (lean) flames at small wave numbers corresponds to situations where the dispersion relation has complex roots, and the disturbances propagate along the surface of the flame as gravity waves, similar to the waves on the surface of the sea.

Figure 5.1.7a shows a side view of a lean propane flame, 10 cm in diameter, propagating downward in a top-hat flow. The flame speed is 9 cm/s, below the stability threshold, and the flame is stable at all wavelengths. Figure 5.1.7b shows a near stoichiometric flame in the same burner. The flame is seen at an angle from underneath. The mixture is diluted with nitrogen gas to reduce to flame speed to the instability threshold (10.1 cm/s), so that the cells are linear in nature. The cell size here is 1.9 cm. Figure 5.1.7c shows a flame far above the instability threshold, the cell shape becomes cusped, and the cells move chaotically.

Figure 5.1.8 shows time-resolved growth of the cellular instability on an initially planar lean propane-air flame [19]. The flame is seen from the side and the apparent thickening arises from deformations of the flame along the line of sight. The growth rates measured directly from this experiment for a range of flame speeds from 11 to 20 cm/s is in good agreement with the predictions of Equations 5.1.5 through 5.1.8.

5.1.2.4 Flame Instability on a Bunsen Flame

We have shown that planar premixed hydrocarbon-air flames are unstable over a range of wavelengths, typically from several centimeter to a few millimeter. The only exception is for slow flames propagating downward,

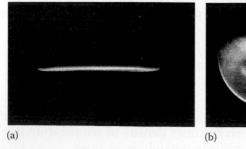
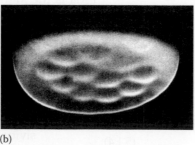
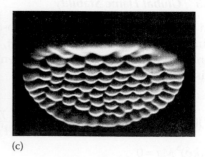

(a)   (b)   (c)

FIGURE 5.1.7
(a) Lean propane flame of 10 cm diameter in a top-hat flow, stabilized only by gravity. The flame is seen from the side. (b) Near stoichiometric propane flame (equivalence ratio 1.04) diluted with excess nitrogen to reduce the flame speed to the stability threshold (10.1 cm/s). The cell size is $\lambda = 1.9$ cm. (c) Near stoichiometric diluted flame, with the flame speed far above threshold. (From Quinard, J., Limites de stabilité et structures cellulaires dans les flammes de prémélange, PhD thesis, Université de Provence, Marseille, December 1984; Clavin, P., *Prog. Energy Combust. Sci.*, 11, 1, 1985. With permission.)

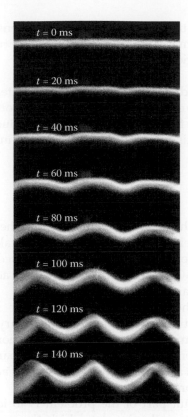

FIGURE 5.1.8
Sequence showing the temporal growth of instability on an initially planar lean propane-air flame. The flame speed is 11.5 cm/s. (From Clanet, C. and Searby, G., *Phys. Rev. Lett.*, 80, 3867, 1998. With permission.)

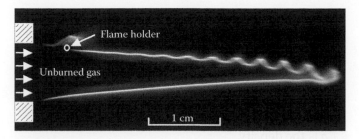

FIGURE 5.1.9
Short exposure image of growth of instability on a propane-air flame enriched with oxygen. The image has been rotated 90°. (From Searby, G., Truffaut, J.M., and Joulin, G., *Phys. Fluids*, 13, 3270, 2001. With permission.)

which can be stabilized by gravity. One may then wonder why small Bunsen flames, such as those typified by the gas burner of a kitchen stove or a domestic gas-fired water-heater, do not develop cellular structures. The reason for the apparent stability of these flames lies in the fact that they are anchored on the rim of the burner, and there is a strong velocity component tangential to the flame surface, see Figure 5.1.2. All structures on the flame surface are convected toward the flame tip at the tangential velocity, and thus have a finite residence time, $\tau = L/U_{//} \equiv h/(U_o \cos^2(\alpha)) \equiv h \tan(\alpha)/(S_L \cos(\alpha))$, where L is the length of the inclined flame, h is the vertical height of the Bunsen flame, and $\alpha = \cos^{-1}(h/L)$ is the half angle of the flame tip. This residence time must be compared with the growth time of the instability, $1/\sigma$. If the residence time is not large when compared with the growth time, then small perturbations at the base of the inclined flame will not have the time to grow to an appreciable amplitude before they are convected out of the flame. This is generally the case.

In some recent experiments, Truffaut et al. have used 2D slot burner to investigate the dynamics of instabilities on inclined flames [20–22]. Figure 5.1.9 shows a short exposure image of a propane flame. The image has been turned 90° clockwise. The mixture has been

enriched with oxygen (28%) to increase the flame speed and the growth rate. The flow velocity is 8.6 m/s and the flame speed is 0.64 m/s. The residence time of perturbations is 5 ms. The growth time of the most unstable wavelength (3.64 mm) is 1.07 ms. The turbulence of the flow is sufficiently low (0.1%) so that the initial amplitude of perturbations excited at the base of the flame is less than 1/10th of the flame thickness. During the residence time, the amplitude of these disturbances grows to scarcely more than the flame thickness, see the lower flame front. However, the other side of the flame is attached to a flame holder and periodically perturbed by an alternating voltage applied between the flame holder and the burner exit. Here, the excitation frequency is 2500 Hz and the amplitude of the induced perturbations is ≈0.07 mm. These perturbations grow and eventually saturate before reaching the tip of the flame. The growth rates of the instability measured in this experiment are again in good agreement with the predictions of Equations 5.1.5 through 5.1.8 [22].

5.1.3 Thermo-Acoustic Instabilities

Unsteady combustion is a strong source of acoustic noise. The emission of sound by gaseous combustion is governed by the classical set of conservation equations:

Mass conservation:

$$\frac{D\rho}{Dt} + \rho \nabla \cdot \mathbf{v} = 0 \qquad (5.1.9)$$

(Inviscid) Momentum conservation:

$$\rho \frac{D\mathbf{v}}{Dt} = -\nabla p \qquad (5.1.10)$$

Energy conservation:

$$\rho C_P \frac{DT}{Dt} = \dot{q} + \frac{Dp}{Dt} + \nabla \cdot (\lambda \nabla T) \qquad (5.1.11)$$

and an equation of state (for simplicity, we will use the ideal gas law):

$$\frac{p}{\rho} = (C_p - C_v)T = \frac{C_v}{C_p}c^2 \qquad (5.1.12)$$

where

D( )/Dt is the Lagrangian time derivative
ρ is the density
p is the pressure
v is the gas velocity
T is the temperature
C_p and C_v are the specific heats, assumed here to be constant for simplicity
$\dot{q}$ is the heat release rate per unit volume
λ is the thermal conductivity
c is the local speed of sound

Using the equation of state (Equation 5.1.12) in the energy Equation 5.1.11, and neglecting the diffusion of heat on the scale of the acoustic wavelength, one finds

$$\frac{Dp}{Dt} = c^2 \frac{Dp}{Dt} + \left(\frac{C_p - C_v}{C_v}\right)\dot{q} \qquad (5.1.13)$$

Equation 5.1.13 shows how heat release acts a volume source. Assuming that the combustion takes place in a uniform medium at rest (Mach $\ll$ 0), and writing for small perturbations, $a = \bar{a} + a'$ ($a = p, \rho,$ **v**), the linearized conservation equations for mass and momentum can be used to eliminate the density in 5.1.13 to obtain a wave equation for the pressure in the presence of local heat release:

$$\frac{\partial^2 p'}{\partial t^2} - c^2 \nabla^2 p' = \left(\frac{C_p - C_v}{C_v}\right)\frac{\partial \dot{q}}{\partial t} \qquad (5.1.14)$$

In the absence of fluctuations in the heat release rate, $\partial \dot{q}/\partial t = 0$, Equation 5.1.14 reduces to the standard wave equation for the acoustic pressure. It can be seen that a fluctuating heat release then acts as a source term for the acoustic pressure.

Combustion-generated noise is a problem in itself. However, if an acoustic wave can interact with the combustion zone, so that the heat release rate is a function of the acoustic pressure, $\dot{q} = f(p')$, then Equation 5.1.14 describes a forced oscillator, whose amplitude can potentially reach a high value. The condition for positive feedback was first stated by Rayleigh [23]:

"If heat be periodically communicated to, and abstracted from a mass of vibrating air, the effect produced will depend on the phase of the vibration at which the transfer takes place. If heat be given to the air at the moment of greatest condensation, or taken from it at the moment of greatest rarefaction, the vibration is encouraged."

In more modern terms, the "Rayleigh criterion" states that positive energy is transferred to the acoustic wave if the pressure fluctuation and heat release fluctuation are in phase. This criterion is usually written in an integral form:

$$\int_v \int_0^{2\pi} p'\dot{q}'dt\,dv > 0 \qquad (5.1.15)$$

where p' is the pressure fluctuation, $\dot{q}'$ is the fluctuation in the heat release rate, and the integral is taken over the volume of combustion, v, and over one acoustic cycle. The system will be globally unstable if the acoustic gain is greater than the acoustic losses.

Despite the fact that thermo-acoustic instabilities have been studied for more than a century, their control and elimination in practical combustion devices is still a problem that is difficult to master, particularly in devices with a high energy density such as aero-engines and rocket propulsion systems [24–28].

The difficult part of the problem is to identify and describe the mechanism by which the acoustic wave modulates the combustion rate. There are many possible mechanisms by which an acoustic wave can influence combustion, and the dominant mechanism varies with the design of the combustion device. Possible coupling mechanisms include

1. Direct sensitivity of the chemical reaction rate to the local pressure [29–34]

2. Oscillations of flame area induced by the acoustic acceleration [35–38]

3. Oscillations of total flame area induced by convective effects [39–42]

4. Periodic oscillations of the equivalence ratio when the fuel is injected as a liquid [43–45]

This list is not exhaustive. A review of the relative strengths of mechanisms 1, 2, and 4 in a simple 1D configuration has been performed by Clanet et al. [46]. They will be discussed in the next sections. Oscillations of total flame area induced by convective effects will be discussed in Chapter 5.2.

5.1.3.1 Pressure Coupling

In the standard ZFK flame model [6], the chemical reaction rate, Ω, is governed by a first-order irreversible one-step Arrhenius law

$$\Omega = A_o \rho Y \exp(-E_a/(RT))$$

where

A_o is the Arrhenius rate prefactor
Y is the mass fraction of the limiting reactant
E_a is the activation energy of the reaction
R is the gas constant

Since the reaction rate is proportional to the density, ρ, it is clear that the heat release rate will increase with pressure. However, since acoustic waves are adiabatic, they are also accompanied by a temperature oscillation

$$\frac{\delta T}{T_o} = \frac{C_p - C_v}{C_p} \frac{\delta p}{p_o}$$

and for large activation energy, the heat release rate is even more sensitive to the temperature oscillation than to the pressure oscillation. This was first noticed by Dunlap [29]. The effect of acoustic pressure on the ZFK flame was solved by Harten et al. in the limit of small gas expansion [30] and with no restriction by Clavin et al. [31]. Equivalent results were also obtained by McIntosh [32]. In the low-frequency limit, they show that the normalized response of the flame is given by

$$\frac{\dot{q}'/\bar{q}}{p'/\bar{p}} = \frac{\beta}{2} \frac{C_p - C_v}{C_p}$$

where the prime indicates the amplitude of oscillation, the bar indicates the mean value, and β is the reduced activation energy, $\beta = E_A(T_b - T_o)/RT_b^2$, T_b is the mean temperature of the burnt gas. For acoustic frequencies comparable with the inverse of the flame transit time, the internal structure of the flame (i.e., the temperature and concentration gradients) does not have time to adjust to the changing boundary conditions. The response of the heat release rate of the flame is found to increase, with a small dependency on the Lewis number [31,32]:

$$Z(\omega) = \frac{\dot{q}'/\bar{q}}{p'/\bar{p}} = \frac{\beta}{2} \frac{T_b}{T_o} \frac{C_p - C_v}{C_p} \frac{A(\omega)}{B(\omega)} \quad (5.1.16)$$

with

$$A(\omega) = [n(\omega) - (T_b - T_o)/T_b][n(\omega) - 1]n(\omega)$$

$$B(\omega) = [n(\omega) - 1]n^2(\omega) - \frac{\beta}{2}(Le - 1)[1 - n(\omega) + 2i\omega\tau_t]$$

$$n(\omega) = \{1 + 4i\omega\tau_t\}^{\frac{1}{2}}$$

where

Le is the Lewis number
ω is the angular frequency
$\tau_t = D_{th}/S_L^2$ is the flame transit time

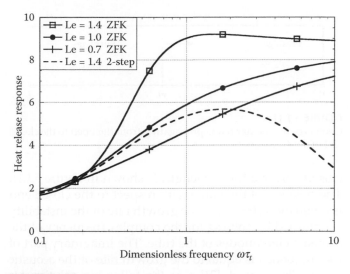

FIGURE 5.1.10
Semilog plot of the real part of heat release response, $\mathrm{Re}[(\dot{q}'/\bar{q})],(\rho'/\bar{\rho})$ of a ZFK flame to acoustic pressure oscillations [31]. Dotted line shows response of a simple two-step flame [47].

The response function in Equation 5.1.16 has a different normalization to the transfer function defined by Clavin et al. [31]. Here $Z = (Z_{Clavin}/M)(C_v/C_p)$, where $M = S_L/c$ is the Mach number of the flame.

A plot of the real part of the relative heat release response for three Lewis numbers is shown in Figure 5.1.10. This plot was calculated for a reduced activation energy $\beta = 10$ and a burnt gas temperature of 1800 K, representative of a lean hydrocarbon-air flame. Note that the order of magnitude of the relative response of the flame is only a little more than unity. This is a relatively weak response. For example, a sound pressure level of 120 dB corresponds to a relative pressure oscillation $p'/\bar{p} = 2 \times 10^{-4}$, so the fluctuation in the heat release rate will be of the same order of magnitude.

Recent experimental measurements by Wangher et al. [48] of the response of perfectly planar premixed methane and propane flames to acoustic pressure oscillations show that the order of magnitude of this analysis is correct, but the increase in response is not observed. The reason for this disagreement between experiment and analytical theory seems to arise from the over-simplifying assumption of an irreversible one-step Arrhenius law for the chemical reaction rate. As a first step to providing an understanding, Clavin and Searby [47] have investigated the response of a simple two-step chain-branching reaction. Although this model is still not realistic, the results confirm that multistep chemical kinetics substantially modify the unsteady pressure response of premixed flames, as shown by the dotted line in Figure 5.1.10.

The growth rate of the instability depends on the relative geometry of the flame front and the combustion chamber. Here, we give the results for the simple geometry of a flame propagating from the open to the

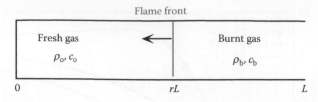

FIGURE 5.1.11
Geometry of a planar flame propagating from the open to the closed end of a tube of length L.

closed end of a tube of length L, shown in Figure 5.1.11. The position of the flame with respect to the closed end is given by rL, $0 < r < 1$. The growth rate of the instability is obtained by solving for the complex frequency of the acoustic eigenmodes of the tube. The imaginary part of the frequency is equal to the growth rate of the acoustic wave. Clavin et al. [31] give the following solution for the growth rate, σ, in the geometry of Figure 5.1.11

$$\sigma = \frac{S_L}{L} \frac{C_p}{C_v} \text{Re}[Z(\omega)] F(r, \omega_n) \qquad (5.1.17)$$

S_L is the laminar flame velocity, the function $Z(\omega)$ is the heat response function Equation 5.1.16, whose real part is plotted in Figure 5.1.10. The function $F(r, \omega_n)$ is a dimensionless acoustic structure factor that depends only on the resonant frequency, ω_n, the relative position, r, of the flame, and the density ratio ρ_b/ρ_o.

$$F(r, \omega_n) = \frac{1}{r(1 + \tan^2(rX_n)) + \frac{\rho_b}{\rho_o}(1-r)}$$
$$\times \frac{1}{(1 + \tan^2((1-r)\frac{c_o}{c_b} X_n)) \tan^2(rX_n)} \qquad (5.1.18)$$

where $X_n = \omega_n L / c_o$ are the dimensionless resonant frequencies ω_n of the tube with the flame at a distance rL from the closed end. If the gain of the instability is small, the X_n are the frequencies of the free eigenmodes, given by:

$$\frac{\rho_b c_b}{\rho_o c_o} \tan(rX_n) \tan\left((1-r)\frac{c_o}{c_b} X_n\right) = 1 \qquad (5.1.19)$$

$F(r, \omega_n)$ represents a normalized square of the acoustic pressure of mode n at the position of the flame front. It is plotted in Figure 5.1.12 for the first two acoustic modes of the tube. This function goes to zero at the open end of the tube, which is a pressure node. For the fundamental mode of the tube, the gain remains small until the flame has traveled at least halfway down the tube.

We can now estimate the order of magnitude of the acoustic growth rate expected for this mechanism. Consider a lean hydrocarbon-air flame, $S_L \approx 0.3 \text{ m/s}$. The ratio of the specific heats of the mixture is $C_p/C_v = 1.4$.

In a tube 1 m long [49], fundamental frequency is close to 100 Hz, the reduced frequency is less than unity, so $\text{Re}[Z] \approx 3$ and we can take $F \approx 0.5$. From Equation 5.17, it can be seen that the expected growth rate is

$$\sigma = 0.63 \text{ s}^{-1} \qquad (5.1.20)$$

This is a very small growth rate. The gain per acoustic cycle is 0.63%, much smaller than the acoustic losses, which are typically ≈2%–3% per cycle in this configuration. Therefore, this mechanism is not a strong source of thermo-acoustic instability. At maximum, the gain could be a little higher than the acoustic losses for reduced frequencies greater than unity. In real experiments in a half-open tube [49], the growth rate is found to be very much higher; obviously another mechanism is at work.

5.1.3.2 Acceleration Coupling

On the scale of the acoustic wavelength, a flame front is an interface separating two fluids of different densities. The flame front will thus react to gravity or to an imposed acceleration field. It is for this reason that a downward propagating flame can be stabilized by gravity, see Section 5.1.2.3. If the downward propagating flame is above the instability threshold, cellular structures will appear at the unstable wavelengths, see Figure 5.1.7.

In the presence of an acoustic *velocity* field, $u_a(t) = u'_a \cos(\omega_a t)$, the flame front is subjected to an oscillating acceleration $-\omega_a u'_a \sin(\omega_a t)$. When this acceleration is oriented toward the burnt gas, then the amplitude of the cells will tend to decrease. When it is oriented toward the unburnt gas, the amplitude of the cells will tend to

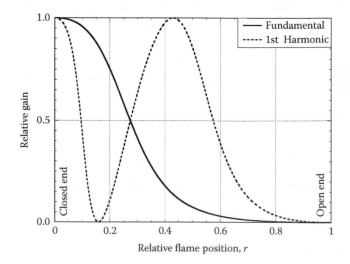

FIGURE 5.1.12
Acoustic structure functions for pressure coupling in an open-closed tube, for the fundamental and first harmonic.

increase. The acoustic field can thus modulate the total surface area of the flame, which in turn modulates the instantaneous heat release rate. This is another possible coupling mechanism between the acoustic wave and the heat release rate. This mechanism was first recognized by Rauschenbakh [50]. Pelcé and Rochwerger [38] have calculated the response function for this mechanism. They performed a linear analysis for a thin flame in the regime of small amplitude sinusoidal cells, $ak \ll 1$, where a is the amplitude of the sinusoidal wrinkling and $k = 2\pi/\lambda$ is the wave number. This regime is fulfilled only for flames just above the stability threshold. Clanet et al. [46] extended these calculations to include the effect of temperature-dependent diffusivities. They also used a heuristic approximation to apply this analysis to cellular flames far from threshold. Pelcé and Rochwerger define a response function, Tr, of the heat release rate to the acoustic velocity at the flame front, u'_a:

$$\text{Tr} = \frac{\dot{q}'/\bar{\dot{q}}}{u_a/S_L} \quad (5.1.21)$$

For the geometry of Figure 5.1.11, the growth rate of the instability is found to be [38]

$$\sigma = \frac{c}{L}\text{Im}[\text{Tr}]G(r, \omega_n) \quad (5.1.22)$$

Anticipating that the functions Tr and G will be of order unity, it is immediately obvious that the growth rate in Equation 5.1.22 is greater than that of the pressure coupling mechanism Equation 5.1.17 by a factor c/S_L (the inverse of the Mach number of the flame). The response function, Tr, is given by [46]:

$$\text{Tr} = \frac{(ak)^2}{2}\left(\frac{T_b - T_o}{T_o}\right)\frac{-i\omega\tau_t D(k\delta)}{-(\omega\tau_t)^2 A(k\delta) + i\omega\tau_t B(k\delta) + C(k\delta)} \quad (5.1.23)$$

where the coefficients A, B, and C are those of the dispersion relation, Equations 5.1.5. The coefficient $D(k\delta)$ is given by

$$D(k\delta) = \gamma k\delta[1 - k\delta(Ma - J/\gamma)] \quad (5.1.24)$$

Ma is defined by Equation 5.1.8 and J is defined by Equation 5.1.7. Typical curves for the imaginary part of the transfer function, Im[Tr], are plotted in Figure 5.1.13. These curves are calculated for a flame speed of 0.3 m/s, the other parameters in the coefficients A, B, C, and D are appropriate for a lean methane flame. The response is shown for three typical dimensionless wave numbers, $k\delta = 0.01$, 0.03, and 0.1, which correspond to dimensional

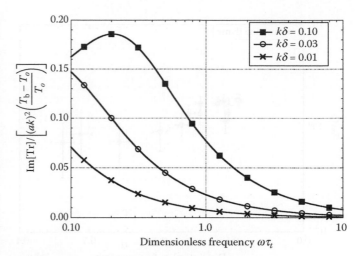

FIGURE 5.1.13
Semilog plot of the frequency-dependent part of heat release response, Im $[(\dot{q}'/\bar{\dot{q}})/(\rho'/\bar{\rho})/(ak)^2(T_b-T_o)]$ of a ZFK flame to acousity velocity oscilations [46], for three dimensionless wave numbers.

wavelengths $\lambda = 4.4$, 1.4, and 0.44 cm, respectively. The dimensionless frequency $\omega\tau_t = 1$ corresponds to a dimensional frequency of 642 Hz.

The acoustic structure function $G(r, \omega_n)$ is given by

$$G(r, \omega_n) = F(r, \omega_n)\tan\left(r\omega_n\frac{L}{c_o}\right) \quad (5.1.25)$$

where the resonant frequencies ω_n are given by Equation 5.1.19. $G(r, \omega_n)$ represents a normalized product of the acoustic pressure times the acoustic velocity at the flame front. It goes to zero at the pressure nodes and also at the velocity nodes. It is plotted in Figure 5.1.14 for a density ratio of 7.

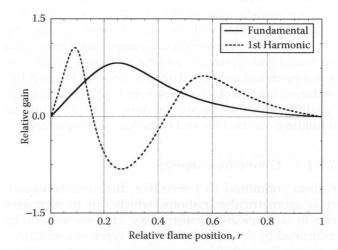

FIGURE 5.1.14
Acoustic structure function for acceleration coupling in an open-closed tube, for the fundamental resonance and the first harmonic.

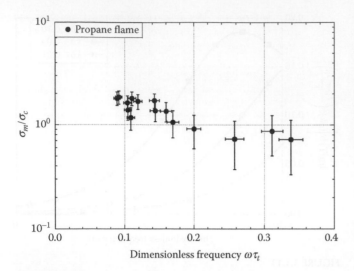

FIGURE 5.1.15
Comparison between the measured growth rate, σ_m, and the growth rate calculated from Equation 5.1.22, σ_c, for propane-air flames in a 1 m tube according to Ref. [46].

We can now estimate the order of magnitude of the acoustic growth rate expected for this mechanism. For a lean methane-air flame with a typical laminar flame velocity $S_L \approx 0.3$ m/s, the burnt gas temperature is $T_b \approx$ 2095 K. For saturated Darrieus–Landau cells, the aspect ratio $a/\lambda \approx 0.15$ (see Figure 5.1.8), and the observed wavelength λ is about 4 cm [46] so $(ak)^2 \approx 1$ and $kd \approx 0.01$. In the same 1 m tube as before, the reduced fundamental frequency is $\omega\tau_t = 0.16$, and so $\text{Im}[\text{Tr}(k\delta, \omega\tau_t)] \approx 0.21$, and we can take $F \approx 0.5$. From Equation 5.1.22, the expected growth rate is found to be:

$$\sigma = 35\ \text{s}^{-1} \qquad (5.1.26)$$

This is a strong instability. The corresponding growth rate per acoustic cycle is 35% at 100 Hz. This estimation is quite compatible with the growth rates measured by Clanet et al. [46]. The results of a comparison between the measured growth rate, σ_m, and the calculated growth rate, σ_c, for premixed propane flames is shown in Figure 5.1.15 (redrawn from [46]). The agreement is surprisingly good, considering that the theory was derived for 2D sinusoidal wrinkling, and that the real flame had 3D cusped cells.

5.1.3.3 Convective Coupling

Flames submitted to convective disturbances experience geometrical variations, which can in turn give rise to heat release unsteadiness. This process can be examined by considering different types of interactions between incident velocity or equivalence ratio modulations and combustion. The flame dynamics resulting from these interactions give rise to sound radiation and

again this eventually feeds energy in the acoustic modes of the system. A feedback loop is established between the flow, the combustion process, and the resonant acoustic modes, and under certain conditions, the perturbations are amplified leading to combustion oscillations. This process designated in the literature as combustion instabilities or thermo-acoustic instabilities observed in many practical devices has been the subject of considerable attention because it perturbs the normal operation of the system and in extreme cases leads to failure. The dynamics of perturbed flames and its relation with thermo-acoustic instabilities is the subject of the next chapter.

References

1. G. Darrieus. Propagation d'un front de flamme. Unpublished work presented at La Technique Moderne (1938), and at Le Congrès de Mécanique Appliquée (1945), 1938.
2. L. Landau. On the theory of slow combustion. *Acta Physicochimica URSS*, 19:77–85, 1944.
3. P. Clavin and F.A. Williams. Theory of premixed-flame propagation in large-scale turbulence. *Journal of Fluid Mechanics*, 90(pt 3):589–604, 1979.
4. F.A. Williams. *Combustion Theory*. 2nd ed., Benjamin/Cummings, Menlo Park, CA, 1985.
5. Ya.B. Zel'dovich, G.I. Barenblatt, V.B. Librovich, and G.M. Makhviladze. *The Mathematical Theory of Combustion and Explosions*. Plenum, New York, 1985.
6. Ya.B. Zel'dovich and D.A Frank-Kamenetskii. A theory of thermal flame propagation. *Acta Physicochimica URSS*, IX:341–350, 1938.
7. G.H. Markstein. Instability phenomena in combustion waves. *Proceedings of the Combustion Institute*, 4:44–59, 1952.
8. P. Clavin and F.A. Williams. Effects of molecular diffusion and of thermal expansion on the structure and dynamics of premixed flames in turbulent flows of large scale and low intensity. *Journal of Fluid Mechanics*, 116:251–282, 1982.
9. S.G. Davis, J. Quinard, and G. Searby. Markstein numbers in counterflow, methane- and propane-air flames: A computational study. *Combustion and Flame*, 130:123–136, 2002.
10. G.H. Markstein. *Nonsteady Flame Propagation*. Pergamon, New York, 1964.
11. G. Searby and P. Clavin. Weakly turbulent wrinkled flames in premixed gases. *Combustion Science and Technology*, 46:167–193, 1986.
12. P. Clavin. Dynamic behaviour of premixed flame fronts in laminar and turbulent flows. *Progress in Energy and Combustion Science*, 11:1–59, 1985.
13. M. Matalon and B.J. Matkowsky. Flames as gas dynamic discontinuities. *Journal of Fluid Mechanics*, 124:239–259, 1982.
14. M.L. Frankel and G.I. Sivashinsky. On the effects due to thermal expansion and Lewis number in spherical flame propagation. *Combustion Science and Technology*, 31:131–138, 1983.

15. P. Clavin and P. Garcia. The influence of the temperature dependence of diffusivities on the dynamics of flame fronts. *Journal de Mécanique Théorique et Appliquée*, 2(2):245–263, 1983.

16. J. Quinard, G. Searby, and L. Boyer. Stability limits and critical size of structures in premixed flames. *Progress in Astronautics and Aeronautics*, 95:129–141, 1985.

17. G. Searby and J. Quinard. Direct and indirect measurements of Markstein numbers of premixed flames. *Combustion and Flame*, 82(3–4):298–311, 1990.

18. J. Quinard. Limites de stabilité et structures cellulaires dans les flammes de prémélange. PhD thesis, Université de Provence, Marseille, December 1984.

19. C. Clanet and G. Searby. First experimental study of the Darrieus-Landau instability. *Physical Review Letters*, 80(17):3867–3870, 1998.

20. G. Searby, J.M. Truffaut, and G. Joulin. Comparison of experiments and a non-linear model for spatially developing flame instability. *Physics of Fluids*, 13:3270–3276, 2001.

21. J.M. Truffaut. Etude expérimentale de l'origine du bruit émis par les flammes de chalumeaux. University thesis, Université d'Aix-Marseille I, 1998.

22. J.M. Truffaut and G. Searby. Experimental study of the Darrieus-Landau instability on an inverted-'V' flame and measurement of the Markstein number. *Combustion Science and Technology*, 149:35–52, 1999.

23. J.W.S. Rayleigh. The explanation of certain acoustical phenomena. *Nature*, 18:319–321, 1878.

24. L. Crocco and S. Cheng. *Theory of Combustion Instability in Liquid Propellant Rocket Motors*. Butterworths, London, 1956.

25. D.T. Harrje and F.H. Reardon. Liquid propellant rocket combustion instability. Technical Report SP-194, NASA, Washington, DC, 1972.

26. F.E.C. Culick. Combustion instabilities in liquid-fueled propulsion systems. an overview. *AGARD Conference Proceedings Combustion Instabilities in Liquid Fuelled Propulsion Systems*, 450, pp. 1.1–1.73. NATO, 1988.

27. V. Yang and A. Anderson. Liquid rocket engine combustion instability, volume 169 of *Progress in Astronautics and Aeronautics*. AIAA, Washington DC, 1995.

28. W. Krebs, P. Flohr, B. Prade, and S. Hoffmann. Thermoacoustic stability chart for high intense gas turbine combustion systems. *Combustion Science and Technology*, 174:99–128, 2002.

29. R.A. Dunlap. Resonance of flames in a parallel-walled combustion chamber. Technical Report Project MX833, Report UMM-43, Aeronautical Research Center. University of Michigan, 1950.

30. A. van Harten, A.K. Kapila, and B. J. Matkowsky. Acoustic coupling of flames. *SIAM Journal on Applied Mathematics*, 44(5):982–995, 1984. doi: 10.1137/0144069. URL http://link.aip.org/link/?SMM/44/982/1.

31. P. Clavin, P. Pelcé, and L. He. One-dimensional vibratory instability of planar flames propagating in tubes. *Journal of Fluid Mechanics*, 216:299–322, 1990.

32. A.C. McIntosh. Pressure disturbances of different length scales interacting with conventional flames. *Combustion Science and Technology*, 75:287–309, 1991.

33. A.C. McIntosh. The linearised response of the mass burning rate of a premixed flame to rapid pressure changes. *Combustion Science and Technology*, 91:329–346, 1993.

34. A.C. McIntosh. Deflagration fronts and compressibility. *Philosphical Transactions of the Royal Society of London A*, 357:3523–3538, 1999.

35. A.A. Putnam and R.D. Williams. Organ pipe oscillations in a flame filled tube. *Proceedings of the Combustion Institute*, 4:556–575, 1952.

36. G.H. Markstein. Flames as amplifiers of fluid mechanical disturbances. In *Proceeding of the sixth National congress for Applied Mechanics*, Cambridge, MA, pp. 11–33, 1970.

37. G. Searby and D. Rochwerger. A parametric acoustic instability in premixed flames. *Journal of Fluid Mechanics*, 231:529–543, 1991.

38. P. Pelcé and D. Rochwerger. Vibratory instability of cellular flames propagating in tubes. *Journal of Fluid Mechanics*, 239:293–307, 1992.

39. T. Poinsot, A. Trouvé, D. Veynante, S. Candel, and E. Esposito. Vortex driven acoustically coupled combustion instabilities. *Journal of Fluid Mechanics*, 177:265–292, 1987.

40. D. Durox, T. Schuller, and S. Candel. Self-induced instability of premixed jet flame impinging on a plate. *Proceedings of the Combustion Institute*, 29:69–75, 2002.

41. S. Ducruix, D. Durox, and S. Candel. Theoretical and experimental determination of the transfer function of a laminar premixed flame. *Proceedings of the Combustion Institute*, 28:765–773, 2000.

42. T. Schuller, D. Durox, and S. Candel. Self-induced combustion oscillations of laminar premixed flames stabilized on annular burners. *Combustion and Flame*, 135:525–537, 2003.

43. P. Clavin and J. Sun. Theory of acoustic instabilities of planar flames propagating in sprays or particle-laden gases. *Combustion Science and Technology*, 78:265–288, 1991.

44. J.D. Buckmaster and P. Clavin. An acoustic instability theory for particle-cloud flames. *Proceedings of the Combustion Institute*, 24:29–36, 1992.

45. T.C. Lieuwen, Y. Neumeier, and B.T. Zinn. The role of unmixedness and chemical kinetics in driving combustion instabilities in lean premixed combustors. *Combustion Science and Technology*, 135:193–211, 1998.

46. C. Clanet, G. Searby, and P. Clavin. Primary acoustic instability of flames propagating in tubes : Cases of spray and premixed gas combustion. *Journal of Fluid Mechanics*, 385:157–197, 1999.

47. P. Clavin and G. Searby. Unsteady response of chain-branching premixed-flames to pressure waves. *Combustion Theory and Modelling*, 12(3):545–567, 2008.

48. A. Wangher, G. Searby, and J. Quinard. Experimental investigation of the unsteady response of a flame front to pressure waves. *Combustion and Flame*, 154(1-2):310–318, 2008.

49. G. Searby. Acoustic instability in premixed flames. *Combustion Science and Technology*, 81:221–231, 1992.

50. B.V. Rauschenbakh. *Vibrational Combustion*. Fizmatgiz, Mir, Moscow, 1961.

5.2 Perturbed Flame Dynamics and Thermo-Acoustic Instabilities

Sébastien Candel, Daniel Durox, and Thierry Schuller

5.2.1 Introduction

Flame dynamics is intimately related to combustion instability and noise radiation. In this chapter, relationships between these different processes are described by making use of systematic experiments in which laminar flames respond to incident perturbations. The response to incoming disturbances is examined and expressions of the radiated pressure are compared with the measurements of heat release rate in the flame. The data indicate that flame dynamics determines the radiation of sound from flames. Links between combustion noise and combustion instabilities are drawn on this basis. These two aspects, usually treated separately, appear as manifestations of the same dynamical process.

It is known that under normal conditions, combustion processes generate broadband incoherent noise characterized by radiation of a "combustion roar." Under unstable operation, a feedback is established between sound sources in the flame and perturbations propagating in the flow [1]. Sound radiation becomes coherent as it is tuned on one of the resonance frequencies of the system. The process generally involves a periodic disturbance of the flow inducing a cyclic motion of the flame and unsteady rates of heat release. Combustion instabilities have been extensively investigated mainly in relation with the development of high-performance combustion systems like liquid rocket engines [2–4] and gas turbine combustors [5,6], but they also occur in industrial processes and domestic boilers [7]. The sound level under unstable operation reaches excessive amplitudes heat fluxes to the walls are enhanced, and vibrations induce structural fatigue affecting the integrity of the system and often leading to failure [8]. Intensification of combustion resulting from acoustically coupled instability is illustrated in Figure 5.2.1, which shows a view of a multiple inlet premixed combustor under stable and unstable operations. The oscillation enhances in this case the luminosity, indicating that the flame is more compact, and that combustion takes place near the chamber backwall augmenting the heat flux to this boundary [9].

Combustion dynamics phenomena have important practical consequences and their prediction, alleviation, and reduction constitute technological challenges [10–12]. This requires an understanding of (1) the driving mechanisms that feed energy into the wave motion and (2) the acoustic coupling mechanisms that close the feedback loop.

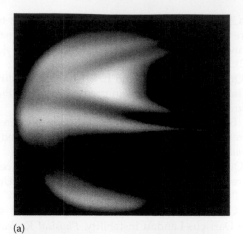

(a)

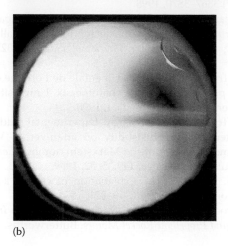

(b)

FIGURE 5.2.1
Light emission from a multiple inlet combustor under (a) stable and (b) unstable operation. (From Poinsot, T.J., Trouvé, A.C., Veynante, D.P., Candel, S.M., and Esposito, E.J., *J. Fluid Mech.*, 177, 265, 1987. With permission.)

The analysis of combustion dynamics is then intimately linked to an understanding of perturbed flame dynamics, the subsequent generation of unsteady rates of heat release, and the associated radiation of sound and resulting acoustic feedback. In practical configurations, the resonance loop involves the flow, the combustion process, and the acoustic modes of the system as represented schematically in Figure 5.2.2.

Many items in this coupled process have been extensively explored in recent years [13–23]. The numerical simulation of combustion instabilities has also progressed quite remarkably and this topic is covered in Ref. [12] and

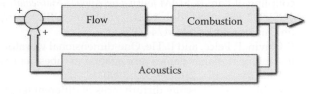

FIGURE 5.2.2
Block diagram of processes involved in thermo-acoustic instabilities.

reviewed in Ref. [24]. Systematic experiments discussed in this chapter serve to identify some of the mechanisms driving instabilities. It is shown that rapid changes in flame surface area constitute a powerful mechanism, which can be at the origin of processes driving instability. This chapter begins with a short review of combustion noise theory. The experimental setup described in the Section 5.2.3 is used in Section 5.2.4 to explore various types of flame dynamics. In these experiments, an external modulation is imposed to allow conditional (phase-locked) analysis of the motion and associated sound pressure field. Visualizations of flame dynamics described in Section 5.2.4 are interpreted in Section 5.2.5 by a time-trace analysis of heat release and radiated pressure field. This serves to identify the processes giving rise to sound radiation from the flame, link sound production and heat release perturbations, and extract common features and differences between the various configurations. While the main part of this chapter is concerned with acoustically perturbed flames, there are other possibilities. It is known, for example, that some of the combustion instabilities observed in gas turbines are due to interactions between equivalence ratio perturbations and combustion [25]. Composition of the fresh mixture is modulated by pressure perturbations propagating to the injection manifold. The composition waves induced in this process then travel to the flame and induce heat release perturbations. This aspect is briefly considered in Section 5.2.6. Numerical simulations are used in this case to illustrate a process that is not easily demonstrated in experiments.

5.2.2 Sound Radiation from Flames and Combustion Acoustics

A classical investigation of combustion noise was carried out by Thomas and Williams [26]. The principle of this clever experiment was to fill soap bubbles with a reactive mixture and record the pressure field radiated by the burning of each isolated bubble. It was found that the pressure signal could be described as the sound generated by a monopole source of strength $d\Delta V/dt$, where ΔV represented the volume increase due to thermal expansion of the gases crossing the reactive front. In this early model, the far-field pressure signal p' was expressed in terms of the volume acceleration induced by nonsteady combustion:

$$p'(r,t) = \frac{\rho_0}{4\pi r} \frac{d^2 \Delta V}{dt^2} \qquad (5.2.1)$$

where
  ρ_0 is the far-field air density
  r designates the distance separating the compact flame and the observation point

According to Equation 5.2.1, combustion noise results from the fluid expansion determined by the heat release rate. It was argued that because this source is isotropic, there is no preferential direction of radiation. These experiments carried out in the laminar case support some earlier theoretical investigations of turbulent combustion noise. In the theory established by Bragg [27], it was postulated that a turbulent flame behaves as a net monopole radiator. An experimental validation of this theory was carried out by Hurle et al. [28]. Starting from Equation 5.2.1 and assuming that a turbulent flame is acoustically equivalent to a distribution of monopole sources of sound with different strengths and frequencies distributed throughout the reaction zone [29], it was possible to express the far-field sound pressure p' radiated by the turbulent flame as a function of the volumetric rate of consumption of reactants q by the flame [28,30]:

$$p'(r,t) = \frac{\rho_0}{4\pi r}\left(\frac{\rho_u}{\rho_b} - 1\right)\left[\frac{dq}{dt}\right]_{t-\tau} \qquad (5.2.2)$$

where
  ρ_u/ρ_b designates the volumetric expansion ratio of burnt to unburnt gases
  τ is the acoustic propagation time from the combustion region to the measurement point r

In Equation 5.2.2, it is assumed that acoustic wavelengths λ are large when compared with any of the characteristic scales of the flow ($\lambda \gg L$) and that the measurement point r is far from the source region ($r \gg \lambda$). The previous expression provides the sound pressure in the far-field for a compact source, but it can be used indifferently for premixed or nonpremixed flames [30].

The previous expression can also be derived by starting from the wave equation for the pressure in the presence of a distribution of heat release:

$$\nabla^2 p' - \frac{1}{\overline{c}^2}\frac{\partial^2 p'}{\partial t^2} = -\frac{1}{c_0^2}\frac{\partial}{\partial t}\left[(\gamma - 1)\dot{Q}'\right] \qquad (5.2.3)$$

where
  $\dot{Q}'$ is the rate of heat release per unit volume
  γ designates the specific heat ratio
  c_0 is the speed of sound in the ambient medium surrounding the flame.

Assuming that combustion takes place in an unconfined domain, the radiated pressure is given by

$$p'(r,t) = \frac{\gamma-1}{4\pi c_0^2}\int_v \frac{1}{|r-r_0|}\frac{\partial}{\partial t}\dot{Q}'\left(r_0, t - \frac{|r-r_0|}{c_0}\right)dV(r_0) \qquad (5.2.4)$$

If the observation point is in the far-field and if the source region is compact, the previous expression becomes

$$p'(r,t) = \frac{\gamma-1}{4\pi c_0^2 r}\frac{\partial}{\partial t}\int_v \dot{Q}'\left(r_0, t - \frac{r}{c_0}\right)dV(r_0) \quad (5.2.5)$$

This expression is close to that derived by Strahle [31,32]. It can be shown to yield the classical Equation 5.2.2 in the case of a premixed flame. Assuming that the flame is isobaric, one has $\rho_u/\rho_b = T_b/T_u$. Using $\rho_0 c_0^2 = \gamma\bar{p}$ and the fact that $c_p(T_b - T_u)$ represents the heat released per unit mass of premixed reactants, one finds that

$$\rho_0\left(\frac{\rho_u}{\rho_b}-1\right)q = \rho_0\left(\frac{T_b}{T_u}-1\right)q = \frac{\gamma-1}{c_0^2}\int\dot{Q}'dV \quad (5.2.6)$$

and Equations 5.2.2 and 5.2.5 exactly match. The last result is, however, more general than the classical expression because it applies to any type of flame (premixed, nonpremixed, or partially premixed).

In the premixed case and for lean conditions (equivalence ratio less than 1), the volumetric rate of reactants consumption q can be estimated from the light emission intensity I of excited radicals like C_2^* or CH^* [28,33] and OH^* [34] in the reaction zone. This can be used effectively to measure the volumetric rate of reactants consumption:

$$q = kI \quad (5.2.7)$$

The coefficient k depends on the fuel, free radical observed, combustion regime, flame shape and type, and experimental setup. This relation is useful to study combustion noise for a homogeneous reactive mixture submitted to flow perturbations. Concerning inhomogeneous mixtures, the relation between the radiated pressure and the flame chemiluminescence is not as straightforward as in the premixed case and one cannot deduce the rate of reaction from emission from excited free radicals. Considering only premixed flames, the linearity coefficient k is determined from separate experiments by plotting the mean emission intensity I versus the mean flow rate, all other parameters remaining fixed. Combining Equations 5.2.2 and 5.2.7, the radiated far-field sound pressure field can be linked to the light emitted from the turbulent combustion region:

$$p'(r,t) = \frac{\rho_0}{4\pi r}\left(\frac{\rho_u}{\rho_b}-1\right)k\left[\frac{dI}{dt}\right]_{t-\tau} \quad (5.2.8)$$

This expression was checked in many early experiments on premixed flames with a fixed equivalence ratio. Light

intensity emitted by free radical is often used to get an estimate of the heat release rate [28]. Assuming that fuel and oxidizer react in stoichiometric proportions, it was also successfully applied to turbulent diffusion flames by keeping a constant global mixture ratio [30].

Considering the case of premixed flames, it is noted by Thomas and Williams [26] that sound radiation can be related to the rate of change of the flame surface area by assuming that the burning velocity is constant. Similarly, Ref. [35] and later Ref. [36] indicate that in the wrinkled flame regime, the rate of chemical conversion is directly linked to the flame surface area $A(t)$. For a mixture of fresh reactants at a constant equivalence ratio, the pressure field is directly linked to the instantaneous flame surface:

$$p'(r,t) = \frac{\rho_0}{4\pi r}\left(\frac{\rho_u}{\rho_b}-1\right)\left[\frac{d(S_L A)}{dt}\right]_{t-\tau} \quad (5.2.9)$$

Assuming a constant laminar burning velocity, one may extract the pressure signal from a direct measurement of the wrinkled flame surface area. Alternatively, this can be accomplished by measuring light emission from free radicals in the flame, deducing from this measurement the heat release rate and determining the resulting pressure field from Equation 5.2.8. Such direct comparisons are carried out in the next sections for a variety of flame configurations. At this point, one should mention that the laminar burning velocity is not constant but depends on the local strain rate and curvature. It is shown theoretically that these effects act in combination in the form of flame stretch [37–39] and that this may induce some additional radiation of sound.

5.2.3 Experimental Setup

The experimental setup sketched in Figure 5.2.3 comprises a burner with a $d = 22$ mm nozzle exit diameter and a driver unit (loudspeaker) fixed at its base. The burner body is a cylindrical tube of 65 mm inner diameter containing a set of grids and a honeycomb followed by a convergent nozzle with an area contraction ratio of $\sigma = 9{:}1$.

This system produces a steady laminar flow with a flat velocity profile at the burner exit for mean flow velocities up to 5 m/s. Velocity fluctuations at the burner outlet are reduced to low levels as $v_{rms}/\bar{v} < 0.01$ on the central axis for free jet injection conditions. The burner is fed with a mixture of methane and air. Experiments-described in what follows are carried out at fixed equivalence ratios. Flow perturbations are produced by the loudspeaker driven by an amplifier, which is fed by a sinusoidal signal synthesizer. Velocity perturbations measured by laser doppler velocimetry (LDV) on the burner symmetry axis above the nozzle exit plane are also purely sinusoidal and their spectral

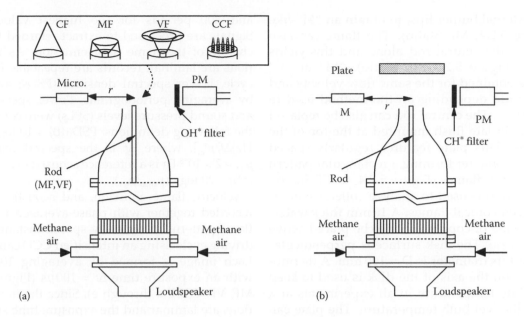

FIGURE 5.2.3
(a) Experimental setup used for conical flame (CF), fountain flame (MF), inverted conical flame (VF), and multipoint injection conical flame (CCF). (b) Experimental setup for flame plate interactions. PM: Photomultiplier equipped with a CH* filter. M: Microphone at $r = 24 - 30$ cm away from the burner axis.

density features a single peak at the driving frequency with a very low harmonic level.

For the range of flow velocities and the two equivalence ratios considered in what follows, the flames are naturally anchored on the burner lips and take a conical shape (Figure 5.2.4, CF Statio). The burner

can also be equipped with a 2 mm diameter cylindrical rod placed inside the convergent unit and centered on the axis providing an additional anchoring region near the burner axis. It is then possible to stabilize the flame on the burner lip to produce a conical flame (CF) or simultaneously on the central rod

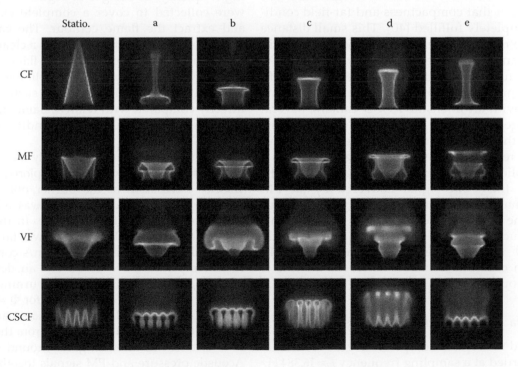

FIGURE 5.2.4
Instantaneous flame images. Statio.: Stationary shapes. a–e: Images taken at equi-spaced instants during a cycle of excitation. CF, conical flame; MF, "M"-shaped flame; VF, "V"-shaped flame; CSCF, collection of small conical flames.

and on the external burner lips, to obtain an "M"-like shape (Figure 5.2.4, MF Statio). The flame can also be attached to the central rod alone and this yields a "V"-shape (Figure 5.2.4, VF Statio). "M" or "V" flames can be obtained for the same flow velocity and equivalence ratio depending on the method used to ignite the system. The central rod can also be replaced by a perforated plate flush mounted at the top of the burner nozzle. This plate features regularly spaced holes of 2 mm diameter forming a rectangular pattern of 25 small conical flames (Figure 5.2.4, CSCF Statio). This configuration is used to analyze collective interactions between conical flames. A 10 mm thick water-cooled disk made of copper can also be placed above the burner nozzle. The disk surface is perpendicular to the axis and its diameter is $D = 100$ mm. A thermocouple placed on the axis of the disk is used to keep a constant plate temperature in all experiments at a value above the wet bulb temperature. The plate can be moved vertically above the burner.

All the flames described previously also feature self-sustained oscillations in the absence of external forcing, but these autonomous dynamical regimes will not be studied in what follows. During external modulation of the flow, the flame oscillates and responds by emitting noise. Radiated sound is measured with the microphone M placed at a distance $T_\infty = 24$–30 cm away from the burner axis depending on the case studied. This distance is not very large when compared with the typical dimension of the reactive region and one cannot consider that compactness and far-field conditions are completely fulfilled [40]. This small distance is adopted to minimize the level of reflections on solid boundaries and on the neighboring equipments. As will be shown later, the observation distance is sufficient for the present purpose. The motion of the flame is recorded with an intensified CCD camera. Several phase-averaged snapshots are gathered to examine the evolution of the flame during a complete cycle of oscillation. Heat release fluctuations are measured with a photomultiplier PM equipped with a CH* or an OH* narrowband filter. The signal delivered by the PM is proportional to the light emitted I by free radicals present in the flame reaction zone. For homogeneous mixtures, fluctuations of this current are directly proportional to variations of heat release [40,41]. This relation ceases to be valid when the mixture is inhomogeneous.

5.2.3.1 Data Acquisition and Processing

LDV, PM, and microphone output voltages are simultaneously recorded at a sampling frequency $f_a = 16,384$ Hz during a period of 2 s. It is thus possible to record at least 100 periods for the lowest driving frequency $f_e = 50$ Hz

and 800 periods for the highest value $f_e = 400$ Hz. Signals are processed to extract retarded time rates of change of the flame light emission. As flow interactions are laminar, records are repeatable from cycle to cycle. Power spectral densities (PSDs) are calculated by averaging periodograms. Power spectral densities and sound pressure levels (SPLs) were computed using the following definitions: PSD(dB) = $10 \log_{10}$ [PSD(Pa²/Hz)$\Delta f/p_{ref}^2$], where Δf is the spectral resolution and $p_{ref} = 2 \times 10^{-5}$ Pa is a reference acoustic pressure and SPL (dB) = $20 \log_{10} (p_{rms}/p_{ref})$.

Velocity, flame emission, and acoustic signals were recorded together with phase-averaged images of the flame patterns at regularly spaced instants during the driving cycle using an intensified CCD camera (ICCD). Each image is formed by averaging 100 snapshots with an exposure time $\Delta t = 100$ µs (Figure 5.2.4—CF, MF, VF, CSCF a through e). Since the flow configurations are laminar and the exposure time small enough when compared with the driving period $T_e = 1/f_e$ ($T_e = 2.5$–20 ms), each phase-averaged image can be considered to be an instantaneous flame pattern in the driving cycle. Cycle–to–cycle reproducibility can also be assessed by examining the image sharpness. Except for some blurring on the flame periphery of the stationary "V"-flame and neighboring inner reaction zones of the collection of small conical flames, other phase-averaged images are perfectly sharp and one may easily follow the flame contour. For each driving frequency, 21 images at regularly spaced phases were collected to cover a complete excitation cycle and extract the flame contour. The camera line of sight crosses a row of holes to get a clear image in the CSCF case. Except for this case, flame surface areas are determined by assuming cylindrical symmetry. The relative fluctuation of flame surface area can be directly compared with relative flame light variations and noise radiation. Operating conditions are summarized in Table 5.2.1.

It is not possible to obtain exactly identical flow conditions for the configurations explored. The level of velocity fluctuation at the burner outlet also differs in the various cases. This level was adjusted to get an acceptable signal-to-noise ratio. In the results presented here, the specific heat ratio was taken as equal to $\gamma = 1.4$, the sound speed $c_0 = 343$ m/s corresponds to a room temperature $T = 293$ K. The air density is taken equal to $\rho_\infty = 1.205$ kg/m. Laminar burning velocities are taken from Ref. [42], $S_L = 0.38$ m/s for $\Phi = 1.11$ and $S_L = 0.34$ m/s for $\Phi = 0.95$. The volumetric expansion ratio $E = \rho_u/\rho_b = T_b/T_u = 7.4$ was estimated from the temperature ratio of burnt to unburnt gases around stoichiometry. Acoustic pressure and PM signals together with flame surface area time-traces extracted from the images are analyzed in what follows.

TABLE 5.2.1

Summary of Operating Conditions

	Φ	f_e (Hz)	SPL (dB)	$\bar{A}$ (m²)	A_{rms} (m²)	v (m/s)	v_{rms} (m/s)
CF	1.11	50	73	1.8×10^{-3}	1.9×10^{-4}	1.7	0.8
MF	1.11	200	92	2.1×10^{-3}	4.0×10^{-4}	2.3	0.4
VF	1.11	100	96	2.1×10^{-3}	9.0×10^{-4}	2.3	0.6
CSCF	0.95	400	96	—	—	5.0	0.9

Note: Φ, equivalence ratio; f_e, modulation frequency; SPL, sound pressure level; $\bar{A}$, mean flame surface area; A_{rms}, fluctuation of flame area; $\bar{v}$, mean velocity at the burner outlet; and v_{rms}, imposed velocity fluctuation level.

5.2.4 Perturbed Flame Dynamics

In the absence of modulation, the flames spread in a stable fashion. Except for the "V"-flame, there is no visible motion of the front. In the "V"-flame case, one observes a small fluctuation of the flame limited to its end region. This can be attributed to interactions with small-scale vortices convected in the mixing layer between the premixed jet and the surrounding air. Without combustion, the sound level is well below 65 dB in the laboratory. In the "V"-flame case, the sound level is slightly louder with an overall value of 65 dB. The sound levels measured in the absence of external modulation are relatively low and the flames are quiet. The situation is quite different in the presence of external modulation. The flames respond with a periodic motion accompanied by a strong noise emission. A qualitative analysis combining a description of the flame patterns at successive phases over a cycle of modulation and the corresponding sound spectra is now carried out to understand the dominant processes leading to sound emission.

5.2.4.1 Conical Flame Dynamics

Under low-frequency excitation, the flame front is wrinkled by velocity modulations (Fig. 5.2.5). The number of undulations is directly linked to frequency. This is true as far as the frequency remains low (in this experiment, between 30 and 400 Hz). The flame deformation is created by hydrodynamic perturbations initiated at the base of the flame and convected along the front. When the velocity modulation amplitude is low, the undulations are sinusoidal and weakly damped as they proceed to the top of the flame. When the modulation amplitude is augmented, a toroidal vortex is generated at the burner outlet and the flame front rolls over the vortex near the burner base. Consumption is fast enough to suppress further winding by the structure as it is convected away from the outlet. This yields a cusp formed toward burnt gases. This process requires some duration and it is obtained when the flame extends over a sufficient axial distance. If the acoustic modulation level remain low (typically: $v'/v < 20\%$),

the flames oscillate at the modulation frequency. The wave crests move at a velocity $v \cos \alpha$.

5.2.4.2 Flame–Wall Interactions

The flame–plate interaction (FP) is investigated at an equivalence ratio $\Phi = 0.95$, a mixture flow velocity $\bar{v} = 1.20$ m/s, an rms modulation level fixed to $v' = 0.36$ m/s, and a modulation frequency $f = 200$ Hz. The velocity perturbation at the burner outlet wrinkles the flame front at the burner lips (Figure 5.2.6a, image at the top of the column), and this perturbation is convected by the flow toward the plate (Figure 5.2.6a, middle images). Two wrinkles following each other are visible in this case. As these perturbations approach the boundary, the flame surface extends along the plate and it increases smoothly up to a maximum value corresponding to the third image in the column (Figure 5.2.6a). In the next image (Figure 5.2.6a, last in the column), a large portion of the flame front has disappeared and the available

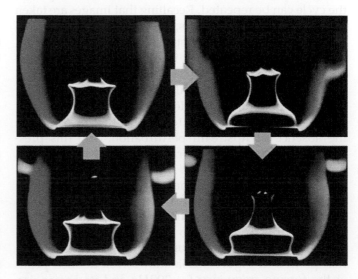

FIGURE 5.2.5

Methane-air premixed flame modulated by acoustic perturbations. Four color Schlieren visualization. $f = 75$ Hz, $\omega_* = 15$, $v'/\bar{v} = 0.2$, and $\Phi = 0.95$.

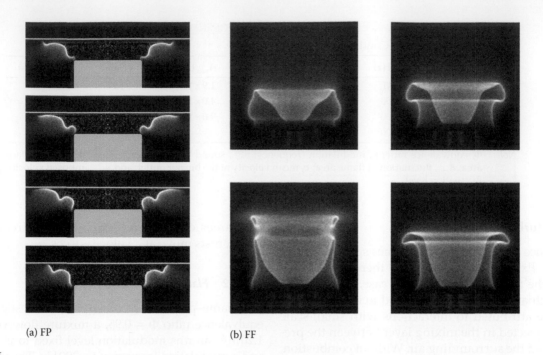

(a) FP   (b) FF

FIGURE 5.2.6

Cyclic flame motions during (a) flame–plate interaction, $\Phi = 0.95$, $v = 1.20\,\text{m/s}$, $f = 200\,\text{Hz}$, and $v' = 0.36\,\text{m/s}$ (a cycle begins with the top image); and (b) flame–flame interaction, $\Phi = 1.13$, $v = 1.71\,\text{m/s}$, $f = 150\,\text{Hz}$, and $v' = 0.50\,\text{m/s}$ (a cycle begins with the top left image). (From Candel, S., Durox, D., and Schuller, T., Flame interactions as a source of noise and combustion instabilities, AIAA paper 2004–2928, *10th AIAA/CEAS Aeroacoustics Conference*, Manchester, U.K., May 2004. With permission.)

flame surface area is reduced with respect to that available at the previous instant. This indicates that a sudden reduction of flame surface area has occurred accompanied by a large negative rate of change of heat release. The flame is extinguished by thermal losses at the cold boundary, which is water cooled and maintained at $T \sim 300\,\text{K}$ in its center [40,48]. The flame in the last image in Figure 5.2.6a has nearly retrieved its initial shape and the cycle can be repeated. Recalling that images are taken at regularly spaced instants, this suggests that the rate of change of heat release takes higher values during phases of flame destruction in the cycle than during the smooth increase in the flame surface area. The motion is periodic at a frequency equal to the forcing frequency f.

The overall noise recorded by microphone M now reaches 84 dB, i.e., about 20 dB above that measured in the absence of upstream perturbation. The rms pressure fluctuation in the laboratory is more than 10 times greater when the flame is modulated. Repeating this experiment, but without the plate or without combustion, the noise level falls to a value close to sound level without modulation (~60 dB), implying that the noise source is directly related to the flame–plate interaction [40]. The noise spectrum corresponding to this interaction plotted in Figure 5.2.8a shows well-defined peaks at the forcing frequency $f = 200\,\text{Hz}$ and its harmonics. The harmonic power content is nonnegligible and for some cases can exceed that concentrated at the fundamental forcing frequency. In some cases not described here, peaks at half the fundamental frequency are also

visible [40]. If pressure fluctuations have the same period as velocity fluctuations at the burner outlet, the latter are purely harmonic, whereas pressure fluctuations have a richer harmonic content. This suggests a strong nonlinear process in the transfer of energy between upstream velocity fluctuations and radiated sound.

These data indicate that thermal losses during unsteady flame–wall interactions constitute an intense source of combustion noise. This is exemplified in other cases where extinctions result from large coherent structures impacting on solid boundaries, or when a turbulent flame is stabilized close to a wall and impinges on the boundary. However, in many cases, the flame is stabilized away from the boundaries and this mechanism may not be operational.

5.2.4.3 Flame–Flame Interactions and Mutual Annihilation of Flame Area

The "M"-flame case shows a different kind of flame interaction illustrated in Figure 5.2.4 (MF). The "M"-shape comprises two reactive sheets separated by fresh reactants. This gives rise to flame–flame interactions between neighboring branches of the "M"-shape [41]. The case presented corresponds to an equivalence ratio $\Phi = 1.13$, a mixture flow velocity $\bar{v}/\bar{v} = 1.13\,\text{m/s}$, a modulation level fixed to $v' = 0.50\,\text{m/s}$, and a modulation frequency $f = 150\,\text{Hz}$. The description of the flame motion over a cycle of excitation starts as in the flame–plate interaction. A velocity perturbation is generated at

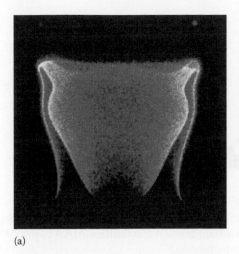
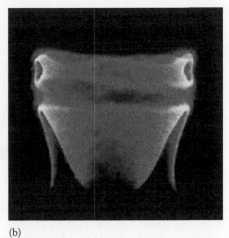

(a)   (b)

FIGURE 5.2.7
False color images of adjacent flame interactions in the FF configuration. A torus of fresh reactants is formed in this case.

the burner lips and produces a deformation of the flame front at the burner base (Figure 5.2.6b, top-left image). The perturbation mostly affects the outer branch of the "M"-flame. It is then convected by the mean flow toward the top of the flame (Figure 5.2.6b, top-right image). As the deformation travels along the flame front, the two branches of the "M" are stretched in the vertical direction and get closer (Figure 5.2.6b, bottom-right image), up to an instant in the cycle where the flame surface area is maximum and two flame elements interact (Figure 5.2.6, bottom-left image). The outcome of this mutual annihilation depends on the spatial position of the first interaction. In some cases, pockets of fresh reactants may be trapped in a flame torus, while in other cases this will not occur [41,48]. The torus is quite visible in the images displayed in Figure 5.2.7.

For some operating conditions not shown here, up to two flame toruses can be produced. During interaction of these flame elements, the reactive front shape is strongly altered. As in the flame–plate situation, after the mutual interaction, the flame quickly retrieves its shape at the beginning of the cycle (Figure 5.2.6, top-left image). The short phase in the cycle during flame surface destruction produces a faster rate of change of flame surface area than the longer phase where flame surface is produced by stretch. The same mechanism operates as in the flame–plate interaction, except that flame surface destruction is produced by mutual annihilation of neighboring sheets.

The overall sound pressure level reaches 91 dB. The pressure spectrum in Figure 5.2.8b is quite similar to that associated with the flame–plate interaction in Figure 5.2.8a. The presence of harmonics of the fundamental frequency indicates that the pressure signal is also periodic with an oscillation frequency corresponding to the flame oscillation frequency, but that the flame response is nonlinear with a rich harmonic content. These energetic harmonics indicate that the

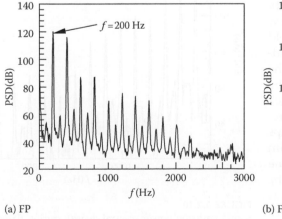
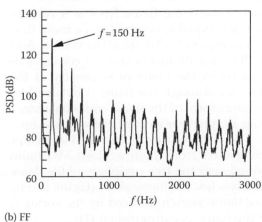

(a) FP   (b) FF

FIGURE 5.2.8
Sound emission power spectral density during (a) flame–plate interaction, $\Phi = 0.95$, $v = 1.20\,\text{m/s}$, $f = 200\,\text{Hz}$, and $v' = 0.36\,\text{m/s}$; and (b) flame–flame interaction, $\Phi = 1.13$, $v = 1.71\,\text{m/s}$, $f = 150\,\text{Hz}$, $v' = 0.50\,\text{m/s}$, and $r_\infty = 0.25\,\text{cm}$.

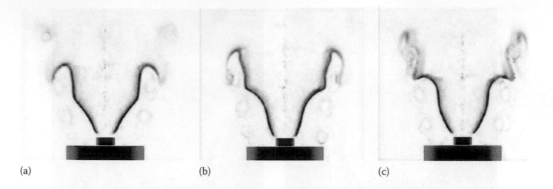

(a)   (b)   (c)

FIGURE 5.2.9
ICF flame motion during cyclic modulation of the flow: $\Phi = 0.8$, $\bar{v} = 1.87 = $ m/s, $f = 150$ Hz, and $v' = 0.15$ m/s. (Adapted from Candel, S., Durox, D., and Schuller, T., Flame interactions as a source of noise and combustion instabilities, AIAA paper 2004–2928, *10th AIAA/ CEAS Aeroacoustics Conference*, Manchester, U.K., May 2004. With permission.)

physical process at the origin of the noise involves a rapid change in the rate of heat release.

As mutual flame annihilation is believed to control and limit flame surface area in turbulent combustion, the previous results suggest that this mechanism could also be a source of intense noise radiation in turbulent combustors.

5.2.4.4 Flame Roll-Up by Vortex Structures

The third mechanism discussed in this chapter involves flame–vortex interactions. This is exemplified in the "V"-flame configuration (VF in Figure 5.2.4). The flame is stabilized on the central rod and extends freely from the edge of this body. Toroidal vortical structures are shed in the outer mixing layer of the jet exhausted by the burner. In this layer, atmospheric air is entrained by the inner jet of premixed reactants. Operating conditions for the case under consideration are $\Phi = 0.8$, $\bar{v} = 1.87$ m/s, $v' = 0.15$ m/s, and a forcing frequency $f = 150$ Hz. When a velocity perturbation is produced at the burner outlet, a vortical structure is shed from the rim and is convected toward the flame front at a velocity that is approximately equal to one-half of the mean flow velocity. Images in Figure 5.2.9 display the vortical field deduced from PIV measurements and the corresponding flame positions in the plane of symmetry of the burner [19] (the slice through the flame is obtained by taking the Abel transform of the flame emission image). Vortex shedding is synchronized by the acoustic forcing frequency. The vortices travel from the burner lips and roll up the reactive front (Figure 5.2.9a). Maximum roll-up is obtained near the free periphery of the flame when the vortex reaches the flame sheet (Figure 5.2.9b). At some points, flame stretch induced by the vortex is so strong that the flame is extinguished (Figure 5.2.9c). A large amount of flame surface vanishes during this interaction over a short period of time.

The overall sound radiated is 83 dB. The case examined in Figure 5.2.10 slightly differs from that documented in the images displayed in Figure 5.2.9. As for the two previous cases, the power spectral density also features harmonic components with a fundamental frequency corresponding to the modulation frequency (Figure 5.2.10), indicating that the mechanism at the origin of the noise production is strongly nonlinear.

5.2.5 Time-Trace Analysis

The qualitative analysis developed in the previous section is now complemented by examining correlations between the sound field and the flame dynamics. This is accomplished by comparing time traces of pressure fluctuation p' and light emission I signals recorded during the various experiments. Results are displayed in Figures 5.2.11 and 5.2.12, respectively, for the various

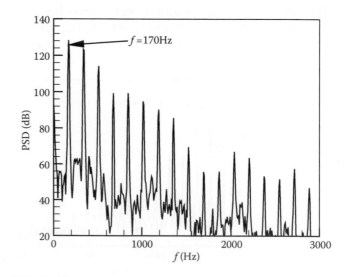

FIGURE 5.2.10
Sound emission power spectral density during flame–vortex interaction (ICF configuration): $\Phi = 0.92$, $v = 2.56$ m/s, $f = 170$ Hz, $v' = 0.30$ m/s, and $r_\infty = 24$ cm.

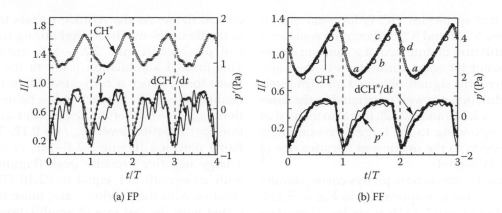

FIGURE 5.2.11

Time history of acoustic pressure and heat-release during (a) flame–plate interaction, $\Phi = 0.95$, $\bar{v} = 1.20\,\text{m/s}$, $f = 200\,\text{Hz}$, and $v' = 0.36\,\text{m/s}$; and (b) flame–flame interaction (FF), $\Phi = 1.13$, $v = 1.71\,\text{m/s}$, $f = 150\,\text{Hz}$, $v' = 0.50\,\text{m/s}$, and $r_\infty = 0.25\,\text{cm}$.

interactions. The relative intensity $I/\bar{I}$ of the CH* emission signal recorded by the PM is plotted at the top of these figures. The pressure fluctuations p recorded by microphone M are also plotted at the bottom of these figures with the time rate of change dCH*/dt. This latter signal is obtained by differentiating the CH* signal.

A delay equal to the acoustic time lag $\tau = r/c_0$ is also included to account for wave propagation over the distance r between the flame and the microphone M. The resulting signal is then scaled to get a signal dCH*/dt(t) with the same amplitude as the pressure fluctuation $p'(t+\tau)$. Owing to the combustion

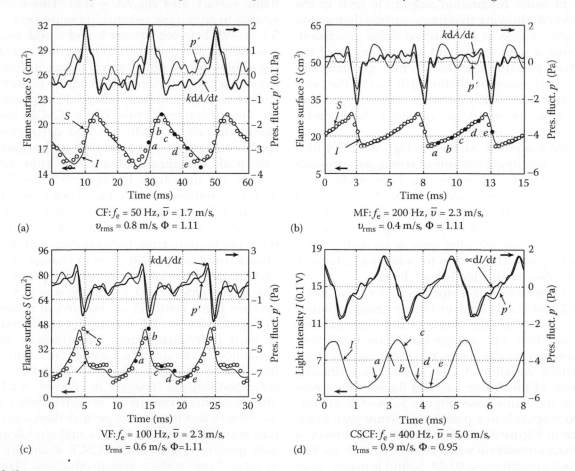

FIGURE 5.2.12

Time traces : OH* light intensity I, flame surface area A, pressure fluctuations p', and computed pressure fluctuations $k\,dA/dt$. Circles indicate extracted flame surface areas A in cm² (S and A are used indifferently to designate the flame surface). Black circles marked a, b, c, d correspond to flame patterns presented in images from Figure 5.2.3.

noise theory, the quantities $dCH^*/dt(t)$ and $p'(t+\tau)$ plotted in Figures 5.2.11 and 5.2.12 should be equal for a homogeneous mixture maintained at a constant equivalence ratio. Except for some differences in the high-frequency content, the signals obtained for high levels of velocity modulation are well correlated in the cases explored (Figures 5.2.11 and 5.2.12). It is also found that the phase corresponding to the largest pressure fluctuation corresponds to the maximum negative rate of change of the CH^* signal.

Concerning the FF interaction, phases corresponding to the conditioned images appearing in Figure 5.2.6b are also plotted in Figure 5.2.11b as circles. These data were obtained by directly measuring the relative flame surface area fluctuations $A/\bar{A}$ using the snapshots of Figure 5.2.6b and by assuming an axisymmetric configuration. During a cycle of oscillation, the flame light emission level increases slowly when the flame surface increases up to its maximum value. It then drops suddenly to its minimum value on a timescale shorter than that corresponding to flame surface increase. The maximum rate of decrease of the CH^* signal always fixes the pressure fluctuation amplitude level in the cases explored, showing that flame surface destruction is the major process in combustion noise generation. When coupled with a resonant acoustic mode of the burner, these pressure fluctuations may generate self-sustained combustion oscillations [19,41,43]. The same relation (Equation 5.2.2) between far-field pressure fluctuations and heat release variations holds under self-sustained oscillations.

Time trace analysis is pursued in Figure 5.2.12, which also aims at identifying instants of strong noise production. This is done by comparing the flame surface area extracted from the flame contours of the phase-averaged images at different instants in the driving cycle and the PM signal. For the sake of simplicity, light emission signals are superposed with the same scale on the flame surface area plots. This serves to point out instants of maximum noise production in each image.

The first interaction corresponds to a conical flame (Figure 5.2.4—CF, a through e). Under strong harmonic forcing $v_{rms}/\bar{v}= 0.47$, the shape of the flame exhibits periodic wrinkles with a large bulge convected from the burner lips to the flame tip inducing moderate flame surface variation $A_{rms}/\bar{A} = 0.11$. Nevertheless, the creation of this bulge leads to flame surface production at the flame base (Figure 5.2.4—CF, a). This instant corresponds to a positive pressure peak in the time trace in Figure 5.2.4—CF, but the sound level of 73 dB remains moderate when compared with the laboratory background noise 60 dB. Sound is mainly associated with phases of flame surface production. The sequence of images in Figure 5.2.4—MF corresponds to an "M"-flame submitted to the same harmonic forcing. The response differs significantly from that of the

conical flame. As the wrinkle travels toward the top of the flame, elements of neighboring fronts approach and pinch-off (Figure 5.2.4—MF, instants c and d). This mutual annihilation leads to the creation of a flame torus, which is convected downstream. The net result of this cyclic interaction is a flame surface variation $A_{rms}/\bar{A} = 0.19$ of the same order as the incoming flow perturbation level $v_{rms}/v= 0.17$. The instant of flame pinching (Figure 5.2.4—MF, e) corresponds to a large negative pressure peak (Figure 5.2.12—MF), with an overall SPL equal to 92 dB (Table 5.2.1). In contrast with the previous case, noise is mainly associated with the fast rate of annihilation of flame surface area. This interaction produces a large sound level even for relatively low levels of modulation. In the cycle presented in Figure 5.2.4—VF, the roll-up at the flame periphery is due to a toroidal vortex shed from the burner lips generated by the flow perturbation at the burner outlet $v_{rms}/v = 0.26$. As flame elements are rolled up at the front periphery, they are stretched, mutually interact (Figure 5.2.4—VF, b and c), and contribute to a large rate of destruction of the flame surface area $A_{rms}/\bar{A} = 0.43$. This event corresponds to an intense negative pressure peak in Figure 5.2.12—VF between instant b and c and results in a total SPL radiated of 96 dB. The last interaction involving a collection of small conical flames differs to some extent from the previous cases because phases of flame surface creation and destruction take approximatively the same duration (Figure 5.2.12—CSCF). Instants of maximum rate of production and destruction correspond, respectively, to a positive and negative pressure peak of approximatively the same amplitude in pressure time trace in Figure 5.2.12—CSCF. This observation differs from some of our previous expectations [44] that mechanisms leading to flame surface destruction were the main contributors to the overall noise production. Here, both creation and destruction phases result in opposite rates of change of the flame surface area associated with intense pressure peaks. The flame surface area evolution was not extracted because flame contours are blurred in many images in Figure 5.2.4—CSCF due to the combined motion of each individual flame. Nevertheless, the flame dynamics can be described using flame emission signals. Letters in Figure 5.2.12—CSCF indicate instants corresponding to images in Figure 5.2.4—CSCF. The mean flow velocity at the plate holes outlet is $v = 5.0$ m/s and the relative perturbation level $v_{rms}/v= 0.16$ remains moderate. Flames first stretch and elongate quickly (Figure 5.2.4—CSCF a and b), leading to large flame surface area production. Flames then suddenly break up and release small spherical pockets of fresh reactions (Figure 5.2.4—CSCF c and d). It is interesting to note that for about the same flow perturbation level $v_{rms}=0.8$–0.9 m/s, the SPL radiated by the collection of small conical flames reaches 96 dB, whereas the single conical flame remains relatively quiet (CF, SPL = 73 dB, Table 5.2.1). It should also be mentioned that this result was obtained for a reduced mass flow rate = 0.29 g/s for the CSCF configuration to be compared with = 0.59 g/s consumed by the single CF. The ensemble of small conical flames is a more compact configuration than the single conical flame leading to higher flame surface densities and to stronger noise production.

Further quantitative comparisons are given in what follows. Flame surface area A extracted from phase-averaged images and indicated as circles in cm^2 are presented in Figure 5.2.12 together with flame light intensity I displayed as solid lines. Emission signals I were rescaled using Equation 5.2.9 to obtain quantities comparable with the flame surface area A. Both signals perfectly coincide during all phases of the driving cycle indicating that the flame dynamics can be well retrieved by examining OH* radical light emission during flame interactions with convective waves (Figure 5.2.12—CF), mutual annihilation of flame elements (Figure 5.2.12—MF), or flame vortex interactions (Figure 5.2.12—VF). Time traces reveal strong nonlinearities during short phases of the driving cycles leading to either large rates of surface production (CF) or surface destruction (MF, VF). The pressure p' radiated in the laboratory measured by the microphone can be compared with the theoretical pressure fluctuation computed from the retarded rate of change of the flame surface area dA/dt, or equivalently from dI/dt. Results are plotted at the top of Figure 5.2.12. Pressure peak positions, amplitudes, and rms levels are correctly retrieved for the conical, "M" and

"V"-shaped flames submitted to convective waves (Figure 5.2.12—CF, MF, and VF). It should be noted that these results obtained for a microphone located in the near-field of the flame still obey the classical relation derived for far-field conditions.

5.2.6 Interactions with Equivalence Ratio Perturbations

Previous data have concerned unconfined flame configurations driven by velocity perturbations. These cases are less dependent on the geometry because sound generation is not modified by reflection from boundaries. It is also easier to examine unconfined flames with optical techniques. However, in many applications, combustion takes place in confined environments and sound radiation takes place from the combustor inlet or exhaust sections. The presence of boundaries has two main effects:

- Boundaries modify the reflection response of the system
- Boundaries modify the flame dynamics by changing the geometry of the flow in the system

The first item is easily treated by considering the eigenmodes of the system and expanding the pressure field on a basis formed by these modes. The second item is less well documented but is clearly important. The presence of boundaries not only modifies the structure of the mean flow but also influences the flame dynamics. This is demonstrated in a set of recent experiments in which the lateral confinement was varied systematically [45].

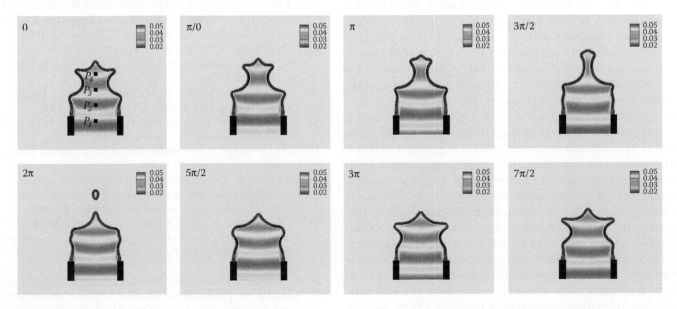

FIGURE 5.2.13
Dynamics of a methane-air conical flame submitted to a convective wave of equivalence ratio perturbations, $f = 175$ Hz, and $\phi/\bar{\phi} = 0.1$. The mass fraction perturbation is shown on the color scale. The flame is represented by temperature isocontours. Two cycles of modulation are displayed.

In addition to velocity perturbations, it is important to examine flames perturbed by other types of disturbances like composition fluctuations. These can result from the response of the injection system to pressure waves radiated by combustion and traveling in the upstream manifold. In premixed configurations, this gives rise to equivalence ratio fluctuations, which are convected downstream and interact with the flame [46]. This process is important in practice but difficult to study experimentally. It is here illustrated by numerical simulations carried out by imposing a convective wave to a conical flame [47]. Results shown in Figure 5.2.13 indicate that the flame is wrinkled by this perturbation and that this gives rise to heat release fluctuations. It is also interesting to note that the flame motion takes place with a period that equals twice that of the imposed modulation in equivalence ratio.

5.2.7 Conclusions

Analysis of perturbed flames carried out in this chapter indicates that rapid changes of heat release constitute a source of combustion noise, a mechanism that can effectively drive combustion instabilities. In premixed flames, the rapid changes in heat release are associated with variations in flame surface area dA/dt. These are produced, for example, when the flame interacts with convective vortices. The flame–vortex interaction induces a roll up of the reactive sheet followed by mutual annihilation of reactive elements resulting in rapid destruction of flame surface area. Rapid changes in flame surface area can also result from interactions between adjacent flame sheets, which consume the intervening reactant or from the collective motion of multiple flames synchronous pinching and pocket formation. Experiments indicate that these processes induce large positive or negative rates of change of heat release rate $\partial \dot{Q}/\partial t$, which constitute sources of noise radiation. When these sources are synchronized by coherent perturbations propagating inside the combustion chamber, they feed energy into the motion and this can lead to the growth of oscillation in the system. Rates of change of heat release are often found to reach maximum values during phases of flame surface destruction, implying that flame surface dissipation is the dominant mechanism of combustion noise production in combustors operating in a continuous mode (gas turbine combustors, industrial boilers, furnaces, etc.). This is so because processes of flame annihilation are usually faster than those generating flame surface. This is, however, not always the case and intense radiation can result from ignition, flame stretch, and collective dynamics. In automotive engines, for example, where combustion noise mainly originates from the violent ignition of the reactants introduced in the cylinder, the fast rate of production of flame surface area is the dominant mechanism. From these experimental data, it seems possible to reduce the level of driving by minimizing the rate of change of heat release resulting from the response of flames to incident perturbations.

References

1. F.A. Williams. *Combustion Theory*. The Benjamin/Cummings Publishing Company, Inc., California, 1985.
2. L. Crocco. Aspects of combustion instability in liquid propellant rocket motors. part 1. *J. Am. Rocket Soc.*, 21:163–178, 1951.
3. L. Crocco and S.I. Cheng. *Theory of Combustion Instability in Liquid Propellant Rocket Motors*. AGARDograph number 8, Butterworths Science Publication, London, 1956.
4. M. Barrère and F.A. Williams. Comparison of combustion instabilities found in various types of combustion chambers. *Proc. Combust. Inst.*, 12:169–181, 1969.
5. T.C. Lieuwen and V. Yang, eds. *Combustion Instabilities in Gas Turbine Engines: Operational Experience, Fundamental Mechanisms, and Modeling*. Progress in Astronautics and Aeronautics, Vol. 210, AIAA, 2005.
6. S. Candel. Combustion instabilities coupled by pressure waves and their active control. *Proc. Combust. Inst.*, 24:1277–1296, 1992.
7. A.A. Putnam. *Combustion Driven Oscillations in Industry*. Elsevier, New York, 1971.
8. F.E.C. Culick and V. Yang. Overview of combustion instabilities in liquid-propellant rocket engines. *Liquid Rocket Engine Combustion Instability*. Progress in Astronautics and Aeronautics, Vol. 169, pp. 3–37, Chapter 1, AIAA, 1995.
9. T.J. Poinsot, A.C. Trouvé, D.P. Veynante, S.M. Candel, and E.J. Esposito. Vortex driven acoustically coupled combustion instabilities. *J. Fluid Mech.*, 177:265–292, 1987.
10. K. McManus, T. Poinsot, and S. Candel. A review of active control of combustion instabilities. *Prog. Energ. Combust. Sci.*, 19:1–29, 1993.
11. S. Candel. Combustion dynamics and control: Progress and challenges. *Proc. Combust. Inst.*, 29:1–28, 2002.
12. T. Poinsot and V. Veynante. *Theoretical and Numerical Combustion*. Edwards, Philadelphia, 2001.
13. A.P. Dowling. A kinematic model of a ducted flame. *J. Fluid Mech.*, 394:51–72, 1999.
14. S. Ducruix, D. Durox, and S. Candel. Theoretical and experimental determination of the transfer function of a laminar premixed flame. *Proc. Combust. Inst.*, 28:765–773, 2000.
15. Y. Huang and V. Yang. Bifurcation of flame structure in a lean-premixed swirl-stabilized combustor: Transition from stable to unstable flame. *Combust. Flame*, 136(3):383–389, 2004.
16. T. Schuller, S. Ducruix, D. Durox, and S. Candel. Modeling tools for the prediction of premixed flame transfer functions. *Proc. Combust. Inst.*, 29:107–113, 2002.
17. T. Schuller, D. Durox, and S. Candel. A unified model for the prediction of flame transfer functions: Comparison

between conical and v-flames dynamics. *Combust. Flame*, 134:21–34, 2003.

18. T. Lieuwen. Nonlinear kinematic response of premixed flames to harmonic velocity disturbances. *Proc. Combust. Inst.*, 30:1725–1732, 2005.

19. D. Durox, T. Schuller, and S. Candel. Combustion dynamics of inverted conical flames. *Proc. Combust. Inst.*, 30:1717–1724, 2005.

20. R. Balachandran, B.O. Ayoola, C.F. Kaminski, A.P. Dowling, and E. Mastorakos. Experimental investigation of the non linear response of turbulent premixed flames to imposed inlet velocity oscillations. *Combust. Flame*, 143:37–55, 2005.

21. N. Noiray, D. Durox, T. Schuller, and S. Candel. Self-induced instabilities of premixed flames in a multiple self-induced instabilities of premixed flames in a multiple injection configuration. *Combust. Flame*, 145:435–446, 2006.

22. V.N. Kornilov, K.R.A.M. Schreel, and L.P.H. de Goey. Experimental assessment of the acoustic response of laminar premixed bunsen flames. *Proc. Combust. Inst.*, 31:1239–1246, 2007.

23. N. Noiray, D. Durox, T. Schuller, and S. Candel. A unified framework for nonlinear combustion instability analysis based on the flame describing function. *J. Fluid Mech.*, 2008 (In press).

24. S. Candel, A.-L. Birbaud, F. Richecoeur, S. Ducruix, and C. Nottin. Computational flame dynamics (invited lecture). In *Second ECCOMAS Thematic Conference on Computational Combustion*, Delft, The Netherlands, July 2007.

25. T. Lieuwen and B.T. Zinn. The role of equivalence ratio fluctuations in driving combustion instabilities in low nox, gas turbines. *Proc. Combust. Inst.*, 27:1809–1816, 1998.

26. A. Thomas and G.T. Williams. Flame noise: Sound emission from spark-ignited bubbles of combustible gas. *Proc. R. Soc. Lond. A*, 294:449–466, 1966.

27. S.L. Bragg. Combustion noise. *J. Inst. Fuel*, 36:12–16, 1963.

28. I.R. Hurle, R.B. Price, T.M. Sudgen, and A. Thomas. Sound emission from open turbulent flames. *Proc. R. Soc. Lond. A*, 303:409–427, 1968.

29. T.J.B. Smith and J.K. Kilham. Noise generated by open turbulent flame. *J. Acoust. Soc. Am.*, 35:715–724, 1963.

30. R.B. Price, I.R. Hurle, and T.M. Sudgen. Optical studies of the generation of noise in turbulent flames. *Proc. Combust. Inst.*, 12:1093–1102, 1968.

31. W.C. Strahle. On combustion generated noise. *J. Fluid Mech.*, 49:399–414, 1971.

32. W.C. Strahle. Combustion noise. *Prog. Energ. Combust. Sci.*, 4:157–176, 1978.

33. B.N. Shivashankara, W.C. Strahle, and J.C. Handley. Evaluation of combustion noise scaling laws by an optical technique. *AIAA J.*, 13:623–627, 1975.

34. M. Katsuki, Y. Mizutani, M. Chikami, and Kittaka T. Sound emission from a turbulent flame. *Proc. Combust. Inst.*, 21:1543–1550, 1986.

35. D.I. Abugov and O.I. Obrezkov. Acoustic noise in turbulent flames. *Combust. Explosions Shock Waves*, 14:606–612, 1978.

36. P. Clavin and E.D. Siggia. Turbulent premixed flames and sound generation. *Combust. Sci. Technol.*, 78:147–155, 1991.

37. G. Markstein. *Non Steady Flame Propagation*. Pergamon Press, Elmsford, NY, 1964.

38. P. Pelcé and P. Clavin. Influence of hydrodynamics and diffusion upon the stability limits of laminar premixed flames. *J. Fluid Mech.*, 124:219–237, 1982.

39. M. Matalon and B.J. Matkowsky. Flames in fluids: Their interaction and stability. *Combust. Sci. Technol.*, 34:295–316, 1983.

40. T. Schuller, D. Durox, and S. Candel. Dynamics of and noise radiated by a perturbed impinging premixed jet flame. *Combust. Flame*, 128:88–110, 2002.

41. T. Schuller, D. Durox, and S. Candel. Self-induced combustion oscillation of flames stabilized on annular burners. *Combust. Flame*, 135:525–537, 2003.

42. C.M. Vagelopoulos, F.N. Egolfopoulos, and C.K. Law. Further considerations on the determination of laminar flame speeds with the counterflow twin-flame technique. *Proc. Combust. Inst.*, 25:1341–1347, 1994.

43. D. Durox, T. Schuller, and S. Candel. Self-sustained oscillations of a premixed impinging jet flame on a plate. *Proc. Combust. Inst.*, 29:69–75, 2002.

44. S. Candel, D. Durox, and T. Schuller. Flame interactions as a source of noise and combustion instabilities. AIAA Paper 2004–2928, 2004.

45. A.L. Birbaud, D. Durox, S. Ducruix, and S. Candel. Dynamics of confined premixed flames submitted to upstream acoustic modulations. *Proc. Combust. Inst.*, 31(1):1257–1265, 2007.

46. J.H. Cho and T. Lieuwen. Laminar premixed flame response to equivalence ratio oscillations. *Combust. Flame*, 140:116–129, 2005.

47. A.L. Birbaud. Dynamique d'interactions sources des instabilités de combustion, PhD Thesis, Ecole Centrale Paris, Chatenay-Malabry, 2006.

48. S. Candel, D. Durox, T. Schuller. Flame interactions as a source of noise and combustion instabilities, AIAA paper 2004–2928, *10th AIAA/CEAS Aeroacoustics Conference*, Manchester, U.K., May 2004.

5.3 Tulip Flames: The Shape of Deflagrations in Closed Tubes

Derek Dunn-Rankin

5.3.1 Introduction

The propagation of premixed flames in closed vessels has been a subject of combustion research since its inception as a defined field of study in the late 1800s, when Mallard and LeChatelier [1] explored the behavior of explosions in the tunnels of coal mines. In the early decades of the twentieth century, experimenters used streak cameras to monitor the progress of premixed flame fronts propagating in tubes and channels without recording the detailed shape of the flame. Sustained interest in measuring flame speeds photographically and in observing the transition of deflagrations to detonations, however, produced advances in observation techniques that allowed flame shape monitoring during combustion. For example, Mason and Wheeler [2] presented an instantaneous image of a flame propagating in a tube, and demonstrated it to be unsymmetrical. Using this image, these researchers pointed out that flame shape complicates the concept of a flame's uniform movement (or normal propagation speed). Arguably, the most remarkable photographic study of flame-shape changes in closed vessels (even to this day) was published by Ellis using a then newly developed rotating shutter camera [3] that captured a stroboscopic time history of flame propagation in closed vessels. With this camera, he gathered images of CO/O_2 flames in spheres, cubes, odd-shaped cylinders, and cylindrical closed tubes [4]. The work included a variety of mixture compositions, ignition locations, and even flames propagating into each other. All of these closed-vessel studies produced very interesting visual results, and one such result, the very distinct cusp-shaped flame that occurs in the closed tubes, has come to be known as the "tulip flame" (though Ellis never referred to it with this name). An example of the tulip-flame formation, adapted from Ellis and Wheeler [5], is shown in Figure 5.3.1. The stroboscopic sequence shows that the flame grows out as a fairly symmetrical hemisphere from ignition at one endwall. As it approaches the sidewalls of the tube, however, the flame elongates into a dome shape (also referred to as a "finger," by Clanet and Searby [6]). Then, as the elongated elements burn out near the sidewall, the flame rapidly flattens and turns convex toward the unburned gas. Generally, the flame maintains this convex cusp (or tulip) shape through the rest of its propagation.

5.3.2 Historical Description of the Tulip

The dramatic dynamics of the flame-shape change shown in Figure 5.3.1, along with its proposed relationship to the flame instability and flame-generated flow, has periodically sparked an interest in its study. Before reviewing this flame-shape transition phenomenon, it will be useful to trace the history of the "tulip" name and distinguish this particular flame shape from the myriad of others with which it is often equated or confused.

In 1957, a flame propagating in a long tube under conditions resulting in a deflagration to detonation transition (DDT) was given the name "tulip" by Salamandra et al. [7]. This term was subsequently commonly applied in detonation studies to describe this typical shape [8,9]. Figure 5.3.2 shows a few

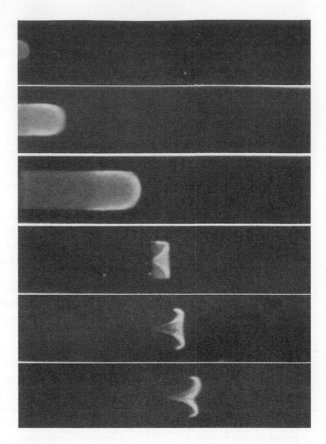

FIGURE 5.3.1
Rotating camera images of a CO/O_2 flame undergoing the inversion from the hemispherical cap flame to an inverted shape that is now considered a tulip or perhaps more accurately a "two-lip" flame. The flame propagates in a 20.3 cm long closed cylindrical tube of 2.5 cm diameter. (Adapted from Ellis, O.C. de C. and Wheeler, R.V., *J. Chem. Soc.*, 2, 3215, 1928.)

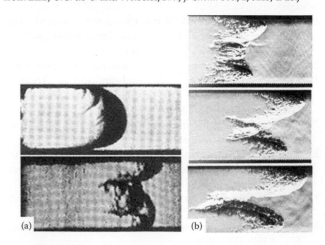

FIGURE 5.3.2
Tulip flames in the context of deflagration transitioning to detonations: (a) the flame shape first called as a "tulip" flame by Salamandra et al. [7] in the sentence describing the slowing-down stage of the flame, "The meniscus-shaped flame front becomes tulip-shaped." The chamber had square cross sections (36.5 mm on the side) and was approximately 1 m long. The mixture is H_2/O_2. (b) Similar images from more recent experiments [10], where the flame is also considered "tulip-shaped." More recent images are also referred to as tulip-shaped.

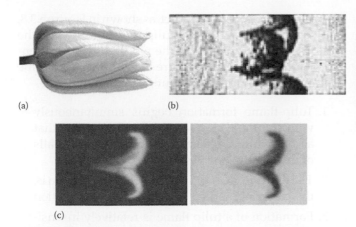

(a)

(b)

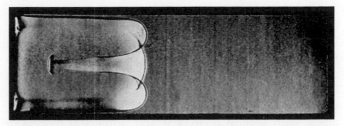

(c)

FIGURE 5.3.3
Comparison between tulips. (a) Image of an actual tulip flower that has been rotated and sized for comparison; (b) the tulip shape noted by Salamandra et al. [7] in flames on their transition to detonation; and (c) the inverted flame shape identified by Ellis and Wheeler [5] in closed tubes that is now being called a tulip flame. The image to the right is simply a negative of that to its left.

examples of these DDT-related tulip flames and Figure 5.3.3 compares this flame with a rotated and sized image of an actual tulip flower to show the reasonableness of this name association. It is important to notice that the Ellis-type flame shown at the bottom of the figure does not resemble the tulip to a great extent, and might even be better described as a "two-lip" flame.

However, despite this misnomer, when interest in the closed-tube flame of the Ellis-type resurfaced in the mid-1980s, the researchers involved were also aware of the DDT tulips and the name transferred to the cusped laminar flame transition [11–14]. An example of a flame image from this era is shown in Figure 5.3.4, and the "tulip" name is now used routinely to

FIGURE 5.3.5
The flame shape created when a weak shock (pressure ratio 1.10) interacts with it from the right. The image shows a flame funnel, 2.5 ms after the interaction begins; the flame is a stoichiometric butane/air mixture. (Adapted from Markstein, G.H., *Nonsteady Flame Propagation*, AGARDograph, Pergamon Press, New York, 1964.)

describe the cusped flame-shape changes that occur during the propagation in relatively short closed tubes [15–21].

To further complicate the situation, the tulip name has also been applied to flames interacting with pressure waves (Lee [8] describing the work of Markstein, [22] as shown in Figure 5.3.5), flames propagating in open tubes [6,23,24], flames in narrow channels where heat loss to the walls is important (e.g., [25,26], and even to flame propagating in a spinning tube [27]. The downward-propagating limit flames, shown in Figure 5.3.6, from Jarosinski et al. [28] might also be candidates for the tulip name, but so far they have not been termed so.

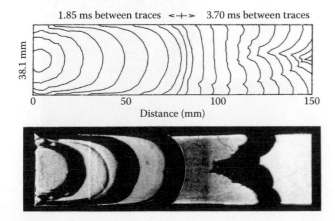

1.85 ms between traces <–+–> 3.70 ms between traces

FIGURE 5.3.4
Flame shape images and traces extracted from the high-speed schlieren movie (5000 frames/s) of a stoichiometric methane/air flame going through a tulip inversion while propagating in a square cross-section (38.1 mm on the side) closed tube.

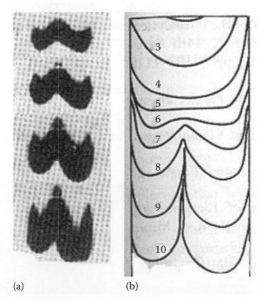

(a)   (b)

FIGURE 5.3.6
A lean methane/air flame propagating downward in a tube closed above. The tube is of 51 mm diameter. (a) Self-light images of the flame; (b) traces of the flame shape at 55 frames/s. (Adapted from Jarosinski, J., Strehlow, R.A., and Azarbarzin, A., *Nineteenth Symposium (International) on Combustion*, The Combustion Institute, Pittsburgh, pp. 1549–1555, 1982.)

As briefly demonstrated above, the combustion literature is replete with flame shapes that suggest the tulip, including those of Ellis and Wheeler [5], Gúenoche and Jouy [29], Jeung et al. [30], and more recently in the papers of Clanet and Searby [6], Matalon and Metzener [19], and Dunn-Rankin and Sawyer [12,31]. Among these papers, there are numerous explanations for the tulip flame, along with several analytical and numerical models of the process. Based on the variety of studies, explanations, and models of the tulip flame, it is likely that the shape does not arise from a single phenomenon, but can occur from different processes that may depend on the system under study. Hence, a single dominant mechanism does not explain all the experimental evidences available in the literature. If, however, we limit our consideration to the short closed-tube condition highlighted by Ellis, we can more easily evaluate the flame/flow/instability condition responsible for this particular form of the tulip flame.

5.3.3 Tulip Flames in Relatively Short Closed Tubes

For the most dramatic flame-inversion behavior to occur, the deflagration is ignited at one end of a tube, closed at both the ends, and filled with a near-stoichiometric mixture of fuel and air. The tube should have a length to transverse dimension ratio of between 4 and 20. The absolute value of the transverse dimension should be large enough (>25 mm) to obviate the wall-heat transfer and boundary-layer effects, but small enough (<100 mm) to prevent large buoyant or dynamic distortions of the flame sheet. The tulip flame would still occur outside these geometric limits; however, the sharpness of the transition would degrade. Figure 5.3.7 is an example drawn from the work of Ellis and Wheeler [5] illustrating the appropriate tube conditions and the sequence of flame shapes that result after ignition at one end.

As mentioned earlier, the tulip flame has been observed under a wide variety of conditions, suggesting that it is a very robust phenomenon. In fact, as shown in Figure 5.3.8, Zhou et al. [32] discovered a tulip flame even when the closed tubes are curved. On the whole, the experimental observations in the literature have shown that in the relatively short closed-tube configuration:

1. Tulip-flame formation begins simultaneously with the rapid decrease in the flame area that accompanies the flame quench at the sidewalls of the combustion vessel

2. The more pronounced the flame-area reduction is, the more pronounced is the tulip-flame transition

3. Formation of a tulip flame is relatively insensitive to mass loss and endwall geometry

4. Formation of a tulip flame is relatively insensitive to mixture composition

Taken together, these observations indicate that the basic tulip-flame formation is a remarkably robust phenomenon that depends somewhat on the overall geometry of the combustion vessel. There is little doubt that the growth of the cusp represents a Darrieus–Landau instability [33–35] that is stabilized by the thermo-diffusive transport at the small scales, but an interesting lingering

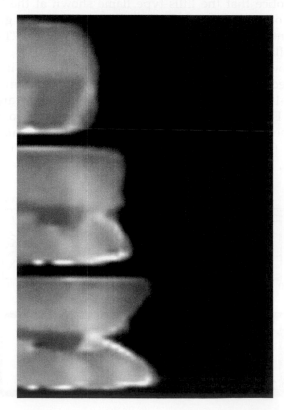

FIGURE 5.3.8
Self-light image of the formation of a tulip flame in a closed tube that is curving. The curve is off to the right and not visible in these photographs, but the classical twin-lobe formation occurs as if the tube was straight. These images are similar to that reported by Zhou et al. [32]. (Courtesy of A. Sobesiak.)

FIGURE 5.3.7
A classic self-light stroboscopic image of a premixed flame undergoing a tulip inversion in a closed tube. There is an interval of 4.1 ms between the images of a water vapor saturated CO/O_2 flame arranged to have a flame speed comparable with that of a stoichiometric methane/air flame. The tube is 2.5 cm in diameter and 20.3 cm long. (Adapted from Ellis, O.C. de C. and Wheeler, R.V., *J. Chem. Soc.*, 2, 3215, 1928.)

question is what provides the dominant trigger for the instability in a way that most often leads to the large twin lobes.

5.3.4 Triggering the Tulip-Flame Instability

Early attempts to explain the trigger for the tulip flame focused on the pressure wave/flame interactions. This was a natural consequence of the well-documented vibratory behavior of flames seen in the very first streak images recorded [2], and the images of Markstein (like that in Figure 5.3.5) showed how pressure waves can radically disturb the flame front. However, numerical studies showed that tulip flames occur even when pressure waves are absent (e.g., [11]), suggesting that a flame/flow interaction might be important. The advent of laser Doppler velocimetry (LDV) provided a method by which this flame flow interaction could be observed quantitatively. In the mid-1980s, Starke and Roth [14], Jeung et al. [30], and Dunn-Rankin [36] made velocity measurements during the tulip formation. This new quantitative data also encouraged a series of numerical simulations of the tulip phenomenon (e.g., [16]), which periodically resurfaces (e.g., [17]). Figures 5.3.9 and 5.3.10 show

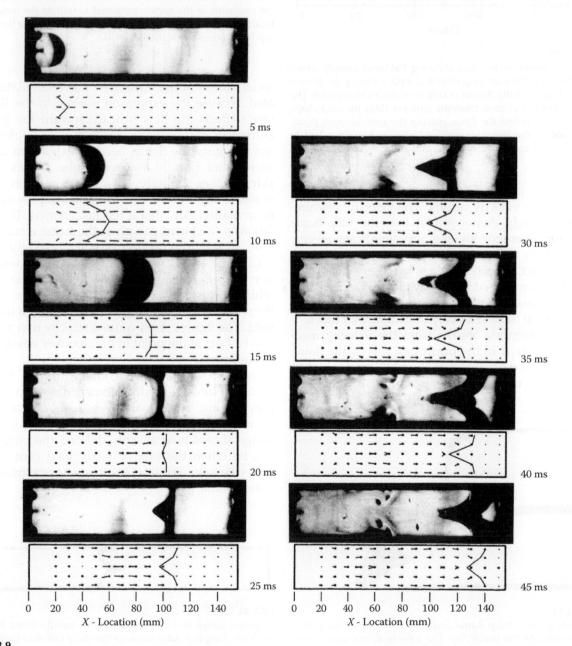

FIGURE 5.3.9
Velocity vectors of the gas flow measured using laser Doppler anemometry inside a closed chamber during the formation of a tulip flame. Images of the flame are also shown, though the velocity measurements required many repeated runs, hence, the image is only representative. The chamber has square cross sections of 38.1 mm on the side. The traces in the velocity fields are the flame locations based on velocity data dropout. The vorticity generated as the flame changes shape appears clearly in the velocity vectors.

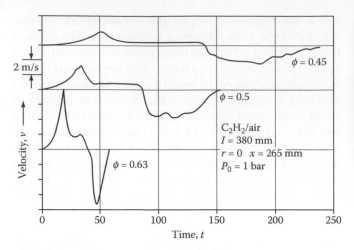

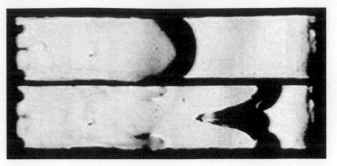

FIGURE 5.3.12
Similar to Figure 5.3.11, another schlieren image of a tulip flame and the remnants of the vortex, proposed to initiate the instability. The location of the vortex suggests that it forms just as the hemispherical cap flame burns out at the sidewalls.

FIGURE 5.3.10
Laser Doppler anemometry data showing the axial velocity along the centerline of a 380 mm long closed chamber during the formation of acetylene/air tulip flames of different equivalence ratios. The velocity is measured 265mm from the ignition; thus, the tulip shape is already formed before the flame reaches the measurement point. This work shows the behavior similar to the results described in Figure 5.3.9. (Adapted from Starke, R. and Roth, P., *Combust. Flame*, 66, 249, 1986.)

samples of these velocity results. The velocity measurements conclusively demonstrated that the tulip-flame phenomenon does not involve any significant reverse flow in the bulk unburnt gas or any squish flow along the walls. In fact, the unburnt gas behaves simply as if it were being compressed by a leaky piston. The measurements also showed that the boundary-layer effects are not significant in these relatively large diameter tubes.

The most intriguing results from the velocity measurements occurred in the burnt gas, just as the flame that is initially convex towards the unburnt gas begins to flatten. The curved flame produces a noticeable

recirculation in the burnt gas. The recirculation is created because the expansion normal to the flame surface deflects the flow upward toward the centerline. During its early growth phase, this recirculation does not influence the flame, because expansion drives the flame front rapidly down the tube, but as the flame skirt reaches the sidewalls, the expansion decreases dramatically and the flame remains in proximity to the recirculation flow that it has just created. This recirculation in the burnt gas gives the flame a small deflection, which triggers the broader instability. Figures 5.3.11 and 5.3.12 show the schlieren images from two tulip-flame formation events. In these cases, the schlieren system was sufficiently sensitive to capture the residual thermal gradients associated with the flow field. The figures show quite distinctly that a recirculation cell sits at the point in the chamber where the tulip flame began.

Figure 5.3.13 shows an image from a high-speed schlieren movie of a stoichiometric methane/air flame propagating in a square cross-section tube (38.1mm on a side) of approximately 300mm in length. The image shows how sharply defined the cusped flame can be when the tube is fairly long (but not open) and when the ignition source is approximately two-dimensional.

FIGURE 5.3.11
Schlieren image of a tulip flame and the remnants of the vortex, proposed to initiate the instability. The tube is acrylic with square cross section of 38.1mm on the side. The mixture is stoichiometric methane/air.

FIGURE 5.3.13
Schlieren image of a very clear tulip or two-lip flame formed in a moderate length, square cross section tube (300mm long × 38.1mm on the side) from a stoichiometric methane/air mixture, with a line igniter across the left wall.

5.3.5 Concluding Remarks

The tulip or "two-lip" flame is an interesting example of how flame/flow interaction can produce changes in flame shape. The phenomenon of a tulip flame, in all of its guises, occurs in many combustion configurations, and the cause of the tulip can depend, at least partly, on the particular system under study. In the specific case of flames propagating in fairly short, moderately narrow tubes, it appears that the tulip flame occurs when a recirculation generated by the curved flame front suddenly finds itself spinning near the flat flame sheet. The recirculation produces the initial trigger for a Landau–Darrieus instability that subsequently grows to the entire tulip.

References

1. Mallard, E. and Le Chatelier, H.L., Recherches experimentales et theoriques sur la combustion des melanges gaseux explosifs, Series 4. *Annales des Mines*, 8, 274–618, 1883.
2. Mason, W. and Wheeler, R.V., The propagation of flame in mixtures of methane and air. Part II. Vertical propagation, *Journal of the Chemical Society Transactions*, 117, 1227–1237, 1920.
3. Ellis, O.C. de C. and Robinson, H.A, new method of flame analysis, *Journal of the Chemical Society*, 127, 760–767, 1925.
4. Ellis, O.C. de C., Flame movement in gaseous explosive mixtures, *Fuel in Science and Practice*, 7(5): 195–205; 7(6): 245–252; 7(7): 300–304; 7(8): 336–344; 7(9): 408–415; 7(10): 449–454; 7(11): 502–508, 1928.
5. Ellis, O.C. de C. and Wheeler, R.V., Explosions in closed cylinders. Part III. The manner of movement of flame, *Journal of the Chemical Society*, Part 2, 3215–3218, 1928.
6. Clanet, C. and Searby, G., On the 'tulip flame' phenomenon, *Combustion and Flame*, 105, 225–238, 1996.
7. Salamandra, G.D., Bazhenova, T.V., and Naboko, I.M., Formation of detonation wave during combustion of gas in combustion tube, *Seventh Symposium (International) on Combustion*, Butterworths, London, pp. 851–855, 1959.
8. Lee, J.H.S., Initiation of gaseous detonation, *Annual Review of Physical Chemistry*, 28, 75–104, 1977.
9. Urtiew, P.A. and Oppenheim, A.K., Experimental observations of the transition to detonation in an explosive gas, *Proceedings of the Royal Society of London. Series A.* 295(1440), 13–28, 1966.
10. Kuznetsov, M., Alekseev, V., Matsukov, I., and Dorofeev, S., DDT in a smooth tube filled with a hydrogen–oxygen mixture, *Shock Waves*, 14, 205–215, 2005.
11. Dunn-Rankin, D., Barr, P.K., and Sawyer, R.F., Numerical and experimental study of "Tulip" flame formation in a closed vessel, *Twenty-First Symposium (International) on Combustion*, The Combustion Institute, Pittsburgh, pp. 1291–1301, 1986.

12. Dunn-Rankin, D. and Sawyer, R.F., Interaction of a laminar flame with its self-generated flow during constant volume combustion, *AIAA Volume from Tenth ICDERS*, 115–130, 1985.
13. Rotman, D.A. and Oppenheim, A.K., Aerothermodynamic properties of stretched flames in enclosures, *Twenty-First Symposium (International) on Combustion*, The Combustion Institute, Pittsburgh, pp. 1303–1312, 1986.
14. Starke, R. and Roth, P., An experimental investigation of flame behavior during cylindrical vessel explosions, *Combustion and Flame*, 66, 249–259, 1986.
15. Cloutman, L.D., Numerical simulation of turbulent premixed combustion, Western States Section/Japanese Section of the Combustion Institute JointMeeting, Fall, Honolulu, November 22–25; Lawrence Livermore National Laboratory Report UCRL-96680, 1987.
16. Gonzalez, M., Borghi, R., and Saouab, A., Interaction of a flame front with its self-generated flow in an enclosure: The "tulip flame" phenomenon, *Combustion and Flame*, 88, 201–220, 1992.
17. Marra, F.S. and Continillo, G., Numerical study of premixed laminar flame propagation in a closed tube with a full Navier-Stokes approach, *Twenty-Sixth Symposium (International) on Combustion*, The Combustion Institute, Pittsburgh, pp. 907–913, 1996.
18. Matalon, M. and McGreevy, J.L., The initial development of a tulip flame, *Twenty-Fifth Symposium (International) on Combustion*, The Combustion Institute, Pittsburgh, pp. 1407–1413, 1994.
19. Matalon, M. and Metzener, P., The propagation of premixed flames in closed tubes, *Journal of Fluid Mechanics*, 336, 31–50, 1997.
20. N'Konga, B., Fernandez, G., Guillard, H., and Larrouturou, B., Numerical investigations of the tulip flame instability—comparisons with experimental results, *Combustion Science and Technology*, 87, 69–89, 1992.
21. Starke, R. and Roth, P., An experimental investigation of flame behavior during explosions in cylindrical enclosures with obstacles, *Combustion and Flame*, 75, 111–121, 1989.
22. Markstein, G.H., Experimental studies of flame-front instability, in *Nonsteady Flame Propagation*, AGARDograph, G.H. Markstein, ed., Pergamon Press, New York, 1964.
23. Chomiak, J. and Zhou, G., A numerical study of large amplitude baroclinic instabilities of flames, *Twenty-Sixth Symposium (International) on Combustion*, The Combustion Institute, Pittsburgh, pp. 883–889, 1996.
24. Dold, J.W. and Joulin, G., An evolution equation modeling inversion of tulip flames, *Combustion and Flame*, 100, 450–456, 1995.
25. Hackert, C.L., Ellzey, J.L., and Ezekoye, O.A., Effect of thermal boundary conditions on flame shape and quenching in ducts, *Combustion and Flame*, 112, 73–84, 1998.
26. Song, Z.B., Ding, X.W., Yu, J.L., and Chen, Y.Z., Propagation and quenching of premixed flames in narrow channels, *Combustion, Explosion, and Shock Waves*, 42, 268–276, 2006.
27. Sakai, Y. and Ishizuka, S., The phenomena of flame propagation in a rotating tube, *Twenty-Sixth Symposium*

(International) on Combustion, The Combustion Institute, Pittsburgh, pp. 847–853, 1996.

28. Jarosinski, J., Strehlow, R.A., and Azarbarzin, A., The mechanisms of lean limit extinguishment of an upward and downward propagating flame in a standard flammability tube, *Nineteenth Symposium (International) on Combustion*, The Combustion Institute, Pittsburgh, pp. 1549–1555, 1982.

29. Gúenoche, H. and Jouy, M., Changes in the shape of flames propagating in tubes, *Fourth Symposium (International) on Combustion*, Williams and Wilkins, Baltimore, pp. 403–407, 1953.

30. Jeung, I., Cho, K., and Jeong, K., Role of flame generated flow in the formation of tulip flame, paper AIAA 89–0492, 27th AIAA Aerospace Sciences Meeting, Reno, Nevada, January 9–12, 1989.

31. Dunn-Rankin, D. and Sawyer, R.F., Tulip flames: Changes in shape of premixed flames propagating in closed tubes, *Experiments in Fluids*, 24, 130–140, 1998.

32. Zhou, B., Sobiesiak, A., and Quan, P., Flame behavior and flame-induced flow in a closed rectangular duct with a 90 degrees bend, *International Journal of Thermal Sciences*, 45, 457–474, 2006.

33. Darrieus, G., Propagation d'un Front de Flamme: Essai de Theórie des Vitesses Anomales de Deflagration par Development Spontane de la Turbulence, unpublished manuscript of a paper presented at La Technique Moderne, 1938, and Le Congrés de Mechanique Appliquée, Paris, 1945, 1938.

34. Kadowaki, S. and Hasegawa, T., Numerical simulation of dynamics of premixed flames: Flame instability and vortex–flame interaction, *Progress in Energy and Combustion Science*, 31, 193–241, 2005.

35. Landau, L., On the theory of slow combustion, *Acta Physicochimica URSS*, 19, 77–85, 1944.

36. Dunn-Rankin, D., The interaction between a laminar flame and its self-generated flow, Ph.D. Dissertation, University of California, Berkeley, 1985.

6

Different Methods of Flame Quenching

CONTENTS

6.1 Flame Propagation in Narrow Channels and Mechanism of Its Quenching

Artur Gutkowski and Jozef Jarosinski

6.1.1 Introduction

The problem of flame quenching by a wall was first studied nearly two centuries ago by Sir Humphry Davy when he was involved in preventing explosions in coal mines. He used an experimental approach that took him several months to solve the problem of the mining safety lamp. He then demonstrated its construction to the public [1]. After this spectacular achievement, flame quenching by walls was neglected by science for more than a 100 years. It was only in 1918 that Payman and Wheeller published their work on the propagation of flames through small diameter tubes [2]. Extensive experimental studies of flame quenching were also undertaken by Holm [3], who was the first to introduce the concept of quenching distance or quenching diameter. However, the most important contribution to the theory of flame quenching by walls was made by Zel'dovich [4,5]. He demonstrated that at the quenching limit, $S_{L,lim}$, the laminar burning velocity, and the maximum value of the limit flame temperature, $(T_{b,max})_{lim}$ were related to their respective values under adiabatic conditions, S_L°, and T_b°, through the relations

$$\frac{S_{L,lim}}{S_L^\circ} = e^{-\frac{1}{2}} = 0.61 \qquad (6.1.1)$$

$$\Delta T_{b,lim} \equiv \left[T_b^\circ - \left(T_{b,max} \right)_{lim} \right] = \frac{b}{\left(S_{L,lim} \right)^2} = \frac{\left(T_b^\circ \right)^2 R}{E} \qquad (6.1.2)$$

where
   b a parameter characterizing heat losses,
   R the universal gas constant,
   E the activation energy.

In his theory, Zel'dovich also showed that the Peclet number, given by the expression

$$\mathrm{Pe} = \frac{S_L^\circ D_Q}{a} \qquad (6.1.3)$$

remained constant at the flame quenching limit, where D_Q is the quenching diameter and a is the thermal diffusivity. His theory, included in his later book [6], has stood the test of time.

Unfortunately, his work was published during the World War II, in the former Soviet Union, and for a long time remained almost unknown to the wider community studying combustion.

As the quenching distance for flames propagating in different mixtures was often important for industrial applications, experimental methods to measure this quantity were developed. The most frequently used ones were

1. Burner method

2. Tube method

3. Flange electrode method

The first method was employed at the atmospheric pressure [3,7] or low pressures [8,9], while the second was mainly carried out at reduced pressures [10–12]. The third method was developed and explored by Blanc et al. [13] and Lewis and von Elbe [14], also employed by other researchers [15], and is still in use in some laboratories.

Some of the early studies were devoted to assessing the effect of walls on the quenching distance. It was found that the nature of the wall material hardly affected the quenching distance [7,8,16].

This is because the heat capacity of a wall of finite thickness is several orders of magnitude higher than that of the hot combustion products. However, some researchers did observe a small effect of the properties of the wall [17] on the quenching distance. This was interpreted in terms of some residual catalytic activity of the wall surface, poisoned by the combustion products from the preceding experiments [18]. With respect to this explanation, the surface of any material moistened through the condensation of the water vapor produced in the reaction is supposed to have very similar, low activity.

However, a number of theories on flame quenching were also developed. Some of these were based on arbitrary assumptions concerning the conditions for quenching [7,11]. Another group of theories were based on the energy conservation equation for the flame, including heat loss [19–21].

In the later years, studies on flame quenching were inspired by different applications and were undertaken sporadically [22–24].

The quenching distance is a very important parameter, characterizing a laminar flame in contact with solid

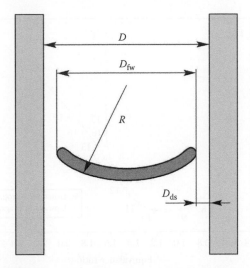

FIGURE 6.1.1
Definitions of flame parameters in channels. D, distance between channel walls effective in flame quenching (quenching distance). D_{fw}, flame width; D_{ds}, dead space; R, radius of curvature of the flame.

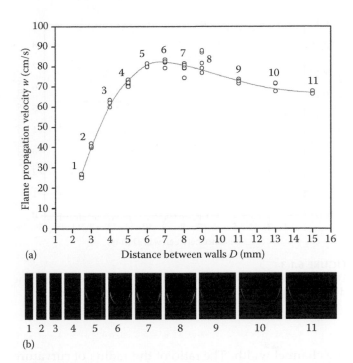

FIGURE 6.1.2
(a) Measured flame propagation velocity as a function of distance between the walls for a flame propagating in stoichiometric propane/air mixture. (b) Pictures of the flame propagating downward in narrow channels. Frame numbers correspond to the numbers of experimental points. (Channel width on the scale in b).

walls. It is directly related to the laminar burning velocity and flame thickness, and brings together the most important processes occurring in a laminar flame front. Flame quenching by a wall is accompanied by a number of interesting phenomena. It is commonly accepted that the quenching distance should be determined for the downward flame propagation. In this context, it is quite interesting to compare the properties of downward and upward propagating flames. Their behavior is similar or different depending on the value of the Lewis number. The limit flame is characterized by quantities, such as "dead space," radius of curvature, length of the zone of hot combustion gases, etc. These quantities are functions of the quenching distance and equivalence ratio, and they all affect quenching. Recognition of their influence on limit flames may facilitate the understanding of the mechanism of flame quenching.

For a while till now, our research group has been involved in studies of the properties of limit flames. Most of the results reported in this chapter were obtained for propane flames, under normal atmospheric conditions, in 300 mm long channels, with a square cross-section. The experimental procedure was described previously [25]. A flame propagating through a stationary mixture in a quenching tube or quenching channel can be characterized by the parameters defined in Figure 6.1.1.

6.1.2 Flame Shape and Propagation Velocity

To understand the mechanism of flame quenching in narrow channels in detail, one should first examine the data of flames in mixtures of constant composition, but in channels of different sizes (Figure 6.1.2). The measured propagation velocities in stoichiometric propane/air mixture are shown in Figure 6.1.2a. For channel widths slightly larger than the quenching distance, the

propagation velocity is lower than the laminar burning velocity. For channel widths over 3 mm, it gradually increases to a maximum value of about $2S_L^o$ for widths of 6–9 mm, falling slowly if the channel width increases further/apart. It can be seen from Figure 6.1.2b that a flame propagating downward is always convex. A convex flame is formed, in spite of the stabilizing effect of gravity. This indicates that a flame front would be unstable. A convex flame can be treated in terms of curved cells, formed after the flat flame loses its stability [6].

The steady states of such systems result from nonlinear hydrodynamic interactions with the gas flow field. For the convex flame, the flame surface area F can be determined from the relation $FS_L = b^2w$, where S_L is the laminar burning velocity, b^2 the cross-section area of the channel, and w is the propagation velocity at the leading point.

The cooling effect of the channel walls on flame parameters is effective for narrow channels. This influence is illustrated in Figure 6.1.3, in the form of the dead-space curve. When the walls are <4 mm apart, the dead space becomes rapidly wider. This is accompanied by falling laminar burning velocity and probably lowering of the local reaction temperature. For wider channels, the propagation velocity w is proportional to the effective flame-front area, which can be readily calculated. On analysis of Figures 6.1.2b and 6.1.3, it is evident that the curvature of the flame is a function of

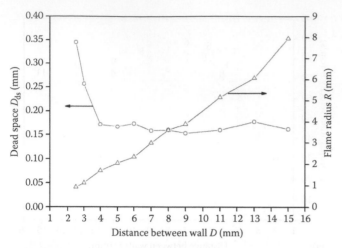

FIGURE 6.1.3
Dead space and the radius of curvature of flame as functions of the distance between channel walls for flames shown in Figure 6.1.2.

the channel width. The ratio of the radius of curvature R to D (defined earlier) is nearly constant and equal to $R/D \approx 0.4$–0.5.

6.1.3 Effect of the Equivalence Ratio on the Quenching Distance

The quenching distance as a function of the equivalence ratio is illustrated by the curve at the top of Figure 6.1.4. To bring out some of the characteristic points, pictures of near-limit flames are also shown in the figure. It can be seen that for lean propane/air mixtures, the quenching distances for flames propagating up and down are practically the same. For rich mixtures, the situation is quite different. In this case, flames propagating in opposite directions are extinguished in the same narrow channel in the mixtures of different compositions. Differences between downward and upward propagation limits are found to increase rapidly with the distance from stoichiometry. The quenching limit curves lie within the flammability limits, but on the rich side, the flammability limits depend on whether the flame propagates up or down.

The effect of natural gravity on flammability limits has been known for a long time. The difference between flammability limits for downward and upward flame propagation was first observed by White [26], for hydrogen/air mixtures. Subsequently, similar effects were also found for other mixtures. For propane flames, the lean flammability limit for both downward and upward propagation was observed to be $\phi = 0.53$. The rich limits were $\phi = 1.64$ for downward and $\phi = 2.62$ for upward propagation. Such wide gap between the flammability limits for rich mixtures is explained in

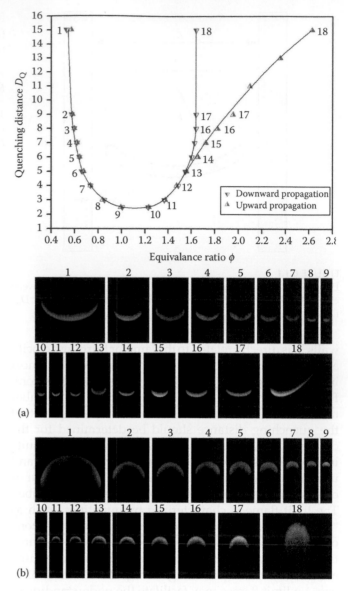

FIGURE 6.1.4
Measured quenching distance as a function of equivalence ratio for propane/air mixture (top), and pictures of (a) downward and (b) upward propagating flames in channels, close to quenching. Channel widths as in the graph. Frame numbers correspond to the numbers of experimental points.

terms of preferential diffusion of the reactant that is deficient (oxygen), in response to the flame stretch of an upward propagating flame (the Lewis number for such flames is <1).

6.1.4 Length of the Zone of Hot Combustion Gases behind the Flame Front

The temperature profile inside a narrow channel was determined using resistance probes with a 10 μm Pt–Ir

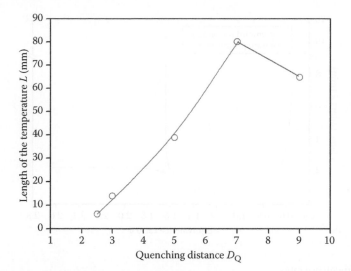

FIGURE 6.1.5
Length of the high-temperature zone behind a flame as a function of quenching distance. Downward propagation in lean propane/air mixture.

sensor, 2.2 mm long. The time constant of the wire was about 5 ms. Measurements in very narrow channels should be treated with caution as they may be affected by errors, owing to the influence of the probe on small volumes of gas. Heat transfer from the hot combustion gases to the walls may be characterized by the length of the high-temperature zone lying just behind the flame front. For up to 80% of their length, the temperature records for this zone were similar. Thus, it was assumed that the effective end of this zone could be placed where the temperature fell to 20% of the maximum temperature rise. The length of the high-temperature zone as a function of the quenching distance is shown in Figure 6.1.5, for downward propagating flames in a lean propane/air mixture (equivalence ratio between 0.53 and 1.00).

It can be observed that the dominant mechanism of heat transfer from the flame to the walls is different for channel widths up to 7 mm as well as for wider channels.

6.1.5 Dead Space

The width of the dead space was determined from the photographs. The results are shown in Figure 6.1.6 as a function of the quenching distance (Figure 6.1.6a) and of the equivalence ratio (Figure 6.1.6b). It can be observed that in channels up to 7 mm wide, the dead space grows linearly with the D_Q, but for rich mixtures the growth is twice as fast as for lean ones. Their values relative to the quenching distance lie near 0.09 for lean and 0.18 for rich mixtures. For lean propane/air mixtures, the dead space for flames propagating upward in the quenching channels is the same as for those propagating down. However, for very rich mixtures, the concept of dead space loses its physical meaning (mechanism of natural self-propagation is replaced by buoyancy). It can be observed that the width of upward propagating rich flames is nearly constant, regardless of the quenching distance (see Figure 6.1.4b).

6.1.6 Radius of Curvature of the Flame

The radius of curvature of flame is shown in Figure 6.1.7 as a function of the quenching distance (Figure 6.1.7a) and of the equivalence ratio (Figure 6.1.7b). The radius was determined from the flame pictures. For lean mixtures, the radius increases linearly with the channel width, both for the downward and upward propagating flames. For rich mixtures and downward propagation, the increase is linear for quenching distances up to $D_Q = 7$ mm, but the increase is not as steep as that of lean mixtures. However, the increase accelerates. For rich

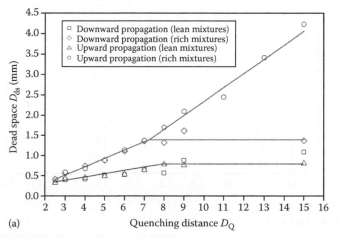

(a)

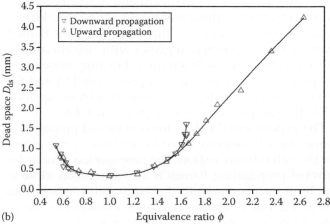

(b)

FIGURE 6.1.6
Dead space as a function of (a) quenching distance and (b) equivalence ratio.

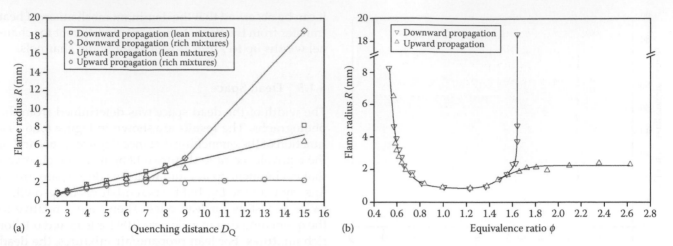

(a)

(b)

FIGURE 6.1.7
Radius of curvature as a function of (a) quenching distance and (b) equivalence ratio.

mixtures and upward flame propagation, the radius of curvature becomes nearly constant at $R = 0.7$–1.2 mm (note that the curvature of those flames is very large).

6.1.7 Burning Velocity Near the Quenching Limit

Convex limit flames propagate in a channel through a stationary mixture, almost without change in shape or structure. As the flame surface is larger than that of a flat flame, they move with a velocity slightly higher than the burning velocity. Taking into account the convex flame surface and its interaction with the walls, it was estimated that the burning velocity of limit flames was approximately 0.9 of the experimentally determined flame propagation velocity. For such flames, the laminar burning velocity is different from that of the adiabatic ones. According to the theory of Zel'dovich et al. [6], the burning velocity under quenching conditions should fall to a limit value of $S_{L,lim} = 0.61 S_L^\circ$.

Flame propagation velocity was measured in quenching channels both for downward and upward propagation. The data from these measurements (with velocities approximately 10% higher than those of limit flames) were plotted on a graph, together with two curves: the most reliable, adiabatic laminar burning velocity S_L° for propane flames determined previously [27] and the limit burning velocity $S_{L,lim}$, calculated from Equation 6.3.1. The comparison is shown in Figure 6.1.8.

The experimental points for downward propagating flames lie satisfactorily close to the limit curve, except for the rich mixture region. The propagation velocity for upward propagating flames is also reasonably close to the limit burning velocity for equivalence ratios from 0.8 to 1.3, but is markedly higher outside this range. The factor responsible for this increase is buoyancy. Preferential diffusion is an additional factor promoting the larger burning velocity increase for rich mixtures.

6.1.8 Peclet Number

The quenching diameter or quenching distance between two plates is related to the properties of the flammable mixture by means of the Peclet number, expressed by Equation 6.1.3. The Peclet number is usually determined for downward propagating flames. Typically, the adiabatic laminar burning velocity and thermal diffusivity at the temperature of the unburned mixture are used in the calculations. For wedge-shaped channels, the limit Peclet number was observed to be Pe = 42 [28]. However, in this study, careful measurements in the rectangular channels showed slightly higher values, Pe = 51. If the quenching distance is compared with the entire flame thickness δ_L (see the definition in Ref. [29]), then their

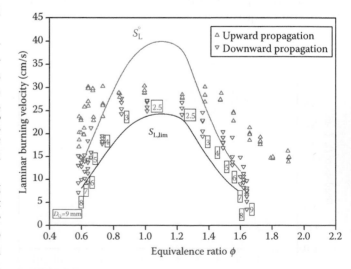

FIGURE 6.1.8
Limit burning velocity as a function of the equivalence ratio for (a) downward propagation (symbol ∇) and (b) upward propagation (symbol $\triangle$).

relation should be approximately constant at $D_Q/\delta_L \approx 2$, over the entire range of mixture compositions [30]. Furthermore, Daou and Matalon [31] also obtained similar results. They used numerical methods to analyze the flame behavior in narrow channels, with and without heat losses. Their calculations showed that flames with heat loss were thicker than those propagating under adiabatic conditions. They also found that the quenching distance was approximately 15 times larger than the characteristic thickness, defined as $\Delta_L = a/S_L$ (for thermal diffusivity at the mean temperature), which corresponds exactly to twice the entire flame thickness, $2\delta_L$. Our experimental results gave $D_Q \approx (15\text{--}17)\Delta_L$. It can be seen from Figure 6.1.4 that the quenching distance curve for downward propagation takes the form of a parabola, with a minimum near stoichiometry. As the product of the laminar burning velocity and the quenching distance is nearly constant for limit flames (see Equation 6.1.3), it is evident that these two quantities are inversely proportional to each other. The constant of proportionality depends on the chemical and transport properties of the mixture, and may be predicted from the existing theory.

For propane flames, the quenching distance (for downward propagating flames) is limited by the distance between the walls of about 10 mm. In larger channels, the flame is quenched at the flammability limits.

6.1.9 Numerical Simulation

To examine the details of the structure of flames in channels under quenching conditions, numerical methods were used. Two-dimensional CFD simulation of a propane flame approaching a channel between parallel plates was carried out using the FLUENT code [25]. The model reproduced the geometry of the real channels investigated experimentally. Close to the quenching limit, the burning velocity, dead space, and radius of curvature of the flames were all close to the experimental values.

The calculated flow and temperature fields for a flame propagating in a channel with the walls separated by $D_Q = 4$ mm is shown in Figures 6.1.9 and 6.1.10.

6.1.10 Behavior and Properties of Flames at the Quenching Limits

On the basis of the observations and results so far presented, it is obvious that the properties of limit flames are very different depending on the width of the quenching channel, the equivalence ratio, and the direction of flame propagation. The reasons for this are detailed in the following four sections.

6.1.10.1 Downward Propagating Lean Limit Flame

Lean limit propane flames propagate under conditions when heat conduction dominates over the molecular

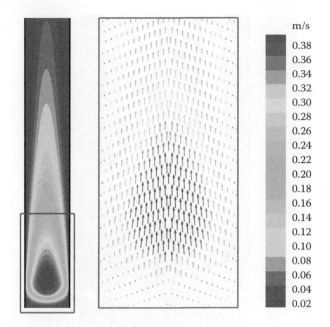

FIGURE 6.1.9
Numerical simulation of the flow field during flame propagation in a channel with $D_Q = 4$ mm ($\phi = 0.73$).

diffusion processes (Le > 1). In comparison with other flames, the dead space, D_{ds}, of such flames is relatively small. It increases linearly with the distance between the channel walls (for D_Q up to 7 mm), but the ratio D_{ds}/D_Q is nearly

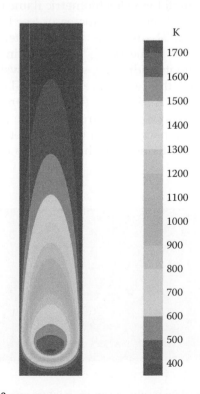

FIGURE 6.1.10
Numerical simulation of the temperature field during flame propagation in a channel with $D_Q = 4$ mm ($\phi = 0.73$).

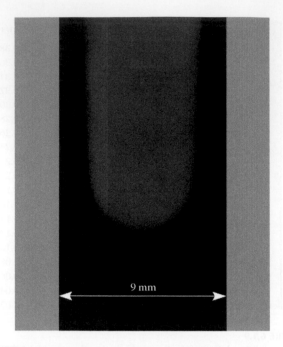

FIGURE 6.1.11
A view of a flame quenched in a channel of 9 mm. Picture recorded by a camera with open shutter ($\phi = 0.57$).

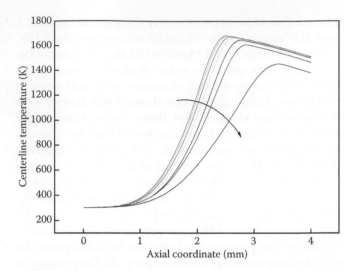

FIGURE 6.1.13
Centerline temperature profiles at various times during flame propagation and extinction in a channel with $D = 4$ mm. Temperature profiles correspond to flame positions in Figure 6.1.12.

constant a value of ≈ 0.1 (see Figure 6.1.6). For $D_Q > 7$ mm, the dead space does not change much. The radius of curvature, very small for a stoichiometric flame ($R \approx 1$ mm), increases several times as D_Q increases (see Figure 6.1.5), but the ratio R/D_Q remains approximately constant ($R/D_Q \approx 0.45$). A flame being quenched in a channel slows down and its width very slowly but systematically decreases, while the dead space grows until the flame is finally quenched. A view of a flame in a 9 mm quenching channel is shown in Figure 6.1.11. The picture was recorded by a camera with an open shutter. Numerical simulation of the flame-quenching process is presented

in Figure 6.1.12. When the flame slows down, the temperature gradients associated with it decrease and finally its characteristic profile is lost (see Figure 6.1.13).

6.1.10.2 Downward Propagating Rich Limit Flame

Rich limit propane flames propagate under conditions when molecular diffusion dominates over heat conduction processes (Le < 1). The dead space, D_{ds}, of these flames is at least twice as large as that of the downward propagating lean limit flames. Also, it grows linearly with increasing distance between the channel walls up to $D_Q = 7$ mm. The ratio D_{ds}/D_Q is nearly constant and is equal to ≈ 0.2 (see Figure 6.1.6). For $D_Q > 7$ mm, the dead space becomes practically constant. The radius of curvature of flame, R, changes with D_Q, similar to the case of limit flames propagating downward, and the ratio R/D_Q also remains nearly constant ($R/D_Q \approx 0.35$).

6.1.10.3 Lean Limit Flames Propagating Upward

The characteristic properties of lean limit flames propagating upward are approximately the same as those of flames propagating down.

6.1.10.4 Rich Limit Flames Propagating Upward

The fact that the quenching limits of upward propagating rich limit flames are much wider in comparison with those moving downward can be explained in terms of preferential diffusion. When the molecular diffusivity of the deficient reactant is higher than the thermal diffusivity of the mixture and when the flame is stretched (upward propagating convex flames

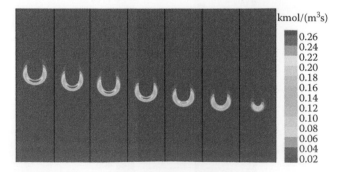

FIGURE 6.1.12
Numerical simulation results of specific mixture consumption rates during flame quenching a channel with $D = 4$ mm. Time between the frames is $\Delta\tau = 0.004$ s ($\phi = 0.729$).

are always stretched), an additional quantity of the deficient reactant are supplied to the surface of such a flame. Also, heat transport from the reaction zone by conduction is less efficient. The important quantities in this context are not the absolute values of the diffusion coefficient and thermal diffusivity, but their ratio (i.e., $Le = a/D_{diff}$). It can be stated that in a rich propane/air mixture, a unit flame surface is supplied with oxygen from a larger volume and the heat from it is transferred to a smaller volume. As a result, the reaction temperature increases. This temperature rise is responsible for the extension of the quenching limit toward richer mixtures. For mixture compositions within the narrow range of $\Phi = 1.64$–1.80, the inner instability phenomena are observed—the propagating flame has the tendency to be cellular. In richer mixtures, the quenching limit curve rises linearly up to the flammability limit and the flame is very stable. However, the flame width and its radius of curvature practically do not change with increasing equivalence ratio, ϕ. Preferential diffusion supplies an oxygen-deficient convex flame surface with more oxygen from the entire volume ahead of it. Furthermore, heat loss by the less-effective conduction ($Le < 1$) is limited to smaller volume. The mixture in the space between the flame and the walls becomes less rich than the average for the mixture. The flame occupies only a small part of the channel cross-section and a significant fraction of the mixture does not participate in the combustion process.

6.1.11 Concluding Remarks

Quenching limits depend on the heat loss to the walls. In quenching channels with wall separation, $D_Q = 7\,mm$, heat is transferred to the wall mainly by conduction. For wider channels, heat transfer can be also influenced by natural convection. There are two sources of heat loss. The first one is located in the preheat zone, and the second one in the zone of hot combustion gases, just behind the flame. The transfer of heat from these zones to the walls lowers the temperature in the reaction zone. Hence, the real flame temperature, T_b, is lower than the adiabatic temperature, T_b°, and the actual propagation velocity, S_L, is lesser than the laminar burning velocity, S_L°, at the adiabatic temperature, T_b°. If heat loss from the flame to the wall increases and reaches a critical value, then the flame temperature and the propagation velocity also take the limit values, $T_{b,lim}$ and $S_{L,lim}$, respectively, in agreement with Equations 6.1.1 and 6.1.2, and the flame is quenched. Thus, a flame can propagate in small tubes or channels only when the difference between the actual flame temperature and the adiabatic temperature is less than $R(T_b^\circ)^2/E$.

Acknowledgments

This work was sponsored by the Marie Curie ToK, project No MTKD-CT-2004-509847, and by the State Committee for Scientific Research, project No 4T12D 035 27. The authors thank Luigi Tecce from Università del Sannio, Benevento, Italy, for his commitments on the numerical simulation, and Elzbieta Bulewicz from the Cracow University of Technology, Poland, for her comments and assistance in the preparation of this chapter.

References

1. Davy, H., On the fire-damp of coal mines and the methods of lighting the mines so as to prevent explosions, *Phil. Trans. Roy. Soc.*, 106: 1, 1816.
2. Payman, W. and Wheeler, R.V., The propagation of flame through tubes of small diameter, *J. Chem. Soc.*, 113: 656, 1918; The propagation of flame through tubes of small diameter, Part II, *ibid.* 115: 36, 1919.
3. Holm, J.M., On the initiation of gaseous explosions by small flames, *Phil. Mag.*, 14: 8, 1932; *ibid.* 15: 329, 1933.
4. Zel'dovich, Ya.B., Theory of limit propagation of slow flame, *Zhur. Eksp. Teor. Fiz.*, 11: 159, 1941.
5. Zel'dovich, Ya.B., *Theory of Combustion and Gas Detonation*, Akad. Nauk SSSR, Moscow, 1944.
6. Zel'dovich, Ya.B., Barenblatt, G.I., Librovich, V.B., and Makhviladze, G.M., *The Mathematical Theory of Combustion and Explosion*, Nauka Publishing House, Moscow, 1980.
7. Friedman, R., The quenching of laminar oxyhydrogen flames by solid surfaces, *Proc. Comb. Inst.*, 3: 110, 1949.
8. Friedman, R. and Johnston, W.C., The wall-quenching of laminar propane flames as a function of pressure, temperature, and air-fuel ratio, *J. Appl. Phys.*, 21: 791, 1950.
9. Anagnostou, E. and Potter, A.E., Jr., Quenching diameters of some fast flames at low pressures, *Combust. Flame*, 3: 453, 1959.
10. Gardner, W.E. and Pugh, A., The propagation of flame in hydrogen-oxygen mixtures, *Trans. Faraday Soc.*, 35: 283, 1939.
11. Simon, D.M., Belles, F.E., and Spakowski, A.E., Investigation and interpretation of the flammability region for some lean hydrocarbon-air mixtures, *Proc. Comb. Inst.*, 4: 126, 1953.
12. Potter, A.E., Jr., and Anagnostou, E., Reaction order in the hydrogen-bromine flame from the pressure dependence of quenching diameter, *Proc. Comb. Inst.*, 7: 347, 1959.
13. Blanc, M.V., Guest, P.G., von Elbe, G., and Lewis, B., Ignition of explosive gas mixtures by electric sparks, *Proc. Comb. Inst.*, 3: 363, 1949.
14. Lewis, B. and von Elbe, G., *Combustion, Flames and Explosions of Gases*, 3rd edition, Academic Press, New York, 1987.
15. Calcote, H.F., Gregory, C.A., Jr., Barnet, C.M., and Gilmer, R.B., Spark ignition. Effect of molecular structure, *Industr. Eng. Chem.*, 44: 2656, 1952.

16. Lewis, B. and von Elbe, G., Stability and structure of burner flames, *J. Chem. Phys.*, 11: 75, 1943.

17. Lafitte, P. and Pannetier, G., The inflammability of mixtures of cyanogen and air; the influence of humidity, *Proc. Comb. Inst.*, 3: 210, 1949.

18. Potter, A.E., Jr., *Progress in Combustion Science and Technology – Volume 1*, Pergamon Press, New York, pp. 145, 1960.

19. von Elbe, G. and Lewis, B., Theory of ignition, quenching and stabilization of flames of nonturbulent gas mixtures, *Proc. Comb. Inst.*, 3: 68, 1949.

20. Mayer, E., A theory of flame propagation limits due to heat loss, *Combust. Flame*, 1: 438, 1957.

21. Spalding, D.B., A theory of flammability limits and flame quenching, *Proc. Roy. Soc.*, A240: 83, 1957.

22. Ferguson, C.R. and Keck, J.C., On laminar flame quenching and its application to spark ignition engines, *Combust. Flame*, 28: 197, 1977.

23. Aly, S.L. and Hermance, C.E., A two-dimensional theory of laminar flame quenching, *Combust. Flame*, 40: 173, 1981.

24. Jarosinski, J., Flame quenching by a cold wall, *Combust. Flame*, 50: 167, 1983.

25. Gutkowski, A., Tecce, L., and Jarosinski, J., Flame quenching by the wall—fundamental characteristics, *J. KONES*, 14(3): 203, 2007.

26. White, A.G., Limits for the propagation of flame in vapour–air mixtures, *J. Chem. Soc.*, 121: 1268, 1922.

27. Vagelopoulos, C.M. and Egolfopoulos, F.N., Direct experimental determination of laminar flame speeds, *Proc. Comb. Inst.*, 27: 513, 1998.

28. Jarosinski, J., Podfilipski, J., and Fodemski, T., Properties of flames propagating in propane-air mixtures near flammability and quenching limits, *Combust. Sci. Tech.*, 174: 167, 2002.

29. Jarosinski, J., The thickness of laminar flames, *Combust. Flame*, 56: 337, 1984.

30. Gutkowski, A., Laminar burning velocity under quenching conditions for propane-air and ethylene-air flames, *Archivum Combustionis*, 26: 163, 2006.

31. Daou, J. and Matalon, M., Influence of conductive heat-losses on the propagation of premixed flames in channels, *Combust. Flame*, 128: 321, 2002.

6.2   Flame Quenching by Turbulence: Criteria of Flame Quenching

Shenqyang S. Shy

The central conclusion of this chapter is that global quenching of premixed flame is characterized by turbulent straining (a turbulent Karlovitz number), equivalence ratio (ϕ), and heat-loss effects. To prove the idea, we shall review the global quenching of premixed flames by turbulence, based on two different reacting systems, "gaseous premixed flames" and "liquid flames." The former was presented at the 29th Combustion Symposium [1], reporting the effects of turbulent straining, equivalence ratio, and radiative heat loss (RHL) on the global quenching of premixed methane/air flames, which will be discussed in detail in this chapter. The latter is an aqueous autocatalytic reaction system that can produce self-propagating fronts with characteristics more closely matching those assumed by flamelet models than gaseous flames, because of the absence of heat-loss, constant density, and the lack of strong non-linearities (exponential dependencies) in the reaction rates. In this chapter, we will only briefly review the global quenching of these self-propagating liquid fronts, and for a detailed description on the treatment of aqueous autocatalytic reactions, front propagation rates, and comparisons with flamelet models, one can refer to the work of Shy et al. [2]. Lastly, the criteria of flame quenching by intense turbulence obtained from both gaseous premixed flames with strong heat losses and liquid flames with negligible heat losses are also presented.

6.2.1   Introduction

Flame quenching by turbulence is of both fundamental and practical importance. Consider a premixed flame propagating through a simplest turbulent flow field that is homogeneous and isotropic. Local quenching of the flame could occur when the external perturbations via aerodynamic stretch or heat losses are strong enough to decrease the reaction rate in the flame to a small value. Many studies of local quenching are available for laminar premixed flames, such as using asymptotic analysis [3], numerical simulations [4,5], and experimental methods [6,7]. The consensus is that quenching by stretch may occur if the flow is nonadiabatic or if the Lewis number (Le) > 1. Much has been learnt about the dynamics of stretched laminar flames [8]. Furthermore, the flame–vortex interactions using direct numerical simulation [9] or experimental approach [10] have further enhanced our understanding on local quenching processes of laminar premixed flames.

However, few studies of global quenching, a complete extinction and not local quenching, of turbulent premixed flames are available [11,12]. This is because flame quenching by turbulence at high Reynolds numbers involves a very wide range of spatio-temporal scales from both turbulence and chemical reactions, which are extremely difficult to measure and model [13,14]. Thus, we designed novel experimental systems to study the flame global quenching by turbulence, including a large cruciform burner for gaseous premixed flames [1] and a vibrating-grids chemical tank for liquid flames [2]. Both systems can be used to generate a large, well-controlled, intense near-isotropic turbulence region to avoid unwanted disturbances from the walls during flame–turbulence interactions.

The concept of turbulent flame stretch was introduced by Karlovitz long ago in [15]. The turbulent Karlovitz number (Ka) can be defined as the ratio of a turbulent strain rate (s) to a characteristic reaction rate (ω), which has been commonly used as a key nondimensional parameter to describe the flame propagation rates and flame quenching by turbulence. For turbulence $s \sim \sqrt{\varepsilon/\nu}$, where the dissipation rate $\varepsilon \sim u'^3/L_I$ and u', L_I, and ν are the rms velocity fluctuation, an integral length scale, and the kinematic viscosity of reactants, respectively, and for premixed flames, $\omega \sim S_L^2/D$, where S_L and D are the laminar burning velocity and the mass diffusivity, respectively. Thus $Ka \equiv (u'/S_L)^2(Sc^2Re_T)^{-0.5}$, where $Re_T \equiv u'L_I/\nu$ and $Sc \equiv \nu/D \approx 1$ for gases or $Sc \approx 600$ for water.

In 1982, Chomiak and Jarosinski [11] studied the quenching phenomena of gaseous laminar flames using intense turbulence and cold walls. They found that upward propagating premixed flames could be quenched by a series of turbulent jets located horizontally and oppositely on two sides of a rectangular duct with a square cross-sectional area, when $Ka_1 = (u'/L_I)(\delta_L/S_L)$ was about 10–20, corresponding to values of Ka varying from 70 to 450. Since, the laminar-flame thickness $\delta_L \sim D/S_L$, $Ka_1 \equiv (u'/S_L)^2(Re_T)^{-1} = Ka(Re_T)^{-0.5}$ taking $Sc = 1$. Note that the difference between Ka and Ka_1 is due to the turbulent length scale, and the Taylor microscale $\lambda \sim L_IRe_T^{-0.5}$ (not L_I) that is used in Ka. Furthermore, Bradley and his coworkers at Leeds [16] also observed global quenching of gaseous premixed flames in a fan-stirred explosion bomb, when $Ka_2 = 0.157(u'/S_L)^2Re_T^{-0.5} = 0.157\,Ka$ was unity corresponding to $Ka = 6.37$ [16]. Bradley [12] further

modified the global-quenching condition to be $Ka_2Le \approx 6$, which was equivalent to $KaLe \approx 38.2$. Clearly, there is no agreement regarding global quenching by turbulence from the above-mentioned two experiments. Though the explosion bomb had an advantage of having high-turbulent intensities with negligible mean velocities, some disadvantages were inevitable owing to the ignition processes that can greatly influence the formation of flame kernel, its subsequent flame development, and thus quenching. Obviously, it is much easier to quench a small flame kernel by turbulence than to quench a fully developed propagating flame. Therefore, the determination of actual global-quenching conditions is extremely difficult to obtain, if possible at all, using the explosion bomb configuration with the spark ignition. To improve or solve such ignition influences and further consider the effect of RHL on global quenching of premixed turbulent flames, a better methodology was proposed [1].

Figure 6.2.1 shows the cruciform burner configuration consisting of a long vertical cylindrical vessel and a large horizontal cylindrical vessel. The vertical vessel of 10 cm diameter provides a downward propagating flame with large surface area. The horizontal vessel equipped with a pair of counter-rotating fans and perforated plates at both the ends can be used to generate a large region of intense near-isotropic turbulence ($\sim15 \times 15 \times 15\,cm^3$) in the core between two perforated plates (23 cm apart) [17–19]. Turbulent intensities in it can be up to 8 m/s with negligible mean velocities when the fan frequency (f) is 170 Hz, values of skewness and flatness are nearly 0 and 3, and corresponding energy spectra have −5/3 slopes.

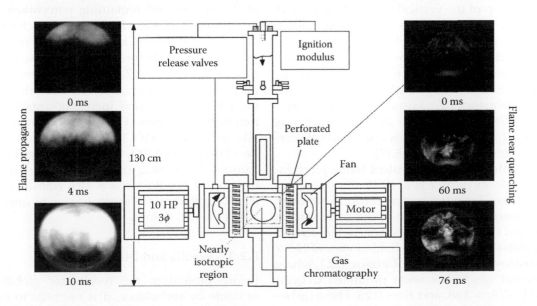

FIGURE 6.2.1
The cruciform burner (central) with two sets of sequential images displaying typical flame propagation (left) and flame near-quenching phenomena (right), where mixtures are methane and air at $\phi = 1.0$ (left) and 0.6 (right) having $u'/S_L = 3.68$ and 69.3, corresponding to $Ka = 0.25$ and 36, respectively. The concentration of the remaining CH_4 after turbulent combustion is measured by the gas chromatography. (From Yang, S.I. and Shy, S.S., *Proc. Combust. Inst.*, 29, 1841, 2002. With permission.)

This novel experimental configuration has several advantages and can be used to obtain the benchmark data for global flame quenching by turbulence. With respect to the effects of RHL on global quenching, several CH_4/air flames with different degrees of RHL, from small (N_2-diluted) to large (CO_2-diluted), were investigated. Each case covered a range of the equivalence ratio (ϕ) with values of u'/S_L, ranging from 0 to about 100, in which high rates of strain are achieved until, ultimately, global quenching of flames occurred.

The following sections describe the experimental methods and dynamics of turbulent premixed flames from propagation to pocket formation to global quench, to determine the global-quenching boundaries on a Ka versus ϕ plot for these CH_4/diluent/air flames. Thus, the critical value of Ka, Ka_c, required to quench these premixed flames globally, as a function of ϕ can be obtained. Based on the behaviors of N_2- and CO_2-diluted flames, the effect of RHL on Ka_c can be determined. For the first time, these results with and without diluents were compared with the previous data [11,12]. Finally, we shall discuss liquid flames [2], which have negligible heat losses and are extremely difficult to quench globally, indicating the importance of heat loss to turbulent flame extinguishment.

6.2.2 Experimental Procedures

Before starting the experiment, the cruciform burner was evacuated and then filled with methane/air mixtures with or without diluting gases at a given ϕ, ranging from 0.60 to 1.45 at one atmosphere. A run began with ignition and simultaneous opening of four large venting valves at the top of the vertical vessel (Figure 6.2.1). A premixed flame with large surface area (at least 10 cm in diameter) was generated, which propagated downwardly through the central uniform region for multitudinous interactions with statistically homogeneous and isotropic turbulence. At any given values of ϕ, the maximum fan frequency (f_{max}) used was 170 Hz, corresponding to $u' = 7.85$ m/s and $Re_T = 24,850$, where L_I was estimated from Taylor's hypothesis and Bradley's correlation for zero mean velocities [16,17].

It has been confirmed that turbulent flame propagation in the cruciform burner is statistically stationary (see Ref. [17]). Taken from the central uniform region of the cruciform burner using a high-speed camcorder, a typical turbulent flame propagation was demonstrated by three sequential images, with a field of view of 11.5×10.0 cm^2, is displayed on the left of Figure 6.2.1, where the experimental conditions were premixed CH_4/air flames at $\phi = 1$, $u'/S_L = 3.68$, and Ka = 0.25. These turbulent flames on propagation were fully developed having large surface, and not just a small flame kernel ignited from the spark electrodes. If the criterion for quenching

such turbulent flames at high Reynolds numbers could be obtained, it should be the actual criterion for global flame quenching by turbulence, and we believe that this was the case in the study by Yang and Shy [1]. In their study [1], the flame–turbulence interactions were not influenced by the ignition source and flame quenching by turbulence was not influenced by unwanted effects of the walls. However, since no image of flames could be shown when global quenching occurred, the best we can do was to show the flame images just prior to global quenching, as presented in Figure 6.2.1 (right), for a very lean CH_4/air mixture case at $\phi = 0.6$ with $u'/S_L = 69.3$ and Ka = 36. As can be seen, the turbulent flames were largely disrupted, becoming fragmented (distributed-like) with pockets or islands propagating randomly. Furthermore, these turbulent distributed flames had very slow overall burning rates that can survive in the central uniform region for a much longer period of time than in case of the typical turbulent flames. Even after 76 ms (Figure 6.2.1), these turbulent distributed flames propagated slowly and randomly, approached the lower vertical vessel, and finally, consumed all the remaining reactants. If the values of Ka could be increased further, these aforementioned distributed-like flames can be globally quenched. Thus, it was a clear-cut demonstration whether the global quench would occur.

We also determined the global quenching of flame via gas chromatography. After a run, the product gases were sampled from the central region of the cruciform burner (Figure 6.2.1) and the remaining CH_4 concentrations as a function of u'/S_L or Ka were measured. Figure 6.2.2 shows a typical example of the variations of the normalized remaining percentage of CH_4 fuel as a function of u'/S_L for both very rich ($\phi = 1.45$) and very lean ($\phi = 0.6$) CH_4/air flames. Furthermore, the values of Ka just before and after global quenching are plotted in Figure 6.2.2. Obviously, there is a transition for global quenching. After the transition, the remaining CH_4 concentration increases drastically for both rich and lean cases. The critical value of Ka of globally quenched, rich CH_4/air flames at $\phi = 1.45$ whose Le ≈ 1.04, is about 9.81. On the other hand, much higher values of Ka_c (> 38.2) are required for global quenching of lean CH_4/air flames at $\phi = 0.6$ whose Le ≈ 0.97. Thus, lean CH_4 flames are much harder to quench globally than rich CH_4 flames.

6.2.3 Results and Discussion

Before presenting the results for global quenching of flame by turbulence, it is essential to first describe and identify the accessible domain of our experimental configuration, limited by the maximum $f = 170$ Hz on a Ka–ϕ plot.

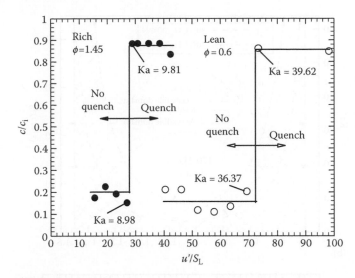

FIGURE 6.2.2

Variation of the normalized remaining percentage of CH_4 fuel (c/c_i) after a run, measured by the gas chromatography, plotted over a very wide range of normalized turbulent intensities ($u'/S_L \approx 10 \sim 100$), where the subscript "i" refers to the initial condition. Both very rich ($\phi = 1.45$; $c_i = 13.2\%$) and very lean ($\phi = 0.6$; $c_i = 5.92\%$) pure methane/air mixtures are investigated, showing critical values of Ka for the transition across which global quench occurs.

6.2.3.1 Accessible Domains

Figure 6.2.3a presents values of the maximum Karlovitz number (Ka_{max}) and S_L for pure CH_4/air flames without any diluents as a function of ϕ, where S_L was obtained from the measurements of Vagelopoulos et al. [20]. The accessible domain marked as the shaded area in Figure 6.2.3a was determined by the curve of Ka_{max}, where the two vertical lines of $\phi = 0.6$ and $\phi = 1.45$ represented the leanest and richest mixtures, respectively, used in the study [1]. Clearly, the accessible domain was very limited, as the values of ϕ were close to stoichiometry, because of larger values of S_L resulting in smaller values of Ka_{max}. Only very lean or very rich CH_4/air flames can experience high enough rates of turbulent straining that are required for global quenching of flame.

To increase the values of Ka_{max} and consequently expand the accessible domain, the values of S_L should be reduced in a subtle way. Hence, various percentages of N_2 and CO_2 diluting gases were blended into CH_4 fuel, as shown in Figure 6.2.3b and c, where values of S_L for N_2- and CO_2-diluted flames were obtained from Stone et al. [21]. With respect to the effect of RHL, CO_2-diluted flames suffer larger heat losses than N_2-diluted flames, because of a higher concentration of CO_2 in the products. This is not the only effect. Probably the lower maximum flame temperature due to higher specific heat of CO_2 are more important. Samaniego and Mantel [22] used a heat-loss coefficient (HL), a ratio of the energy radiated in the flame zone to chemical energy release, for quantifying the RHLs due to the presence of CO_2.

They reported that CO_2-diluted flames had much higher HL, up to four times more, than N_2-diluted flames [22]. For clarity, the accessible and inaccessible domains for pure CH_4/air, CH_4/N_2/air, and CH_4/CO_2/air flames from Figure 6.2.3a, b, and c, respectively, were plotted together on Figure 6.2.3d, where the values of Ka_{max} at both the leanest and richest sides of these diluted flames were also indicated. For example, at $\phi = 0.6$, the values of Ka_{max} increased from about 60 to 320 when CH_4 fuel was diluted with 60% CO_2 (Figure 6.2.3d). Thus, the accessible domain for CH_4/CO_2/air flames can be significantly expanded.

6.2.3.2 Global-Quenching Regimes

We have determined regimes for flame quenching by turbulence on the Ka–ϕ plots for CH_4/air (Figure 6.2.3a), CH_4/N_2/air (Figure 6.2.3b), and CH_4/CO_2/air (Figure 6.2.3c) flames. For any given values of ϕ, hundreds of experiments with the same mixtures but different values of Ka varying from 0 to Ka_{max} ($f = 170\,Hz$) were carried out. Thus, the values of Ka_c for global quenching of these premixed flames can be obtained. Figure 6.2.4a, b, and c shows the variations of Ka_c with ϕ on these accessible domains for CH_4/air, CH_4/N_2/air, and CH_4/CO_2/air flames, respectively. Only a limited range of ϕ on both lean and rich ends can be investigated to identify Ka_c, because of the limitation of the experimental configuration that excluded the region of $Ka_c > Ka_{max}$ (the inaccessible domain). For pure CH_4/air flames (Figure 6.2.4a), it is much harder to globally quench the lean premixed flames for which $Ka_c = 39.6$ when $\phi = 0.6$, than the rich flames for which $Ka_c = 9.81$ when $\phi = 1.45$. It is worth noting that by just slightly increasing the value of ϕ from 0.6 to 0.62, the global quenching of lean CH_4/air flames is not possible even when the value of $Ka = Ka_{max} = 49$ (Figure 6.2.4a). It must be noted that these "no quench" data points, overlapped by the symbol "X" on the Ka_{max} lines are also plotted in Figure 6.2.4a through c. For rich CH_4/air flames, when the value of ϕ decreases from 1.45 to about 1.38, the corresponding values of Ka_c increase from 9.81 to 13.4. Thus, the global-quenching regimes, as indicated by the shaded areas with dashed grid lines, can be determined.

Similar to Figure 6.2.4a, global-quenching regimes and their anticipated curves on the Ka_c–ϕ plot for N_2- and CO_2-diluted flames are presented in Figure 6.2.4b and c, respectively. It can be noted that the values of Ka_c are very sensitive to ϕ. As the values of ϕ gradually approach toward $\phi = 1$ from either the lean or rich side, the values of Ka_c increase drastically. For highly diluted flames, the range of ϕ within the lean and rich flammability limits that can be conducted in the cruciform burner was reduced. For instance, the values of ϕ varied from 0.6 to 0.72 for the leanest mixtures and from 1.45 to

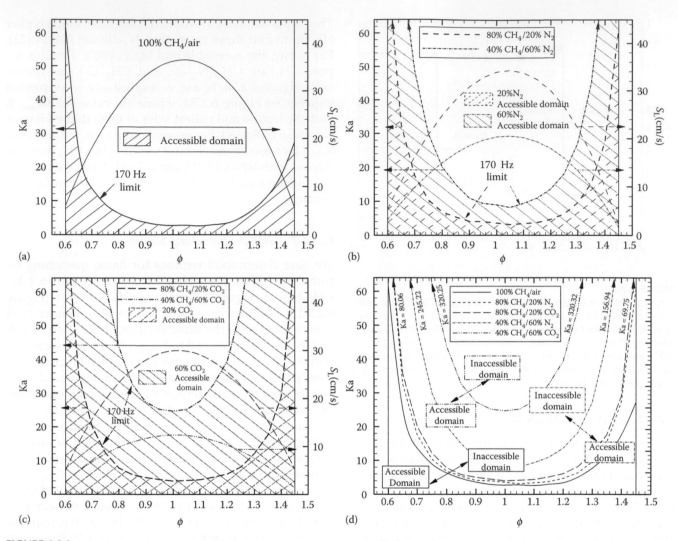

FIGURE 6.2.3
Variations of the maximum Karlovitz number and laminar burning velocities with the equivalence ratio, showing the accessible domain when the maximum f = 170 Hz is operated. (a) CH_4/air mixtures; (b) CH_4 diluted with 20–60% N_2; (c) CH_4 diluted with 20–60% CO_2; and (d) combined plots of these maximum-Ka (f_{max} = 170 Hz) lines from (a–c) for comparison. (From Yang, S.I. and Shy, S.S., *Proc. Combust. Inst.*, 29, 1841, 2002. With permission.)

1.25 for the richest mixtures, when CH_4 fuel was diluted with 60% CO_2 (Figure 6.2.3c). The global-quenching boundary for 60% CO_2-diluted flames, consisting of both real data points (the solid line) and the anticipated curve (the dash line), as shown in Figure 6.2.4c, revealed a complete variation of Ka_c with ϕ, where the maximum Ka_c was assumed to occur near ϕ = 1. Based on the same trend, we predicted the anticipated curves for both pure CH_4/air flames and 60% N_2-diluted flames. These results with both the real (solid lines from Figure 6.2.4a through c) and anticipated (dash lines) data were plotted in Figure 6.2.4d for comparison. However, if we assume that these anticipated curves were accurate, the required Ka_c for global quenching of stoichiometric CH_4/air flames must be as much as $Ka_c \approx 160$ (Figure 6.2.4d). Thus, the vitality of premixed turbulent CH_4/air flames is very impressive.

6.2.3.3 Criteria of Flame Quenching

It is presumed that the global-quenching criteria of premixed flames can be characterized by turbulent straining (effect of Ka), equivalence ratio (effect of ϕ), and heat-loss effects. Based on these aforementioned data, it is obvious that the lean methane flames (Le < 1) are much more difficult to be quenched globally by turbulence than the rich methane flames (Le > 1). This may be explained by the premixed flame structure proposed by Peters [13], for which the premixed flame consisted of a chemically inert preheat zone, a chemically reacting inner layer, and an oxidation layer. Rich methane flames have only the inert preheat layer and the inner layer without the oxidation layers, while the lean methane flames have all the three layers. Since the behavior of the inner layer is responsible for the fuel consumption that

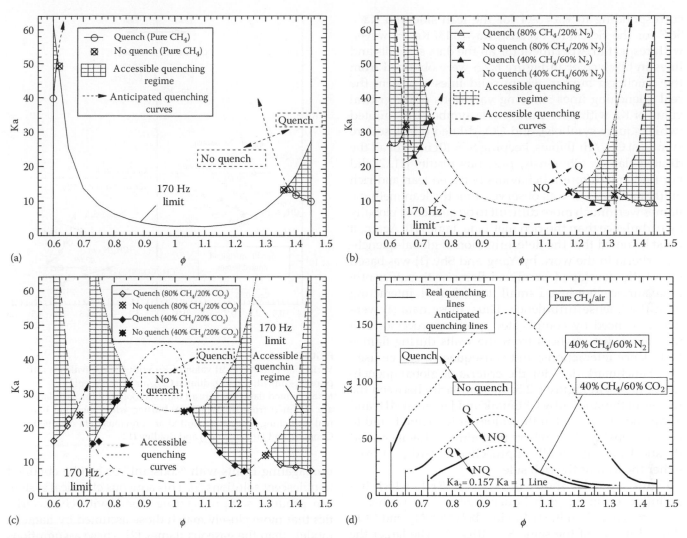

FIGURE 6.2.4
(a–c) Similar to Figure 6.2.3a through c for the accessible domains, but show the critical values of Karlovitz number for global quenching as a function of the equivalence ratio; (d) Values of Ka_c plotted against ϕ for CH_4/air, $CH_4/N_2/air$, and $CH_4/CO_2/air$ flames, respectively, where the solid lines are real quenching lines obtained from the actual data points from (a–c) and the dashed lines are the anticipated quenching lines. (From Yang, S.I. and Shy, S.S., *Proc. Combust. Inst.*, 29, 1841, 2002. With permission.)

can make the reaction process alive or extinct, the oxidation layer is quite vital. Without the oxidation layer, rich methane flames would be easier to be disrupted by intense turbulence than the lean methane. During the fuel consumption in the inner layer, the radicals are depleted by chain-breaking reactions. As pointed out by Seshadri and Peters [23], the rate-determining reaction in the inner layer is very sensitive to the presence of H radicals, and the depletion of H radicals is much more rapid in rich methane flames than in lean methane flames [24,25]. Thus, rich methane flames are more vulnerable to be quenched globally by turbulence than the lean methane flames. The differences are due to the differences in chemical structure of flames.

Figure 6.2.5 reveals the criteria of flame quenching in the Ka–ϕ plot obtained from three different turbulent combustion experiments, which include experiments by (1) Chomiak and Jarosinski [11], (2) Bradley [12], and (3) Yang and Shy [1], using the same premixed methane/air mixtures, where results of (1) and (2) are marked with gray lines or gray symbols. In the case of experiment (1), the available data [11] for the very limited range of ϕ, varying from lean (0.5–0.6) to rich (1.4–1.55) indicated that the global quenching of both lean and rich methane flames occurred when the values of Ka_1 were as large as 10–20, corresponding to the values of Ka ≈ 70–110 for the lean case or 110–200 for the rich case ($L_I = 0.214$ cm), and Ka ≈ 250–450 for both lean and rich cases ($L_I = 0.6$ cm). Again, Ka_1 used the integral length scale, so that $Ka_1 \equiv (u'/S_L)^2 (Re_T)^{-1} = Ka(Re_T)^{-0.5}$. In the case of experiment (2), Bradley reported a criterion for global quenching when $Ka_2 = 1$ [16], which was later modified to $Ka_2 Le = 6$ [12], corresponding to Ka = 39.4 for the lean side and Ka = 36.7 for the rich side, because $Ka_2 = 0.157\,Ka$. However, with respect to experiment (3), all the data symbols and lines in black color in Figure 6.2.5 were obtained from the cruciform burner, where solid lines represent the real quenching lines showing variations of the critical value of Ka with ϕ for pure (circle symbols), N_2-diluted (white square symbols), and CO_2-diluted (black square symbols) CH_4/air flames, keeping $S_L \approx 10\,cm/s$ for all the diluted flames. Obviously, previous results [11,12] did not find any different behaviors between lean and rich methane flames. It was observed that the lean methane flames were much more difficult to be quenched globally by turbulence than the rich methane flames [1]. Again, it must be noted that the determination of global-quenching criteria in the work by Yang and Shy [1] was based on a fully developed premixed flame with sufficiently large surface, not just a small flame kernel, interacting with the intense turbulence. Thus, those data [1] were not influenced by complicated ignition processes and unwanted disturbances from the walls during flame–turbulence interactions, and consequently, can be used as a benchmark data for the criteria of global quenching of flame. In Figure 6.2.5, one similarity between the results of the studies by Chomiak and Jarosinski [11] and Yang and Shy [1] is that the values of Ka_c were found to be very sensitive to ϕ, where the values of Ka_c increased drastically as ϕ gradually approached toward $\phi = 1$ from either the lean or the rich side.

Figure 6.2.5 also shows the effect of RHL, which has an influence on the global quenching of lean methane/air flames based on the behaviors between N_2- and CO_2-diluted flames of the same $S_L \approx 10\,cm/s$. The larger the RHL is, the smaller is the value of Ka_c. For example, $Ka_c = 26.1$ for N_2-diluted flames (small RHL), while $Ka_c = 20.4$ for CO_2-diluted flames (large RHL) when $\phi \approx 0.64$. It is found that for lean methane/air flames of constant S_L, the values of Ka_c increased with ϕ for both N_2- and CO_2-diluted flames, and the difference in the values of Ka_c between these two different diluted flames also increased with ϕ, as shown in Figure 6.2.5. On the other hand, the effects of RHL did not have influence on the global quenching of rich methane/air flames, because $Ka_c \approx 8.4$ for both N_2- and CO_2-diluted flames (values of Ka are in a log plot in Figure 6.2.5).

6.2.3.4 Global Quenching of Liquid Flames

At the 24th Combustion Symposium, Shy et al. [26] introduced an experimental aqueous autocatalytic reaction system to simulate the premixed turbulent combustion in a well-known Taylor–Couette (TC) flow field. By electrochemically initiating this reaction system, the

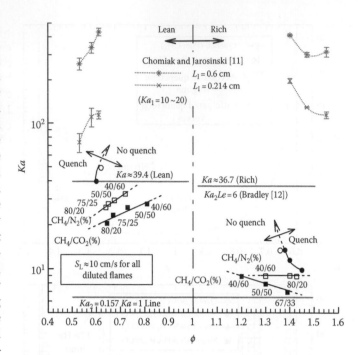

FIGURE 6.2.5
Variations of the critical value of Ka with the equivalence ratio for pure, N_2-diluted, and CO_2-diluted CH_4/air flames keeping $S_L \approx 10\,cm/s$ for all diluted flames, where all data symbols and lines in black color are obtained from the cruciform burner, while the solid lines are real quenching lines. Also plotted are previous data obtained from Chomiak and Jarosinski [11] and Bradley [12].

propagating fronts with "constant" S_L can be obtained in quiescent solutions [2]. This aqueous chemical system can produce self-propagating fronts with characteristics that more closely match those assumed by flamelet models than the gaseous flames [2]. These assumptions commonly included (1) Huygen's propagation, or "thin-flame" model, (2) no heat losses, (3) negligible thermal expansion, (4) constant transport properties, and (5) an idealized turbulence, which is homogeneous and isotropic. Since the aqueous autocatalytic mixture has $Sc \approx 500$ [26], while the $Sc \approx 1$ in gases, Huygen's propagation assumption might be valid at higher values of u'/S_L in the former than in the latter condition for a fixed Re_T. These aqueous propagating fronts have very small exothermicity (typically 1 K), across which the density changes are about 0.02% and the Zel'dovich number (Ze) is close to zero [26], compared with 600% and $Ze \approx 10–20$ in gas combustion. The small exothermicity might render heat diffusion (the Lewis number) and activation temperature (the Zel'dovich number) irrelevant to the propagation mechanism.

In the study by Shy et al. [26], the appropriate chemical solutions were identified and applied to the TC flow in the annulus between two rotating concentric cylinders, with the outer cylinder remaining stationary.

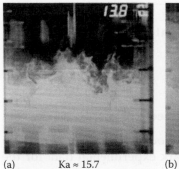 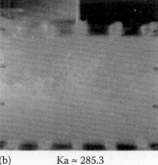

(a)   Ka ≈ 15.7   (b)   Ka ≈ 285.3

FIGURE 6.2.6
Liquid flames propagating in a transparent rectangular tank equipped with a pair of vertically vibrating grids, generating a region of statistically homogeneous and isotropic turbulence, showing instantaneous cross-sectional LIF photographs of the propagating fronts at two different modes: (a) largely wrinkled but still sharp fronts and (b) front broadening just before global quenching.

No global quenching of aqueous propagating fronts was observed, even at values of Ka that were 1000 times greater than those that extinguished gaseous fronts, and thus, it was suggested that without heat loss, the turbulence alone may not be sufficient to quench the flames globally [26]. However, the TC flow had a very narrow annulus gap (<1 cm), limiting the interactions between the disturbances and the wrinkled fronts. Therefore, Shy et al. [2] applied a pair of horizontally oriented and vertically vibrating grids in an aqueous chemical tank ($15 \times 15 \times 30$ cm^3) to generate a large region of statistically homogeneous and isotropic turbulence in the core between the two grids. Figure 6.2.6 shows the two instantaneous cross-sectional LIF photographs of these aqueous propagating fronts in the vibrating-grids turbulence at different values of Ka: (1) Ka ≈ 16 showing largely wrinkled but still sharp fronts and (2) Ka ≈ 285 revealing drastic front broadening just before global quenching. It was observed that the global quenching of liquid flames was possible, provided the value of Ka was as large as 300, a value well beyond the critical value for global quenching of gaseous flames.

6.2.4 Conclusions

This study contributes to our understanding of global-quenching processes of premixed turbulent flames, in which global-quenching regimes are identified on Ka$_c$–ϕ plots. It is much harder to globally quench the lean CH$_4$/air flames by intense turbulence than the rich CH$_4$/air flames, which signifies that the turbulent premixed flames are sensitive to recombination reaction. Values of Ka$_c$ increase largely as ϕ gradually approaches toward $\phi = 1$ from either the lean or rich side, with the maximum

Ka$_c$ occurring possibly near $\phi = 1$. The vitality of premixed turbulent CH$_4$/air flames at $\phi = 1$ is astonishing, because the anticipated value of Ka$_c$ was as high as 160 for the occurrence of global quench. Moreover, global quenching of lean/rich CH$_4$/diluent/air flames ($S_L \approx$ 10 cm/s) was/was not influenced by the RHL, respectively. For lean mixtures, the smaller the RHL was, the larger was the value of Ka$_c$ (>20). Values of Ka$_c$ for both rich N$_2$- and CO$_2$-diluted flames were not much different (Ka$_c \sim 8.3$) with the same $S_L \approx 10$ cm/s in the range of $\phi \approx 1.20$–1.45. Based on the results for liquid flames, it is extremely difficult to globally quench the nearly adiabatic (slightly exothermic, typically 1 K) self-propagating reaction fronts by intense turbulence, revealing the importance of heat loss to flame quenching. Thus, it can be concluded that the global flame quenching by turbulence is characterized by multitudinous interactions among turbulent straining, flame chemistry, and heat losses, as discussed.

References

1. Yang, S.I. and Shy, S.S., Global quenching of premixed CH4/air flames: Effects of turbulent straining, equivalence ratio, and radiative heat loss, *Proc. Combust. Inst.*, 29, 1841, 2002.
2. Shy, S.S., Jang, R.H., and Tang, C.Y., Simulation of turbulent burning velocities using aqueous autocatalytic reactions in a near-homogeneous turbulence, *Combust. Flame*, 105, 54, 1996.
3. Libby, P., Liñán, A., and Williams, F., Strained premixed laminar flames with nonunity Lewis numbers, *Combust. Sci. Technol.*, 34, 257, 1983.
4. Darabiha, N., Candel, S., and Marble, F., The effect of strain rate on a premixed laminar flame, *Combust. Flame*, 64, 203, 1986.
5. Giovangigli, V. and Smooke, M., Extinction of strained premixed laminar flames with complex chemistry, *Combust. Sci. Technol.*, 53, 23, 1987.
6. Ishizuka, S. and Law, C.K., An experimental study on extinction and stability of stretched premixed flames, *Proc. Combust. Inst.*, 19, 327, 1982.
7. Sato, J., Effects of Lewis number on extinction behavior of premixed flames in a stagnation flow, *Proc. Combust. Inst.*, 19, 1541, 1982.
8. Law, C.K., Dynamics of stretched flames, *Proc. Combsut. Inst.*, 22, 1381, 1988.
9. Poinsot, T., Veynante, D., and Candel, S., Quenching processes and premixed turbulent combustion diagrams, *J. Fluid Mech.*, 228, 561, 1991.
10. Roberts, W.L., Driscoll, J.F., Drake, M.C., and Ratcliffe, J.W., OH fluorescence images of the quenching of a premixed flame during an interaction with a vortex, *Proc.*

Combust. Inst., 24, 169, 1992; Roberts, W.L., Driscoll, J.F., Drake, M.C., and Goss, L.P., Images of the quenching of a flame by a vortex—To quantify regimes of turbulent combustion, *Combust. Flame*, 94, 58, 1993.

11. Chomiak, J. and Jarosinski, J., Flame quenching by turbulence, *Combust. Flame*, 48, 241, 1982; Jarosinski, J., Strehlow, R.A., and Azarbarzin, A., The mechanisms of lean limit extinguishment of an upward and downward propagating flame in a standard flammability tube, *Proc. Combust. Inst.*, 19, 1549, 1982.

12. Bradley, D., How fast can we burn, *Proc. Combust. Inst.*, 24, 247, 1992.

13. Peters, N., *Turbulent Combustion*, Cambridge University Press, Cambridge, 2000.

14. Ronney, P.D., Some open issues in premixed turbulent combustion, In: *Modeling in Combustion Science*, J.D. Buckmaster and T. Takeno, Eds., *Lecture Notes in Physics*, Springer-Verlag, Berlin, Vol. 449, p. 3, 1995.

15. Karlovitz, B., Denniston, D.W., Knapschaefer, D.H., and Wells, F.E., Studies on turbulent flames, *Proc. Combust. Inst.*, 4, 613, 1953.

16. Abdel-Gayed, R.G., Bradley, D., and Lawes, M., Turbulent burning velocities: A general correlation in terms of straining rates, *Proc. R. Soc. Lond. A*, 414, 389, 1987.

17. Shy, S.S., I, W.K., and Lin, M.L., A new cruciform burner and its turbulence measurements for premixed turbulent combustion study, *Exp. Thermal Fluid Sci.*, 20, 105, 2000.

18. Shy, S.S., Lin, W.J., and Wei, J.C., An experimental correlation of turbulent burning velocities for premixed turbulent methane-air combustion, *Proc. R. Soc. Lond. A*, 456, 1997, 2000.

19. Shy, S.S., Lin, W.J., and Peng, K.Z., High-intensity turbulent premixed combustion: General correlations of turbulent burning velocities in a new cruciform burner, *Proc. Combust. Inst.*, 28, 561, 2000.

20. Vagelopoulos, C.M., Egolfopoulos, F.N., and Law, C.K., Further considerations on the determination of laminar flame speeds with the counterflow twin-flame technique, *Proc. Combust. Inst.*, 25, 1341, 1994.

21. Stone, R., Clarke, A., and Beckwith, P., Correlations for the laminar-burning velocity of methane/diluent/air mixtures obtained in free-fall experiments, *Combust. Flame*, 114, 546, 1998.

22. Samaniego, J.M. and Mantel, T., Fundamental mechanisms in premixed turbulent flame propagation via flame–vortex interactions: Part I: Experiment, *Combust. Flame*, 118, 537, 1999.

23. Seshadri, K. and Peters, N., The inner structure of methane–air flames, *Combust. Flame* 81:96 1990.

24. William, F., *Combustion Theory*, 2nd ed., Addison-Wesley Publishing Co., New York, 1985.

25. Seshadri, K., Bai, X.S., and Pitsch, H., Asymptotic structure of rich methane-air flames, *Combust. Flame*, 127, 2265, 2001.

26. Shy, S.S., Ronney, P., Buckley, S., and Yahkot, V., Experimental simulation of premixed turbulent combustion using a liquid-phase autocatalytic reaction, *Proc. Combust. Inst.*, 24, 543, 1993.

6.3 Extinction of Counterflow Premixed Flames

Chih-Jen Sung

6.3.1 Motivation and Objectives

The counterflow configuration has been extensively utilized to provide benchmark experimental data for the study of stretched flame phenomena and the modeling of turbulent flames through the concept of laminar flamelets. Global flame properties of a fuel/oxidizer mixture obtained using this configuration, such as laminar flame speed and extinction stretch rate, have also been widely used as target responses for the development, validation, and optimization of a detailed reaction mechanism. In particular, extinction stretch rate represents a kinetics-affected phenomenon and characterizes the interaction between a characteristic flame time and a characteristic flow time. Furthermore, the study of extinction phenomena is of fundamental and practical importance in the field of combustion, and is closely related to the areas of safety, fire suppression, and control of combustion processes.

One significant result from the studies of stretched premixed flames is that the flame temperature and the consequent burning intensity are critically affected by the combined effects of nonequidiffusion and aerodynamic stretch of the mixture (e.g., Refs. [1–7]). These influences can be collectively quantified by a lumped parameter $S \sim (\mathrm{Le}^{-1}-1)\kappa$, where Le is the mixture Lewis number and κ the stretch rate experienced by the flame. Specifically, the flame temperature is increased if $S > 0$, and decreased otherwise. Since Le can be greater or smaller than unity, while κ can be positive or negative, the flame response can reverse its trend when either Le or κ crosses its respective critical value. For instance, in the case of the positively stretched, counterflow flame, with $\kappa > 0$, the burning intensity is increased over the corresponding unstretched, planar, one-dimensional flame for Le < 1 mixtures, but is decreased for Le > 1 mixtures.

It is also well known that there exist different extinction modes in the presence of radiative heat loss (RHL) from the stretched premixed flame (e.g., Refs. [8–13]). When RHL is included, the radiative flames can behave differently from the adiabatic ones, both qualitatively and quantitatively. Figure 6.3.1 shows the computed maximum flame temperature $T_{\max}$ as a function of the stretch rate κ for lean counterflow methane/air flames of equivalence ratio $\phi = 0.455$, with and without RHL. The stretch rate in this case is defined as the negative maximum of the local axial-velocity gradient ahead of the thermal mixing layer. For the lean methane/air flames,

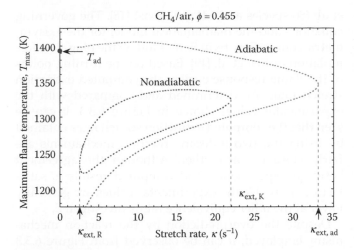

FIGURE 6.3.1
Maximum flame temperature T_{max} as a function of the stretch rate κ for lean methane/air ($\phi = 0.455$) flames, with and without RHL, respectively, indicated as nonadiabatic and adiabatic. Symbol indicates the adiabatic flame temperature T_{ad}.

T_{max} is typically the temperature at the stagnation surface. The reaction mechanism used for methane flame calculations was taken from GRI-Mech 1.2 [14], which consists of 32 species and 177 elementary reaction steps. For the nonadiabatic situations, the RHL was included in the energy equation based on the optically thin assumption, considered to be emitted only from the combustion products: CO_2, H_2O, and CO. The Planck mean absorption coefficients for the radiating species are given as a function of temperature in certain earlier studies [15,16]. The present quasi-one-dimensional formulation also implies negligible lateral conductive heat loss, and the calculations were performed for 1 atmosphere pressure and 300 K upstream temperature.

For the adiabatic condition in which RHL is suppressed, the flame response exhibits the conventional upper and middle branches of the characteristic ignition–extinction curve, with the upper branch representing the physically realistic solutions. It can be noted that the effective Le of this lean methane/air mixture is sub-unity. It can be seen from Figure 6.3.1 that, with increasing stretch rate, T_{max} first increases owing to the nonequidiffusion effects ($S > 0$), and then decreases as the extinction state is approached, owing to incomplete reaction. Furthermore, T_{max} is also expected to degenerate to the adiabatic flame temperature, T_{ad}, when $\kappa = 0$.

On the contrary, the nonadiabatic flames exhibit an isola response, with dual extinction turning points that are designated as $\kappa_{ext,K}$ and $\kappa_{ext,R}$ for the higher and lower values of κ, respectively. Over the upper, stable branch of the isola, it can be observed that, with increasing stretch rate, the flame temperature first increases and then decreases, with the extinction occurring at a

maximum stretch rate, $\kappa_{ext,K}$. Extinction mechanism and response at this limit are similar to those of the adiabatic flame, except for the additional radiative loss that facilitates extinction and renders $\kappa_{ext,K} < \kappa_{ext,ad}$. This extinction state is the stretch-induced extinction limit.

Extinction, however, can also be induced with decreasing stretch rate, because of the progressive increase in the flow time and RHL from the flame. Thus, there is also a minimum stretch rate, $\kappa_{ext,R}$, below which steady burning is again not possible. This limit is known as the radiative loss-induced extinction limit, nevertheless recognizing that the stretch is still operative in influencing the flame response. Consequently, there is only a finite range of the stretch rate over which steady burning is possible. We can further note that while $\kappa_{ext,K}$ is about 66% of $\kappa_{ext,ad}$, their extinction flame temperatures only differ by 23 K and are lower than T_{ad} by <60 K. However, the extinction flame temperature at the lower stretch limit is about 160 K below T_{ad}. Thus, the flame is much weaker at the state of the radiative loss-induced extinction.

This section emphasizes on flame quenching by stretch, as well as highlights and separately discusses the four aspects of counterflow premixed flame extinction limits, including (1) effect of nonequidiffusion, (2) influence of different boundary conditions, (3) effect of pulsating instability, and (4) relationship of the fundamental limit of flammability.

6.3.2 Effect of Nonequidiffusion

When using the counterflow twin-flame configuration to experimentally determine the stretch-induced extinction limits, extinction owing to flame blowoff can be brought about with increasing flow rate, and hence stretch rate. The stretch rate just prior to the onset of extinction can then be measured and identified as the extinction stretch rate using laser Doppler velocimetry (LDV) or digital particle image velocimetry (DPIV). Figure 6.3.2 shows the direct images of the stretched twin n-decane/O_2/N_2 flames right before the occurrence of abrupt extinction at three different equivalence ratios. Here, the molar ratio of $N_2/(N_2 + O_2)$ is 0.84, and the unburned mixture temperature T_u is 400 K.

Experimentally, two modes of extinction, based on the separation between the twin flames are observed. Specifically, the extinction of lean counterflow flames of n-decane/O_2/N_2 mixtures occurs with a finite separation distance, while that of rich flames exhibits a merging of two luminous flamelets. The two distinct extinction modes can be clearly seen in Figure 6.3.2. As discussed earlier, the reactivity of a positively stretched flame with Le smaller (greater) than unity increases (decreases) with the increasing stretch rate. Therefore, the experimental observation is in agreement with the

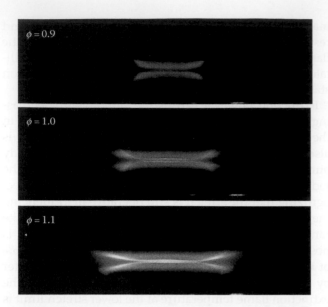

FIGURE 6.3.2
Direct images of near-extinction n-decane/O_2/N_2 flames with unburned mixture temperature $T_u = 400\,K$. The molar ratio of $N_2/(N_2 + O_2)$ is 0.84.

anticipated behavior that the sub-unity Le (Le < 1) counterflow flames, such as rich n-decane/O_2/N_2 mixtures, extinguish in the merged flame mode owing to incomplete reaction, while the Le > 1 counterflow flames, such as lean n-decane/O_2/N_2 mixtures, extinguish by being located at a finite distance away from the stagnation surface owing to the nonequidiffusion effect ($S < 0$).

The measured extinction stretch rates for n-decane/O_2/N_2 mixtures at 400 K preheat temperature as a function of equivalence ratio are shown in Figure 6.3.3. The flame response curves at varying equivalence ratios are also computed using the kinetic mechanisms of Bikas and Peters (67 species and 354 reactions) [17] and Zhao

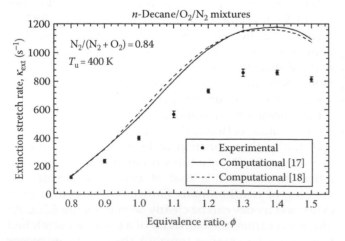

FIGURE 6.3.3
Experimental (symbols) and computed (lines) extinction stretch rates of n-decane/O_2/N_2 mixtures with $T_u = 400\,K$. The molar ratio of $N_2/(N_2 + O_2)$ is 0.84.

et al. (86 species and 641 reactions) [18]. The governing equations and the mathematical model for the axisymmetric counterflow twin flames follow the plug-flow formulation of Kee et al. [19]. Based on the turning points of the flame response curves, the computed extinction stretch rates are determined and compared with the experimental values shown in Figure 6.3.3. It can be seen that the computed extinction stretch rates obtained by using the two n-decane reaction mechanisms are fairly close to each other. Although the agreement between experimental and computed results is satisfactory at $\phi = 0.8$, the experimental values are generally lower at the other equivalence ratios investigated.

Despite the overprediction by the reaction mechanisms employed, it can be observed from Figure 6.3.3 that both the experimental and predicted extinction stretch rates for the n-decane counterflow flames show the same trend, with the extinction stretch rate peaking at $\phi \sim 1.4$. This rich shift is caused by the combined effects of positive stretch and sub-unity Lewis number for rich mixtures, as discussed earlier. It can be further noted that the overprediction of extinction stretch rate could be due to the deficiencies of the combustion chemistry of n-decane as well as the quasi-one-dimensional nature of the counterflow flame modeling. The former is reflected from the fact that the existing reaction mechanisms still cannot well-predict the experimental laminar flame speed data [20]. For the latter, a detailed computational study is necessary to compare the extinction stretch rates predicted by quasi-one-dimensional and two-dimensional modeling. At the global level, both the flame images (cf. Figure 6.3.2) and the DPIV-determined flow fields indicate that the core region of the flame remains fairly one-dimensional, even prior to the abrupt blow-off. However, experimental mapping of major and key minor species contours are needed to confirm the one-dimensionality of the near-extinction flame structure.

6.3.3 Influence of Boundary Conditions

In the numerical modeling of counterflow flames, two types of fluid mechanical description of the strained-flow field have been conventionally used, namely the potential-flow and the plug-flow formulations. By recognizing that the experimental outer-flow field is neither a plug flow nor a potential flow, Kee et al. [19] compared the experimentally determined extinction stretch rate as well as the computed extinction stretch rates obtained with the potential-flow and plug-flow formulations, and demonstrated that, though they differ from each other, the plug-flow results seem to agree better with the experimental data and hence is a better boundary condition. Furthermore, the flame structure remained nearly the same for calculations employing either of the flows as the boundary condition [19].

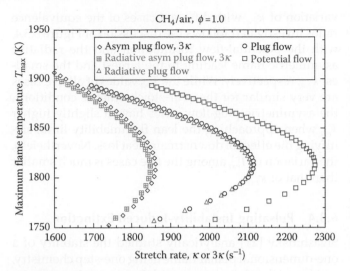

FIGURE 6.3.4
Responses of maximum flame temperature to stretch rate for stoichiometric methane/air flame, employing boundary conditions of potential flow (twin flames), plug flow (twin flames), asymmetric plug-flow (single flame), radiative plug twin-flames, and radiative asymmetric plug single-flame. For asymmetric plug-flow cases, 3κ is plotted along the abscissa axis.

We re-examined the results of Kee et al. [19] with respect to three considerations. First, as the axial-velocity gradient of the outer flow continuously changes for the plug-flow formulation, it is reasonable to expect that there would be considerable uncertainty in determining the effective stretch rate and consequently, in assessing the difference found in the comparison. Second, it is somewhat perplexing that the computed extinction stretch rates of Kee et al. [19] was observed to differ from each other for the two formulations, while the flame structure remained nearly the same. Third, there is no explicit display of the flame structure used in

the comparison in the work of Kee et al. [19], rendering it somewhat difficult to assess the extent of disagreement/agreement obtained from the two formulations.

To scrutinize the sensitivity of the flame structure to the description of the outer-flow field, we compared the flame structure obtained from the two limiting boundary conditions at the extinction state, which can be considered to be the most aerodynamically and kinetically sensitive state of the flame for a given mixture concentration, and demonstrated that they were basically indistinguishable from each other. This result thus suggests that the reported discrepancies in the extinction stretch rates as mentioned in the work by Kee et al. [19] are simply the consequences of the "errors" associated with the evaluation of the velocity gradients.

Figure 6.3.4 compares the upper branch of the S-curve for adiabatic stoichiometric methane/air flames of $T_u = 300\,K$, employing the potential-flow and plug-flow boundary conditions. In this case, the computed maximum axial-velocity gradient ahead of the flame is used to determine the stretch rate, and the turning point of the flame response curve defines the extinction stretch rate κ_{ext}. It can be seen that the potential-flow formulation yields higher κ_{ext} than the plug-flow formulation. However, the maximum temperature at the extinction turning point, T_{ext}, is around 1813 K for both cases, and does not depend on the description of the external flow. Such insensitivity is further demonstrated through the profile comparisons of the velocity, temperature, major species, and important radicals (H, O, and OH), as shown in Figure 6.3.5. It can be observed that despite the differences in the outer flows, these profiles within the thermal mixing layer are essentially indistinguishable. Furthermore, Figure 6.3.5 shows that the local axial-velocity gradient entering the thermal mixing

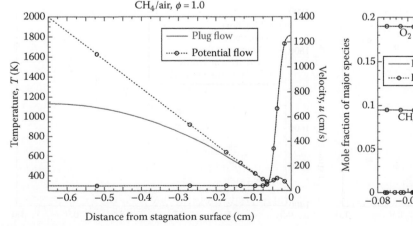
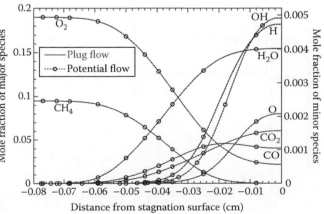

FIGURE 6.3.5
Profile comparison of temperature, velocity, major species (CH_4, O_2, CO, CO_2, and H_2O), and minor species (H, O, and OH) at the extinction state using different outer-flow conditions, for counterflow twin-stoichiometric methane/air flames. For clarity, the symbols do not represent the actual grid distribution employed in the calculation.

layer collates very well for both the outer-flow fields. Consequently, it is reasonable to suggest that the discrepancies between the computed extinction stretch rates of Kee et al. [19] are largely owing to the definition-dependent judgment in evaluating the axial-velocity gradient upstream of the flame. Taking the plug-flow solution as an example, Figure 6.3.5 clearly shows that a significant variation in the value of stretch rate could result depending on how it is evaluated. Therefore, an unambiguous parameter is needed to characterize the extinction limit of a counterflow flame.

Since flame extinction is highly sensitive to downstream heat loss, the twin-flame configuration, which is considered as adiabatic on account of flame symmetry, yields a higher extinction stretch rate for a given fuel/oxidizer mixture than other types of nonadiabatic counterflow/stagnation flames with or without downstream heat loss. Figure 6.3.4 also includes the flame response curve for an asymmetric counterflow configuration, by flowing premixture against nitrogen of the same mixture temperature. It can be seen from Figure 6.3.4 that the extinction stretch rate, κ_{ext}, of the twin-flame case is much higher than the asymmetric case, while the T_{ext} values are basically identical for both cases. Furthermore, the response curves of radiative twin-flame and radiative asymmetric-flame in the optically thin limit are also compared in Figure 6.3.4. Although the radiative flame has a lower κ_{ext} than its nonradiative counterpart, the values of T_{ext} are very similar for this stoichiometric condition.

Figure 6.3.6 further compares κ_{ext} and T_{ext} of lean to stoichiometric methane/air mixtures for all five cases—plug flow, potential flow, asymmetric plug flow, radiative plug-flow, and radiative asymmetric plug-flow. The variation of κ_{ext} with different cases of the equivalence ratios investigated follows the discussion of Figure 6.3.4, with the potential-flow formulation and the radiative asymmetric flame leading to the largest and the smallest κ_{ext}, respectively. While the T_{ext} values of all five cases are very similar for the near-stoichiometric conditions, the asymmetric plug-flow cases have a slightly higher T_{ext} when approaching the lean flammability limit, signifying the effect of downstream heat loss. Nevertheless, the variation of T_{ext} among the five cases is much smaller than that of κ_{ext}.

6.3.4 Pulsating Instability-Induced Extinction

Sivashinsky [21] analytically studied the stability of a one-dimensional planar flame using one-step chemistry, and found that for mixtures with Lewis numbers greater than unity, a self-oscillating regime sets in when the activation energy is sufficiently large, while the cellular instability is promoted for Le < 1 flames and suppressed for Le > 1 flames. The onset of unstable cellular and pulsating modes of flame propagation, predicted for the one-dimensional planar flame, is expected to be modified in the presence of aerodynamic stretch, which is usually present in all practical flames, and is manifested through flow nonuniformity, flame curvature, and flame unsteadiness. Indeed, theories by Sivashinsky et al. [22] and Bechtold and Matalon [23] show that positive stretch, which is associated with, for example, the counterflow flame and the outwardly propagating spherical flame, tends to suppress the cellular instability, while negative stretch promotes it. Regarding pulsating instability, a numerical study [24] on outwardly propagating spherical flames with one-step chemistry

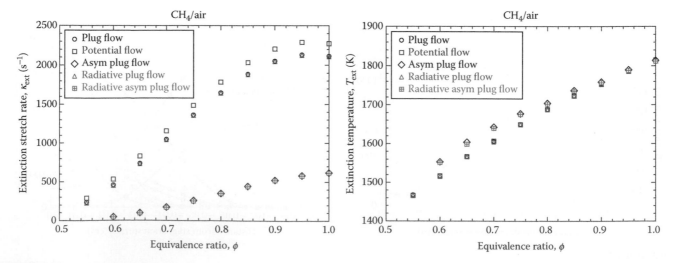

FIGURE 6.3.6
Variations of extinction stretch rate and extinction temperature of methane/air mixtures with equivalence ratio for the five different cases as in Figure 6.3.4.

indicated that oscillation is promoted at small radii, exhibiting strong positive stretch, while it is damped at large radii. Through quasi-steady asymptotic analysis and transient computational simulation with detailed chemistry of the negatively stretched, inwardly propagating spherical flame and the positively stretched counterflow flame in rich hydrogen/air mixtures, Sung et al. [25] demonstrated that positive stretch promotes the onset of flame pulsation, while negative stretch retards it. Thus, in terms of positive/negative nature of the stretch, the influence of stretch on pulsating instability appears to be completely opposite to that on cellular instability.

Since stretch affects the onset of flame pulsation, it should correspondingly affect the state of extinction in the pulsating mode. We shall therefore assess the influence of stretch in modifying the state of extinction. If the extinction turning point of a steady-flame response curve is neutrally stable, the entire upper branch should be dynamically stable; then, the corresponding static-extinction stretch rate is the physical limit.

The inset of Figure 6.3.7 shows the flame response for rich hydrogen/air mixture of $\phi = 7.0$. Since the Lewis number of this mixture is sufficiently greater than unity, it is susceptible to diffusional-thermal pulsating instability. Four flames, denoted by Flames I–IV along the

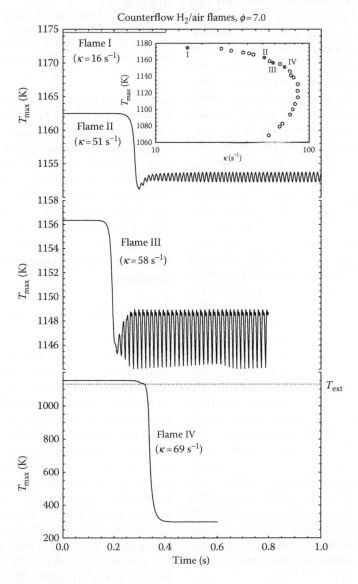

FIGURE 6.3.7
Time variations of maximum flame temperature (T_{max}) for Flames I–IV. The inset shows the steady-state flame response for hydrogen/air mixture of $\phi = 7.0$. Results demonstrate that Flame I is dynamically stable, Flame II is monochromatically oscillatory, Flame III exhibits pulsation with period doubling, and Flame IV is extinguished through pulsation.

upper branch, are used as the initial conditions for the unsteady runs, with Flame IV being the state near the static-extinction turning point. The detailed reaction mechanism used in the calculations for the hydrogen flames was obtained from Kim et al. [26]. Thermal diffusion was also included in the calculation. First, it can be seen from Figure 6.3.7 that Flame I is dynamically stable. As the stretch rate is increased to $51\,s^{-1}$ (Flame II), periodic oscillation develops with frequency of ~67 Hz. Since this equivalence ratio ($\phi = 7.0$) is smaller than the onset value of $\phi = 7.4$ for the one-dimensional, planar, unstretched flame [27], the result demonstrates that positive stretch promotes pulsating instability.

By further increasing the stretch rate to $58\,s^{-1}$ (Flame III), it can be noted in Figure 6.3.7 that not only the oscillation intensifies, but period doubling also develops with a similar frequency of oscillation (~61 Hz) as that of Flame II. It can also be observed that the characteristic flame time of the one-dimensional, planar, $\phi = 7.0$ hydrogen/air flame is around 1.8 ms, based on the adiabatic flame speed of 44 cm/s and the characteristic flame thickness of 0.08 cm. Thus, the period of these intrinsic oscillations is much longer than the corresponding characteristic flame time. As a consequence, the transient flame response is quasi-steady in nature. In particular, the unsteady flame cannot recover once its T_{max} drops below the corresponding T_{ext}. Figure 6.3.7 shows the variation of T_{max} with time for Flame IV. It can be seen that, upon the onset of intrinsic unsteadiness, T_{max} decreases monotonically until the flame extinguishes.

The above results demonstrate that while the one-dimensional, unstretched hydrogen/air flame of $\phi = 7.0$ propagates steadily, upon positive straining its counterflow counterpart could lose stability to pulsation beyond a critical stretch rate. Moreover, the unsteady counterflow flame could also lead to extinction even though the initial stretch rate is smaller than the corresponding steady-state extinction stretch rate. This suggests that pulsating instability reduces the flammable range. Figure 6.3.8 compares the static-extinction stretch rates and the onset stretch rates leading to transient extinction for a range of fuel-rich equivalence ratios. It can be noted that for $\phi = 5.0$, 5.5, and 6.0, the entire upper branches of the steady-state flame response curves are found to be dynamically stable, and hence, the static-extinction turning points are physically realistic limits. However, at $\phi = 6.5$ and 7.0, the steady-state flame response overpredicts the actual, dynamic, extinction stretch rate. The extent of overprediction can be as large as 20% for the $\phi = 7.0$ case. Therefore, for the determination of the actual extinction limits we must consider the occurrence of pulsation.

Furthermore, since the controlling factor for the onset of pulsation is a large Lewis number, the above results have significant implications on the more practically

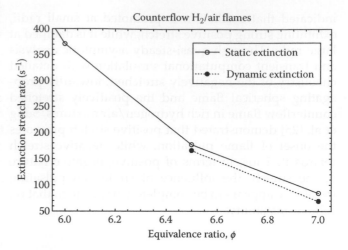

FIGURE 6.3.8
Comparison of static- and dynamic-extinction stretch rates for various hydrogen/air flames. When the equivalence ratio is sufficiently rich, the dynamic-extinction strain rate can be substantially lower than the corresponding static extinction limit.

relevant mixtures, such as those of lean hydrocarbons in air characterized by Le > 1. In particular, a recent study [28] on the one-dimensional, unstretched, lean heptane/air flames demonstrated that the transition from permanent oscillation to extinction is fairly abrupt, which is quite different from that of the rich hydrogen/air flame. As such, for the positively stretched, lean, large hydrocarbon/air flames, the intrinsic oscillation-induced extinction could occur at the onset of pulsating instability, thereby further narrowing the flammable range in stretch rate when compared with the steady-state flame response.

6.3.5 Fundamental Limits of Flammability

A definition of the fundamental limit of a given lean or rich fuel/oxidizer system is the concentration or pressure limit beyond which the steady propagation of the one-dimensional planar flame is not possible. It has been commonly understood that the flammability limit is an inherent property of a given reactant mixture, with little sensitivity to the specific heat-loss mechanism leading to extinction. Identifying and characterizing such a fundamental limit for various fuel types is of primary and practical importance in the performance and safety of combustion devices. A counterflow-based technique was proposed [29] for the experimental determination of these fundamental flammability limits. In this technique [29], extinction stretch rates were first determined for counterflow premixed flames with progressively weaker concentrations. By linearly extrapolating the results to zero-stretch rate, the corresponding mixture concentration was concluded to be that of the fundamental flammability limit. Implicit in such an approach is the assumption that the extinction stretch rate varies linearly with the concentration in the neighborhood of the fundamental flammability limit.

While there are published data on the steady flammability limits for various fuels based on some standard test method [30], recent studies (e.g., Refs. [31,32]) found that such limits may have to be substantially modified under rapid temporal/spatial fluctuations in the turbulent scalar fields. It is, therefore, essential to understand the fundamental characteristics of local quenching for the premixed flames that are situated in the presence of mixture stratification or/and flow straining. Figure 6.3.9 shows two possible counterflow configurations to establish a steady, partially quenched premixed flamefront or the so-called premixed edge flame—a coaxial geometry with an inner jet, for such a study. Unlike the transient premixed edge flame, the steady edge flame facilitates experimentation with high resolution and fidelity.

In the coaxial configuration, such a nonuniform flow field is formed by creating a local gradient in either the flow stretch rate or the equivalence ratio of the mixture. For this purpose, an inner jet located at the center of the counterflow burner is used to independently control the velocity and fuel mixture composition over the central area. The creation of a flame hole, at the center of the flame, is shown in Figure 6.3.9. The annular edge around the hole, where the reacting flow meets the nonreacting flow, is the focus of this study. The use of heated nitrogen is to lessen the downstream heat loss from the flame. Additionally, the conventional twin premixed flames can easily be established by issuing identical reactant composition and velocity from both sides of the burner exits.

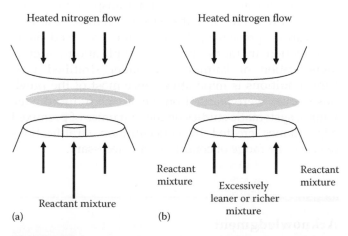

FIGURE 6.3.9
Schematic of premixed edge flames in a counterflow burner: (a) Mean velocity of the inner tube is greater than that of the outer tube, creating a stretch-induced edge flame. (b) Equivalent ratio of the mixture in the inner tube is different from that in the outer tube, creating a stratification-induced edge flame.

Figure 6.3.9a illustrates the edge flame formation by stretch-induced quenching. Given the uniform composition of the reactant mixture, if the nozzle-exit flow through the inner tube is larger than that through the outer tube and the corresponding stretch rate exceeds the extinction limit, then local extinction can occur at the center of the flame. Note that this configuration differs from that used in earlier studies [33–35], in that the present premixed edge flames are formed by varying the inlet conditions, rather than by the nonparallel, tilted slot separation as carried out in earlier studies [33–35].

Similarly, Figure 6.3.9b depicts the situation in which partial quenching of the flame results from unequal composition of the reactant mixtures issued from the inner and outer tubes, while keeping the mean velocities constant. If the equivalence ratio in the inner tube is excessively leaner or richer to exceed a typically flammable range, it would result in local extinction, thereby exhibiting a hole in the center of the premixed flame.

The creation of a steady flame hole was previously carried out by Hou et al. [36]. In their experiments, a steady-annular premixed edge flame was formed by diluting the inner mixture below the flammability limit, for both methane/air and propane/air mixtures. They found that a stable flame hole was established when the outer mixture composition was near stoichiometry. Their focus, however, was on the premixed flame interaction, rather than on the edge-flame formation, extinction, or propagation.

To further demonstrate the feasibility of the above-mentioned coaxial configuration, a simple setup including a stagnation plate and a coaxial burner, as shown in Figure 6.3.10a, was used to establish the axisymmetric counterflow ethylene/air flames under different conditions of the inner and outer tubes. Several samples of flame images indicated the presence of a quasi-one-dimensional flame (Figure 6.3.10b) as well as a stable flame hole, either by dilution of the inner mixture (Figure 6.3.10c) or by producing a local stretch-rate gradient (Figure 6.3.10d).

Using the coaxial configuration, for a given stretch rate and equivalence ratio in the outer tube, the critical lean and rich limits of the inner-core mixture leading to the onset of the hole formation can be systematically identified. The boundaries between the existence of the flame hole and the possible reignition can also be measured. Furthermore, it would be of interest to explore whether the retreat of the flame edge leads to extinction of the surrounding flame.

6.3.6 Concluding Remarks

Localized extinction of the flame surface can readily occur in the turbulent combustion devices, where wrinkled flames interact with turbulent eddies and gas

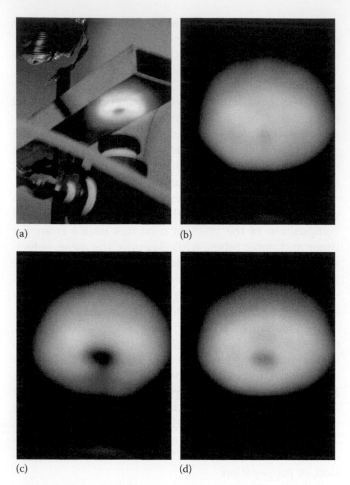

(a)

(b)

(c)

(d)

FIGURE 6.3.10
Photographs of a simple coaxial burner and the resulting sample flame images, viewed diagonally from the bottom. (a) A preliminary setup. Coaxial ethylene/air flames are formed under the following conditions: (b) $U_i = U_o$ and $\phi_i = \phi_o = 0.8$—flat flame; (c) $U_i = U_o$ and $\phi_i = 0$ and $\phi_o = 0.8$—inner flame extinguished owing to inner jet dilution; and (d) $U_i \gg U_o$ and $\phi_i = \phi_o = 0.8$—inner flame extinguished owing to increased local stretch rate. Here, U is the mean exit velocity and ϕ is the equivalence ratio. Subscripts "i" and "o" denote the inner and outer jets, respectively.

pockets. To develop truly predictive models, it is therefore crucial to fundamentally understand the dynamics of various flame-extinction phenomena. The study of extinction processes will also help to develop better description of the fundamental flammability limits of fuel/oxidizer mixtures, allowing improved identification of possible explosion/fire hazards. Within the framework of laminar flamelets, the materials presented in this section covered the four aspects of counterflow flame extinction, with special emphasis on stretch-induced quenching.

First, for nonequidiffusive, positively stretched counterflow flames, results showed that the flame response exhibited opposite behavior when the mixture's effective Lewis number was greater or less than a critical value, which is unity for the flame temperature. These completely opposite trends provide definitive verification of the concept of flame stretch with nonequidiffusion.

Second, the sensitivity of the flame structure and extinction limit to the description of the outer-flow field was examined. Comparison of the flame structure obtained from different boundary conditions at the extinction state suggested that the reported discrepancies in the computed extinction stretch rates were simply the consequences of how the velocity gradient was evaluated. Further investigation is required to identify an unambiguous parameter characterizing the counter-flow-flame extinction limit.

Third, since the pulsating instability observed for the planar unstretched flame was expected to be promoted by positive stretch, pulsation may develop beyond a critical stretch rate smaller than the static extinction limit. As pulsating extinction occurred at a smaller stretch rate than the steady extinction limit, the flame extinguished in the pulsating, instead of the steadily propagating mode, and the flammable range was accordingly narrowed.

Furthermore, characterizing the flammability limits of hydrocarbon combustion would also have a significant impact on the control of the combustion-generated pollutants, such as carbon monoxide and unburnt hydrocarbons, which are the products of incomplete combustion. The amount of these compounds produced at the exit of a combustion device relies strongly on the degree of complete combustion inside the engine, assuming that the product gas through the exhaust path is frozen. The prediction of such pollutants as a result of incomplete combustion depends not only on combustion chemistry, but also on the extinction dynamics of flames involving mixing and transport processes. By recognizing the fact that strong mixture stratification can be present upstream of the premixed flame propagation in many industrial applications, a clear identification of flammability under stratified mixture conditions is imperative. Since most of the previous studies were focused on the stretch-induced edge flames, systematic studies on the dynamics of premixed edge flames created by concentration stratifications over a wide range of conditions are necessary.

Acknowledgment

This chapter was prepared under the sponsorship of the National Aeronautics and Space Administration under Grant No. NNX07AB36Z, with the technical monitoring of Dr. Krishna P. Kundu.

References

1. Sivashinsky, G.I., On a distorted flame front as a hydrodynamic discontinuity, *Acta Astronaut.*, 3, 889, 1976.
2. Clavin, P. and Williams, F.A., Effects of molecular diffusion and of thermal expansion on the structure and dynamics of premixed flames in turbulent flows of large scale and low intensity, *J. Fluid Mech.*, 116, 251, 1982.
3. Matalon, M. and Matkowsky, B.J., Flames as gasdynamic discontinuities, *J. Fluid Mech.*, 124, 239, 1982.
4. Buckmaster, J.D. and Ludford, G.S.S., *Theory of Laminar Flames*, Cambridge University Press, Cambridge, 1982.
5. Williams, F.A., *Combustion Theory*, 2nd ed., Addison-Wesley Publishing Co., Reading, Massachusetts, 1985.
6. Law, C.K., Dynamics of stretched flames, *Proc. Combust. Inst.*, 22, 1381, 1988.
7. Law, C.K. and Sung, C.J., Structure, aerodynamics, and geometry of premixed flamelets, *Prog. Energy Combust. Sci.*, 26, 459, 2000.
8. Platt, J.A. and T'ien, J.S., Flammability of a weakly stretched premixed flame: The effect of radiation loss, Fall Technical Meeting of the Eastern States Section of the Combustion Institute, Orlando, Florida, 1990.
9. Sung, C.J. and Law, C.K., Extinction mechanisms of near-limit premixed flames and extended limits of flammability, *Proc. Combust. Inst.*, 26, 865, 1996.
10. Guo, H., Ju, Y., Maruta, K., Niioka, T., and Liu, F., Radiation extinction limit of counterflow premixed lean methane-air flames, *Combust. Flame*, 109, 639, 1997.
11. Ju, Y., Guo, H., Maruta, K., and Liu, F., On the extinction limit and flammability limit of non-adiabatic stretched methane-air premixed flames, *J. Fluid Mech.*, 342, 315, 1997.
12. Buckmaster, J., The effects of radiation on stretched flames, *Combust. Theory Model.*, 1, 1, 1997.
13. Ju, Y., Guo, H., Maruta, K., and Niioka, T., Flame bifurcations and flammable regions of radiative counterflow premixed flames with general Lewis numbers, *Combust. Flame*, 113, 603, 1998.
14. Frenklach, M., Wang, H., Bowman, C.T., Hanson, R.K., Smith, G.P., Golden, D.M., Gardiner, W.C., and Lissianski, V., GRI-Mech—An optimized detailed chemical reaction mechanism for methane combustion, Report No. GRI-95/0058, 1995.
15. T'ien, C.L., Thermal radiation properties of gases, *Adv. Heat Transfer*, 5, 253, 1967.
16. Hubbard, G.L. and Tien, C.L., Infrared mean absorption coefficients of luminous flames and smoke, *ASME J. Heat Transfer*, 100, 235, 1978.
17. Bikas, G. and Peters, N., Kinetic modelling of *n*-decane combustion and autoignition, *Combust. Flame*, 126, 1456, 2001.
18. Zhao, Z., Li, J., Kazakov, A., and Dryer, F.L., Burning velocities and a high temperature skeletal kinetic model for *n*-decane, *Combust. Sci. Technol.*, 177, 89, 2005.

19. Kee, R.J., Miller, J.A., Evans, G.H., and Dixon-Lewis, G., A computational model of the structure and extinction of strained, opposed flow, premixed methane-air flames, *Proc. Combust. Inst.*, 22, 1479, 1988.
20. Kumar, K. and Sung, C.J., Laminar flame speeds and extinction limits of preheated *n*-decane/O_2/N_2 and *n*-dodecane/O_2/N_2 mixtures, *Combust. Flame*, 151, 209, 2007.
21. Sivashinsky, G.I., Diffusional-thermal theory of cellular flames, *Combust. Sci. Technol.*, 15, 137, 1977.
22. Sivashinsky, G.I., Law, C.K., and Joulin, G., On stability of premixed flames in stagnation-point flow, *Combust. Sci. Technol.*, 28, 155, 1982.
23. Bechtold, J.K. and Matalon, M., Hydrodynamic and diffusion effects on the stability of spherically expanding flames, *Combust. Flame*, 67, 77, 1987.
24. Farmer, J.R. and Ronney, P.D., A numerical study of unsteady nonadiabatic flames, *Combust. Sci. Technol.*, 73, 555, 1990.
25. Sung, C.J., Makino, A., and Law, C.K., On stretch-affected pulsating instability in rich hydrogen/air flames: Asymptotic analysis and computation, *Combust. Flame*, 128, 422, 2002.
26. Kim, T.J., Yetter, R.A., and Dryer, F.L., New results on moist CO oxidation: High pressure, high temperature experiments and comprehensive kinetic modeling, *Proc. Combust. Inst.*, 25, 759, 1994.
27. Christiansen, E.W., Sung, C.J., and Law, C.K., Pulsating instability in the fundamental flammability limit of rich hydrogen/air flames, *Proc. Combust. Inst.*, 27, 555, 1998.
28. Christiansen, E.W., Sung, C.J., and Law, C.K., The role of pulsating instability and global Lewis number on the flammability limit of lean heptane/air flames, *Proc. Combust. Inst.*, 28, 807, 2000.
29. Law, C.K. and Egolfopoulos, F.N., A kinetic criterion of flammability limits: The C-H-O-inert system, *Proc. Combust. Inst.*, 23, 413, 1990.
30. Zabetakis, K.S., Flammability characteristics of combustible gases and vapors, U.S. Department of Mines Bulletin, No. 627, 1965.
31. Marzouk, Y.M., Ghoniem, A.F., and Najm, H.N., Dynamic response of strained premixed flames to equivalence ratio gradients, *Proc. Combust. Inst.*, 28, 1859, 2000.
32. Sankaran, R. and Im, H.G., Dynamic flammability limits of methane-air premixed flames with mixture composition fluctuations, *Proc. Combust. Inst.*, 29, 77, 2002.
33. Liu, J.B. and Ronney, P.D., Premixed edge-flames in spatially varying straining flows, *Combust. Sci. Technol.*, 144, 21, 1999.
34. Kaiser, C., Liu, J.B., and Ronney, P.D., Diffusive-thermal instability of counterflow flames at low lewis number, 38th Aerospace Sciences Meeting and Exhibit, AIAA Paper 2000-0576, 2000.
35. Takita, K., Sado, M., Masuya, G., and Sakaguchi, S., Experimental study of premixed single edge-flame in a counterflow field, *Combust. Flame*, 136, 364, 2004.
36. Hou, S.S., Yang, S.S., Chen, S.J., and Lin, T.H., Interactions for flames in a coaxial flow with a stagnation point, *Combust. Flame*, 132, 58, 2003.

6.4 Flame Propagation in a Rotating Cylindrical Vessel: Mechanism of Flame Quenching

Jerzy Chomiak and Jozef Jarosinski

This chapter presents a physical description of the interaction of flames with fluids in rotating vessels. It covers the interplay of the flame with viscous boundary layers, secondary flows, vorticity, and angular momentum and focuses on the changes in the flame speed and quenching. There is also a short discussion of issues requiring further studies, in particular Coriolis acceleration effects, which remain a totally unknown territory on the map of flame studies.

6.4.1 Introduction

The study of combustion in rotating fluids is of both practical and fundamental interest. Industrial applications not only involve a variety of swirl combustors, but also combustion in different rotating enclosures (cavities), which is important from the point of explosions in electrical motors, centrifugal separators, and turbo machines. The fundamental aspects involve flammability limits, ignition and extinction, flame speed, structure, and stability. The problem is quite wide and can be studied using many approaches. In our study, we would not enter into the realm of mathematical considerations, but following the founding idea of the book, explain the phenomenological account of a flame in a rotating vessel, beginning with ignition and flame development; however, as this problem was already discussed in an earlier study by Chomiak et al. [1], only the case of fully developed flame will be discussed in this chapter.

After ignition at any point in a rotating vessel, the flames driven by centrifugal and Coriolis forces $(F_{cent} = -0.5\rho\nabla|\omega\times r|^2$; $F_{Cor} = 2\rho\omega\times u$, where ρ denotes the density, r the distance from the rotation vector ω, the angular velocity, and u the local velocity) race toward the axis of rotation along a spiral trajectory, develop a cylindrical shape, and propagate toward the periphery. Propagation is in the direction of the centrifugal acceleration, as in the case of a flame propagating downwards in a natural gravity field, but the accelerations are typically of the orders of magnitude higher than the natural acceleration. Thus, propagation and quenching of cylindrical flames* subjected to enormous mass forces is important for rotating vessels. Babkin et al. [3] carried out the first study of such a case. In their work, a closed

vessel of 22.3 cm in diameter and 2.5 cm width was used, filled with lean methane/air mixtures (6.5%–8% methane) at the initial pressures of 0.1, 0.15, and 0.2 MPa, and high rotation speeds (between 565 and 850 1/s) were employed.

After the establishment of a cylindrical flame, continuous reduction of the flame speed relative to an external frame of reference was observed. This reduction was linear in time and quenching occurred in the system when the flame speed was close to zero. The declining flame speed was attributed to heat losses to the walls (reducing the effective expansion ratio in the flame) and quenching was interpreted in terms of the following criterion:

$$\frac{S_L^3}{gk} = b = \text{constant} \qquad (6.4.1)$$

where

S_L is the laminar flame speed
g the centrifugal acceleration
k the thermal diffusivity of the mixture

No physical interpretation of the criterion was provided, but it can be regarded as the ratio of the square of the velocity of a gravity-driven "free fall bubble," of diameter equal to the flame thickness, to the square of the laminar flame speed. This leads to the conclusion that quenching occurs when a flame element quenched at the wall moves ahead of the flame, as observed and as described by Jarosinski et al. [4] (see Fig. 5 in the paper referred to) for downward propagating flames in tubes.

Unfortunately, the degree of scatter of the data is large and for flames in normal gravity ($b = 1.3$) whereas in rotating vessels ($b = 0.02$), i.e., the "constant b" differs by almost two orders of magnitude. Thus, the result is inconclusive. Krivulin et al. [5] provided experimental data supporting Equation 6.4.1 to some extent, based on the observation of flame propagation and quenching in rotating tubes, where the rotation vector was normal to the axis of the tube. However, owing to Coriolis forces, the setup used generated intense secondary flows in the products and thus was not equivalent to a rotating vessel. In addition, the scatter of data was considerable and the "constant b", variable making this contribution inconclusive too.

Subsequently, the problem was investigated by Karpov and Severin [6]. They used closed vessels with a diameter of 10 cm and 10, 5, and 2.5 cm width, initially at atmospheric pressure. The vessels were filled with different lean hydrogen and methane/air mixtures and rotational speeds in the range of 130–420 1/s were employed. They also included data from the study of Babkin et al. [3] in their analysis. Unfortunately, they did not observe the flame itself and measured only the pressure rise in the vessel, which was compared with pressure development in the vessel without rotation, to draw a conclusion with respect to flame speeds and quenching.

---

* The cylindrical flames should be distinguished from the so-called tubular flames generated by the tangential supply of mixture into a cylinder [2], as the latter are subjected to strong axial stretch controlling their behavior and do not interact with the walls.

They concluded that reduction in flame speed and quenching was owing to heat losses to the walls. More specifically, their interpretation of quenching was that it occurred for the following relation between the Peclet number, Pe, and Nusselt number, Nu:

$$Pe^2 \propto Nu \qquad (6.4.2)$$

a relation attributed to Zel'dovich [7].*

Using $Pe = vh/k$ and $Nu \sim \omega r^2/k$, where h is the vessel width, v the flame speed ($v = S_L \varepsilon$, where $\varepsilon > 1$ is the density ratio across the flame), and r the flame radius, a quenching radius

$$r_q = \frac{hv}{7(k\omega)^{1/2}} \qquad (6.4.3)$$

was deduced, where 7 is an empirical constant to give the best fit to the experimental data. A dispersion of the data within 25% was obtained for all the cases investigated, when the chamber width, flame speed, thermal conductivity and rotational speed were changed by a factor 7, 5, 2, and 6.5 respectively.

Further investigations of flame behavior in rotating vessels were reported by Gorczakowski et al. [9], Gorczakowski and Jarosinski [10], and Jarosinski and Gorczakowski [11]. In the study by Gorczakowski et al. [9], a closed vessel, 9 cm in diameter and 10 cm wide, was used to study the combustion of methane/air mixtures at rotational speeds below 628 1/s. The main diagnostic techniques were direct photography and pressure records.

The complicated nature of the results obtained is best illustrated in Figure 6.4.1. It can be clearly seen that in this closed vessel, even with the highest rotational speed, only lean mixtures can be quenched. In some mixtures, before quenching occurs, the propagation exhibits complex features—strongly reduced, passing zero, and even negative flame speeds. In the work of Gorczakowski et al. [9], the reduction in the flame speed was interpreted in terms of heat losses. Quenching was assumed to be identical to that of the downward propagating flames in tubes [4], where combustion products cooled by the walls, under influence of gravity penetrate ahead of the flame, leading to negative flame speeds. However, the experiments show that the negative flame speeds are not a necessary condition for quenching in the rotating flame case.

An interesting feature observed in the above-mentioned investigations, and shown in Figure 6.4.2,

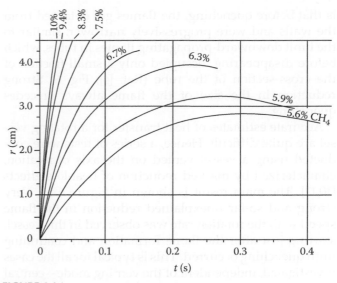

FIGURE 6.4.1
Flame radius changes in time for $\omega = 628\,s^{-1}$ and different methane in air concentrations in the mixture. Closed vessel. (Reproduced from Gorczakowski, A., Zawadzki, A., Jarosinski, J., and Veyssiere, B., *Combust. Flame*, 120, 359, 2000. With permission.)

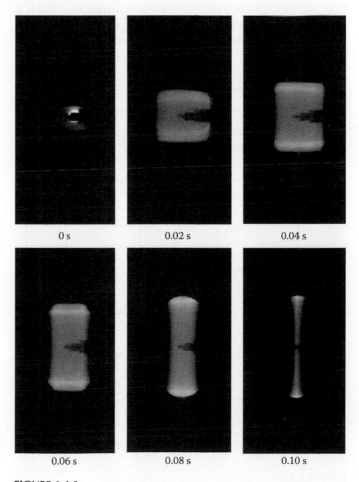

FIGURE 6.4.2
Flame images at different time instants for 3% propane/air mixture at 4400 rpm. Closed vessel.

---

* The above relation is obtained if the constant a obtained from the work of Zel'dovich [7] is assumed to be $a = Nu \frac{k^2}{d^2}$; (which was not suggested in the work of Zel'dovich [7]), where k is the thermal diffusivity and d is the characteristic length. However, Nu/Pe^2 is a parameter in the solution of energy equation for flames with heat losses above certain value for which no solution of the energy equation exist (see, e.g., Ref. [8], p. 108)

is that before quenching, the flames got detached from the walls and were progressively narrower, similar to the limit downward-propagating flames in tubes, which before disappearing, occupied only a small fraction of the cross-section of the tube (Ref. [4], Fig. 6). Strong reduction in the size of the flame always precedes quenching.

Accurate estimates of heat transfer for a rotating vessel are quite difficult. Hence, a series of tests were conducted using a vessel vented on the axis of rotation, characterized by marked reduction of heat-loss effects [10,11]. The main result is shown in Figure 6.4.3. Very strong and so-far unexplained reduction in the flame speed with the rotation rate was observed in this vessel, starting just after the flame formation and continuing until quenching occurred. This is typical for all the cases investigated, independent of the venting mode—central or peripheral venting as well as for closed vessels. The important general result from the study of Gorczakowski and Jarosinski [10] was the identification of the limiting condition for a flame to exist. It is not acceleration, heat losses, position with respect to the walls, or negative flame speed, but a characteristic circumferential (azimuthal) speed, specific for each mixture, which a flame cannot survive. In a recent paper [11], this result was physically interpreted in terms of a critical Froude

number, for which the flame loses its initial barrel shape, owing to the Landau–Darrieus instability, and becomes strictly cylindrical, which is unfavorable for its survival. In this interpretation, comparison is made between the squares of the flame propagation speed and the circumferential velocities and not simply velocities, which is equivalent to the interpretation given by Gorczakowski and Jarosinski [10], who stated that a flame cannot survive a limiting circumferential velocity. Although this conclusion is quite general, the physical reasons for it and for the falling flame speed are unclear and require further explanation. A discussion of a simple experiment will provide us with additional insight into this problem.

6.4.2 Flow Structure in the Vessel during Rotation Transients

This experiment deals with spin-up of a fluid in a container, from rest. Why is this problem important for the combustion of gases in rotational equilibrium with walls and moving as a solid body? The reason is simple—combustion with expansion of the gases causes a strong perturbation of angular speeds in the vessel. This is owing to the radial displacement of the gases caused by changes in their density, and the law of conservation of angular momentum by which differences are induced in the rotation of the displaced gases and walls leading to adjustment transients, best described by the spin-up process.

The experiment discussed in Greenspan's book [12] can be carried out using a very simple apparatus, consisting of a turntable, a light source, and a transparent cylindrical container. The closed container is completely filled with a suspension of a small quantity of aluminum powder along with some detergent in water. The tank is illuminated from the side by a beam passing through a vertical slit and is best viewed at right angle to the light beam. The light reflected by the particles is very sensitive to shearing motion at any point in the tank. Since the particles are usually flat flakes, the motion tends to align them, affecting the intensity of the reflected light, which can serve as an indicator of relative fluid motions. Within a few revolutions from an impulse starting off the rotation, a flow structure appears in the vessel, as shown in Figure 6.4.4. At the start, the boundary layers form at the horizontal surfaces, observed as thin dark ribbons next to the surfaces, through which the motion of the container is passed to the fluid by viscosity. Within these layers, the fluid near the wall is spun-up to a higher angular velocity and is propelled radially outward as in a centrifugal fan. Ekman, who studied the influence of the rotation of the earth on ocean currents, was the first to discuss this effect in 1905, and the boundary layers are now called

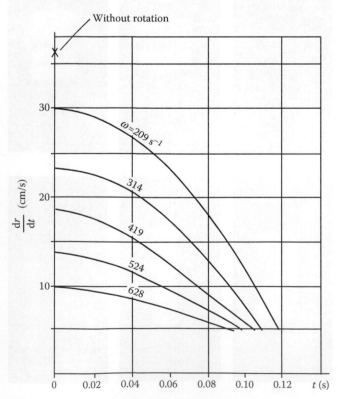

FIGURE 6.4.3
Flame speeds as a function of time for an 8.45% methane/air mixture and different rotation rates. Vessel vented on the axis of rotation.

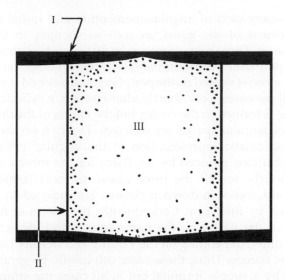

FIGURE 6.4.4
Early-time flow structure during spin-up from rest showing the Ekman layer – I (exaggerated), the light front separating the rotating and non rotating fluids – II, and the quiescent core – III. (From Greenspan, H.P., *The Theory of Rotating Fluids*, Cambridge University Press, 1969. With permission.)

Ekman layers. To compensate for the mass flow in the layers, a small normal flux from the core that remains motionless is required. The fluid from the core is then accumulated on the vessel periphery. The progressively depleted quiescent core provides the much stronger reflected light, since in this region the aluminum particles are randomly oriented, while in the spinning fluid injected by the Ekman layers into the peripheral zone, they are aligned with the flow. Consequently, the two fluids are seen separated by an almost perfect, straight light-front, propagating inward as the flow completes a closed circuit. Thus, the convergence of the fluid into the Ekman layers, together with the constraints of the geometrical configuration produce a radial current into the interior and a global circulation of the fluid. From the above-discussed basic physical picture, it is clear that the flow in the vessel is completely controlled by the Ekman layers. Von Karman [13] first provided a solution for the fluid flow velocities adjacent to a rotating disk, which is the basic element of the flow in the vessel (for a recent review see [14]). He used a similarity assumption for the flow parameters in the form:

$$v_r = r\omega F(z_1); v_\varphi = r\omega G(z_1)$$
$$v_z = \sqrt{\nu\omega}H(z_1); p = \rho\nu\omega P(z_1)$$

(6.4.4)

where

$$z_1 = \sqrt{\frac{\omega}{\nu}}z$$

ν is the kinematic viscosity

v_r is the radial velocity
v_φ is the circumferential (azimuthal) velocity
v_z is the axial velocity toward the wall
p is the pressure

After substituting into the Navier–Stokes and continuity equations and using the following boundary conditions,

$$F = 0 ; G = 1 ; H = 0 \quad \text{for } z_1 = 0$$
$$F = 0 ; G = 0 \quad \text{for } z_1 \to \infty$$

(6.4.5)

the system was solved numerically. The solution is illustrated in Figure 6.4.5, showing the functions F, G, and H. The limiting value of function H, for $z_1 \to \infty$ is -0.886; in other words, the velocity of the flow toward the wall at infinity is

$$v_z(\infty) = 0.866\sqrt{\nu\omega}$$

(6.4.6)

For a cylindrical vessel of width h, the characteristic timescale for the spin-up process, neglecting the effects at the periphery of the vessel is then

$$\tau_s \sim \frac{h}{2v_z} \sim \frac{h}{1.77\sqrt{\nu\omega}}$$

(6.4.7)

This is by orders of magnitude less than the timescale of the diffusive processes, $\tau_d \sim \frac{h^2}{\nu}$.

Owing to the continuity, the axial flow introduces a radial velocity in the vessel, which is given by

$$v_r' = \frac{r}{h}v_z = 0.886\frac{r}{h}\sqrt{\nu\omega}$$

(6.4.8)

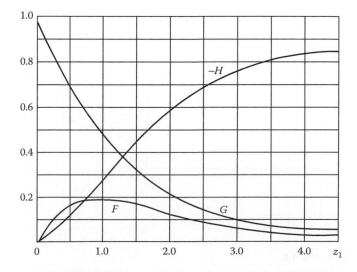

FIGURE 6.4.5
Similarity solutions of the velocity profile functions for the Von Karman problem. (From Von Karman, Th, Z., *Angew. Math. Mech.*, 1, 231, 1921.)

6.4.3 Heat Transfer between the Walls and the Gas in the Vessel

The high velocities in the Ekman layers and the thinness of the layers strongly enhance the heat transfer between the gas and the sidewalls. There exist a variety of well-established analytical and experimental correlations for the heat transfer between gas and a rotating disc (or the wall of the vessel). Cobb and Saunders [15] correlated their experimental investigations of the average laminar heat transfer with the Reynolds number $Re_r = \frac{\omega r^2}{\nu} < 2.4 \times 10^5$ and Prandtl number of 0.72 in the form:

$$Nu_r = 0.36\sqrt{Re_r} \qquad (6.4.9)$$

where the properties of the fluid were taken for a mean temperature between that of the wall and the ambient temperature. Dorfman [16] analytically obtained nearly the same formula (with the coefficient 0.36 replaced by 0.343) and thus Equation 6.4.9 can be taken as a good approximation. It may be noted that the correlation given by Karpov and Severin [5], discussed previously, is quite different, in particular, it strongly overestimates the heat losses. In any case, the heat losses are typically by orders of magnitude larger than that for pure conduction to the walls, when the gas and the vessel are in rotational equilibrium.

6.4.4 Flame Effects on the Flow in the Vessel

A flame in a vessel, as already mentioned, causes a dramatic perturbation of angular speeds of the gas, owing to conservation of angular momentum and radial displacement of the gases, as well as changes in their densities. Three basic cases can be distinguished providing quite different perturbations: a vessel vented on the axis, a vessel vented on the periphery, and a closed vessel. In all the cases, very shortly after ignition, a cylindrical flame is formed in the center, but the effects of the flame on the angular speeds are different. Figure 6.4.6 shows the schematic representation of the angular velocity distributions induced by the flame for the three cases. Obviously, for all the three cases, different transient spin-up and spin-down processes are induced in the vessels by the flame. Consequently, the effects of rotation of the vessel on the flame speeds and extinction will differ, depending on the venting and density ratios in the flames. Thus, these cases can hardly be generalized by a simple formula, but in all cases the strongly enhanced heat transfer is very important.

6.4.5 Flame–Flow Interactions

6.4.5.1 Initial Period

Although the detailed flow-structure in the vessel is unknown and can be predicted by numerical means, only the basic features of the flame–flow interaction can now be depicted. The interaction is shown schematically in Figure 6.4.7. Each of the Ekman layers formed at the sidewalls by the angular velocity perturbation of the flow induces two recirculation cells—one in front of the flame and one behind it, separated by the flame. The recirculation cell in front of the flame is of less importance, as the flow velocities there do not affect the flame,

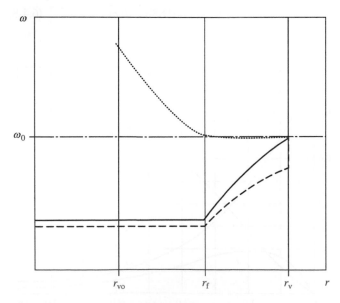

FIGURE 6.4.6
Distribution of angular speeds induced by a flame (schematic); (—) closed vessel; (----) vessel vented at the periphery; (••••) vessel vented at the center. R_{vo} — radius of the venting orifice, r_f — radius of the cylindrical flame, r_v — radius of the vessel, ω_0 — initial angular speed.

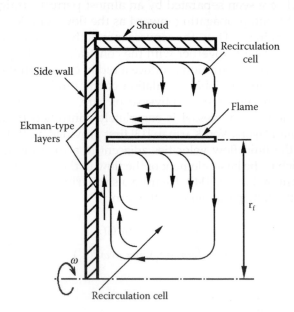

FIGURE 6.4.7
Flame–flow interaction in a rotating vessel, showing the generation by the sidewall of two recirculation cells before and after the flame driven by the Ekman layers.

apart from the minor axial stretch effects. Also, the heat transfer there, even in the closed vessel is of secondary importance, as the temperature difference between the gas and wall is insignificant. However, a very different situation is behind the flame, where the Ekman layer carries products that are strongly cooled down by the wall. When the Ekman flow impinges on the flame, it is unable to penetrate it as the density of combustion products, even after cooling, is still much lesser there than the density of fresh mixture and therefore, the flow is redirected and forced along the flame. The cold annular jet behind the flame effectively reduces the flame speed, even if the temperature of the rest of the products is still unaffected. This explains the observation that the larger the circumferential speed of the vessel at the flame location generating the Ekman flow, the smaller is the flame speed, and the effect is almost instantaneous. This is observed in all the venting cases. Thus, the reduction of flame speed is due to the interaction of the heat losses in the Ekman layers, along with their redirection and penetration far behind the flames. An additional effect is by global heat losses that reduce the effective expansion of gases.

Owing to the large shear in the Ekman layer zone with strong heat losses, the flame is always at a distance from the wall. The distance is small (slightly more than the flame-quenching distance) as long as the flow velocity parallel to the flame induced by the redirection of the Ekman layers is lesser than the propagation speed of the edge flame. However, once the Ekman flow velocity becomes larger, the flame is carried away from the wall and continuously reduced in width until extinction, as observed by Gorczakowski et al. [9], Gorczakowski and Jarosinski [10], and Jarosinski and Gorczakowski [11], which is illustrated in Figure 6.4.2. Thus, the quenching mechanism is now clearly explained and the reason that a given flame cannot survive a certain circumferential speed of the vessel provided. The above-mentioned reasoning is supported by a numerical simulation in a study by Marra [17], from which a snapshot is shown in Figure 6.4.8 illustrating the flow structure close to the wall. A strong axial-annular jet, originating at the wall just behind the flame due to the redirection of the Ekman layer is clearly seen on the left side of the figure, as well as the radial Ekman layer flows in the neighborhood of the flame on the right side.

6.4.5.2 Narrow Flame Period

After detachment of the flame from the wall and reduction of its width, three zones develop in the vessel parallel to each other: a flame and burned gas zone and two zones (adjacent to the sidewalls), where no flame is present and where the gas temperatures are lower than behind the flame. All this happens in a field of very high centrifugal acceleration, which induces a free convection movement of the flame and the product zone behind it toward the

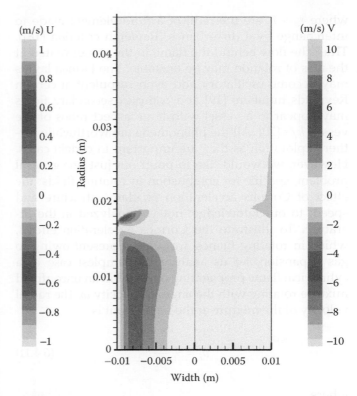

FIGURE 6.4.8
Axial (u_z) left-half and radial (u_r) right-half velocity components at 15.0 ms, $\omega = 314\,\mathrm{s^{-1}}$ for $\phi = 0.879$, methane/air mixture, closed vessel [17].

axis of rotation. The movement reduces the flame speed, relative to an external frame of reference observed in the experiment. The free convection speed is proportional to the square root of the centrifugal acceleration multiplied with the length of the zone behind the flame, in turn proportional to the flame radius. On the whole, the free convection speed reducing the observed flame velocity is proportional to the circumferential speed of the gas at the flame location. This explains the continuous reduction of the flame speed in time and strong dependence of the flame speed on angular velocity, as shown in Figure 6.4.3. It is interesting to note that both the flame-width reduction rate and the flame speed drop are proportional to the circumferential gas speed at the flame location. No wonder that a flame in a certain mixture cannot survive a certain circumferential speed in a rotating vessel, which thus becomes a controlling factor of the rotating flame. However, is still unknown, which of the three physical phenomena is the most important: cooling, reduction of flame width or free convection of the flame zone.

6.4.6 Further Research Issues

The flow in a rotating vessel is prone to several types of instabilities. The rotational flow itself is unstable if

$$\left(r_0 v_0\right)^2 < \left(r_1 v_1\right)^2 \tag{6.4.10}$$

where $r_0 > r_1$ are the radii of a fluid element made to interchange by a disturbance (Rayleigh criterion [18]). Thus, the flow behind the flame in the vessel vented at the axis of rotation may be unstable. The Ekman layers may become oscillatory and even turbulent at certain Reynolds numbers [19] and complex secondary flows may appear in a vessel with large aspect ratios of the vessel, h/r [20]. All the phenomena are worthwhile further exploration as they are important in certain cases. However, we would like to point out just one general problem, specific for combustion in rotating fluids: the effect of Coriolis acceleration on flame structure and speed, to our knowledge, not yet analyzed in the literature. To illustrate the Coriolis acceleration effects, which in rotating flames are always present owing to gas expansion, let us analyze the simplest case of a cylindrical flame propagating outward in an unconfined mixture rotating with the angular velocity ω. The radial velocity of the mixture at the flame front is

$$v_r = S_L \frac{\rho_1}{\rho_2} \tag{6.4.11}$$

where

S_L is the flame speed
ρ_1/ρ_2 is the density ratio across the flame front

Thus, the maximum value of the Coriolis acceleration is at the cold flame boundary, which can be given as

$$a_{cm} = 2\omega v_r \tag{6.4.12}$$

Inside the flame, the acceleration grows from zero to the above value. The timescale for this process is equal to the timescale of the flame

$$\tau = \frac{\delta}{S_L} \tag{6.4.13}$$

where δ is the flame thickness. The acceleration causes sliding of the progress variable iso-surfaces in the flame relative to each other, leading to the tangential velocity of the fresh mixture relative to the burned gases, which is equal to

$$v_t = a_{cm}\tau \tag{6.4.14}$$

where a_{cm} is the mean Coriolis acceleration in the flame. Approximating the temperature distribution in the flame by a linear function, we have

$$v_t = \omega S_L \frac{\rho_1}{\rho_2}\tau \tag{6.4.15}$$

TABLE 6.4.1

Methane/Air Flame Parameters and Rotational Speeds (ω_1) for Which the Tangential and Normal Speeds in an Expanding Flame are Equal

ϕ	0.55	0.63	0.7	0.8	0.9	1
δ [cm]	0.46	0.23	0.175	0.140	0.120	0.1
S_L [cm/s]	12	17.5	22	29	36	43
$T_a - T_1$	1270	1410	1500	1710	1835	1935
ω_1 [s^{-1}]	26.1	76	125.7	207	300	430

Note: Vessel vented at the periphery. ϕ is the equivalence ratio.

leading to a ratio of the tangential velocity to normal velocity equal to

$$\frac{v_t}{v_n} = \omega\tau \tag{6.4.16}$$

In Table 6.4.1, rotational speeds (ω_1) of the system for which $v_t = v_n$ are given for methane/air flames. We may expect that at those rotational speeds, strong effects of rotation on flame structure and speed will be observed. The angular velocities of importance are high, but not very high, considering the rotational motions in fluids. It can be observed that lean flames are more sensitive to rotations than stoichiometric ones. It is presumed that the Coriolis effects present from the very beginning of the flame may provide additional reasons why the flame speeds of the rotating flames are smaller than of nonrotating flames from the start of the experiment (Figure 6.4.3.) Physically, the Coriolis acceleration effects may be due to the deformation of fluid elements in the preheat zone equivalent to stretch. However, this aspect of flame behavior requires further studies. Dramatic flame-shape changes owing to the Coriolis acceleration can be observed for perturbed flames, which is shown in Figure 6.4.9. In this case, venting of the vessel was at the periphery, using just four orifices. This caused square-like deformation of the initial cylindrical shape as the flame approached the rim. The perturbation enhanced by the Coriolis acceleration behind the leading points converted the flame into a multispiral, multilayer structure. The effects will be permanent in the case of turbulent flames in the flamelet regime, leading to strong changes in the flame perturbation structure, important for propagation and not considered in standard analysis.

6.4.7 Conclusions

1. Mechanisms leading to flame speed changes and quenching in rotating vessels (cavities) were discussed.

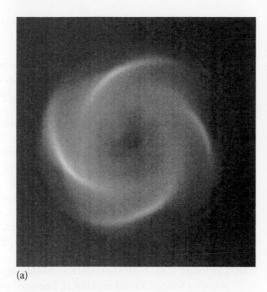

(a)　　　　　　　　　　　　　　　　(b)

FIGURE 6.4.9

Coriolis acceleration effects in a combustion chamber (dia 90 and width 30mm), vented at the periphery through four orifices, having 7.5mm in diameter, 4% propane/air mixture, rotation speed (a) 1000 rpm and (b) 2000 rpm.

2. It is shown that the Ekman layers behind the flame front, generated by the rotational speed changes of the gas owing to expansion, cause flame detachment from the walls and reduction in flame width in the rotating vessels, ultimately causing flame quenching.

3. The mechanism is based on strong cooling of the products adjacent to the walls and injection of the cooled products in the form of an annular jet behind the flame, reducing its width as soon as the speed of the annular jet, proportional to the circumferential speed of the gas at the flame location becomes larger than the propagation speed of the edge flame.

4. After detachment of the flame from the walls, the narrow ever-diminishing hot product zone behind the flame moves owing to the free convection in the centrifugal acceleration field toward the axis of rotation, with a speed scaling with circumferential velocity at the flame location, which reduces the observed flame speed to very low values, and in some cases negative ones.

5. Heat losses may contribute to the process, but do not solely cause the observed flame speed reduction and flame quenching.

6. The secondary flows in the vessels depend on the aspect ratio and venting systems, which does not allow generalizing the results with simple, universal formulas.

7. Owing to the laminar nature of most of the flames in rotating vessels, numerical studies may be quite effective but require resolution of the flow in the Ekman layers and the flame, which is a numerically challenging task.

8. The flames in the rotating vessels represent an interesting case, where the Coriolis effects are strong. However, further studies of the effects on flame propagation and structure are required.

9. True challenges are flows with instabilities and, in particular, the turbulent flows where Coriolis effects change the flame perturbation geometry.

Acknowledgments

This study was sponsored by the Marie Curie TOK project from European 6th FP contract No. MTKD-CT-2004-509847. The authors thank Andrzej Gorczakowski and Francesco Marra for their contributions to the experiments and numerical simulation.

References

1. Chomiak, J., Gorczakowski, A., Parra, T., and Jarosinski, J., Flame kernel growth in a rotating gas, *Combustion Science and Technology*, 180, 391–399, 2008.

2. Ishizuka, S., Characteristics of tubular flames, *Progress in Energy and Combustion Science*, 19, 187–226, 1993; For more recent contributions see Colloquium on tubular flames in *Proceedings of the Combustion Institute*, 31, 1085–1108, 2007.

3. Babkin, V.S., Badalyan, A.M., Borysenko, A.V., and Zamashchikov, V.V., Flame quenching in rotating gas (in Russian), *Fizika Gorenia i Vzryva*, 18, 18–20, 1982.

4. Jarosinski, J., Strehlow, R.A., and Azabrazin, A., The mechanism of lean limit extinguishment of an upward and downward propagating flame in a standard flammability tube, *Proceedings of the Combustion Institute*, 19, 1549–1557, 1982.

5. Krivulin, V.N., Kudriavcev, E.A., Baratov, A.V., Babalian, A.M., and Babkin, V.S., Effects of acceleration on limits of propagation of homogeneous gas flames, (in Russian), *Fizika Gorenia i Vzryva*, 17, 17–52, 1981.

6. Karpov, V.P. and Severin, E.C., Flame propagation in rotating gas (in Russian), *Chemical Physics*, 3, 592–597, 1984.

7. Zel'dovich Ya.B., Theory of limits of propagation of slow flames (in Russian), *Journal of Experimental and Theoretical Physics*, 11(1), 159–169, 1941; Reprinted in Collected Works of Ya.B. Zel'dovich, Nauka, Moscow 1984, 233–246, and the book Theory of Combustion and Explosion, Nauka, Moscow 1981, 271–288.

8. Jarosinski, J., A survey of recent studies on flame extinction, *Progress in Energy and Combustion Science*, 12, 81–116, 1986.

9. Gorczakowski, A., Zawadzki, A., Jarosinski, J., and Veyssiere, B., Combustion mechanism of flame propagation and extinction in a rotating cylindrical vessel, *Combustion and Flame*, 120, 359–371, 2000.

10. Gorczakowski, A. and Jarosinski, J., The phenomena of flame propagation in a cylindrical combustion chamber with swirling mixture, SAE Paper 200001-0195, 2000.

11. Jarosinski, J. and Gorczakowski, A., The mechanism of laminar flame quenching under the action of centrifugal forces, *Combustion Science and Technology*, 178, 1441–1456, 2006.

12. Greenspan, H.P., *The Theory of Rotating Fluids*, Cambridge University Press, Cambridge, 1969, pp. 1–5.

13. Von Karman, Th., Uber laminare und turbulente Reibung, *Zeitschrift für Angewandte Mathematik und Mechanik*, 1, 231–251, 1921.

14. Zandbergen, P.J. and Dijkstro, D., Von Karman swirling flows, *Annual Review of Fluid Mechanics.*, 19, 465–491, 1981.

15. Cobb, E.C. and Saunders, O.A., Heat transfer from a rotating disc, *Proceedings of Royal Society*, 236 (Series A), 343, 1956.

16. Dorfman, L.A., *Hydrodynamic Resistance and the Heat Loss of Rotating Solids*, Olivier and Boyd, London, 1963, pp. 95–97.

17. Marra, F.S., Analysis of premixed flame propagation in a rotating closed vessel by numerical simulation, Mediterranean Combustion Symposium, Monastir, Tunisia, September 2007.

18. Greenspan, H.P., *The Theory of Rotating Fluids*, Cambridge University Press, Cambridge, 1969, pp. 271–275.

19. Ibid, pp. 275–288.

20. Ibid, pp. 293–299.

7

Turbulent Flames

CONTENTS

7.1 Turbulent Premixed Flames: Experimental Studies over the Last Decades

Roland Borghi, Arnaud Mura,
and Alexey A. Burluka

7.1.1 Introduction

There are several industrial devices in which energy conversion occurs through turbulent premixed flames. Gas turbine combustion chambers, afterburners of turbojet engines, and spark-ignited reciprocating engines are the main examples, and they support the research for understanding such turbulent flames. The first research specifically devoted to turbulent premixed flames has been conducted around 1940, by Damköhler [1] in Germany and Shchelkin and others in Russia [2]. Finally, in 1956 the first special session of the (International) Symposium on Combustion devoted to the "structure and propagation of turbulent flames" was held during the sixth edition at the University of Yale (USA).

The first visible example of turbulent premixed flame is the flame above a Bunsen burner, when a flow of premixed reactants is made turbulent within the burner tube because of its large velocity and often with the help of a turbulence-generating grid. Such a rig is very easy to build and it has been widely used to gather information about the structure of turbulent flames with respect to laminar ones, see for instance Chapter VI of Ref. [3]. The apparent thickness of the corresponding turbulent flame is much larger than the thickness of a laminar flame of the same mixture. From the apparent "flame angle", it is possible to deduce roughly a value of the so-called "turbulent flame speed," which is clearly larger than the laminar flame speed. This apparently larger thickness could be simply explained by the fact that turbulence produces some wrinkling and flapping of the classical thin flame sheet, but Damköhler [1] has been the first who suggested that there exist at least two types of turbulent premixed flames. The first one effectively contains a long wrinkled thin and laminar-like flamelet, while chemical reactions take place in a quite thick zone in the other type of flame, which may appear like a very thick flame. Already at the time, several discussions had appeared to clarify whether only two types are possible, to select the most suitable names for them, and to evaluate what are the conditions for their appearances. These discussions were initiated by Karlovitz, Shchelkin, Wohl, Summerfield, Scurlock, and many others. These discussions have later given birth to some synthetic diagrams (see Section 7.4.1 here), very popular even now for representing the different combustion regimes in terms of well-chosen nondimensional ratios built from relevant parameters that characterize both turbulence and laminar flame characteristics.

In the so-called "wrinkled flame regime," the "turbulent flame speed" S_T was expected to be controlled by a characteristic value of the turbulent fluctuations of velocity u', rather than by chemistry and molecular diffusivities. Shchelkin [2] was the first to propose the law $S_T/S_L = (1 + A(u'/S_L)^2)^{1/2}$, where A is a universal constant and S_L the laminar flame velocity of propagation. For the other limiting regime, called "distributed combustion," Summerfield [4] inferred that if the turbulent diffusivity simply replaces the molecular one, then the turbulent flame speed is proportional to the laminar flame speed but multiplied by the square root of the turbulence Reynolds number Re_T.

The knowledge of turbulent premixed flames has improved from this very simple level by following the progress made in experimental and numerical techniques as well as theoretical methods. Much employed in early research, the laboratory Bunsen burners are characterized by relatively low turbulence levels with flow properties that are not constant everywhere in the flame. To alleviate these restrictions, Karpov et al. [5] pioneered as early as in 1959 the studies of turbulent premixed flames initiated by a spark in a more intense turbulence, produced in a fan-stirred quasi-spherical vessel. Other experiments carried out among others by Talantov and his coworkers allowed to determine the so-called turbulent flame speed in a channel of square cross-section with significant levels of turbulence [6].

New experiments devoted to stationary turbulent flames, similar to a Bunsen burner but at a larger scale, have been designed and studied: the so-called V-shaped flame anchored on a wire or rod, see for instance Escudié et al. [7], or a high-velocity flame stabilized, thanks to a bluff body, e.g., Wright and Zukovski [8], or a pilot flame as in the work of Moreau and Borghi [9]. Classical turbulent Bunsen burners have also been used [10,11] as well as propagating unsteady turbulent flames in open flows or enclosures, see Boukhalfa and coworkers [12]. Needless to say that these experimental situations have been studied in much more details by using all the available modern and nonintrusive optical methods providing a deeper insight into the detailed local flame structure. At the same time, numerous theoretical works have been developed on this subject, as well for looking more deeply into the phenomena that are expected to play at small scales than for trying to build realistic models for the prediction of the flames in the different regimes.

At the beginning of the twenty-first century, it is interesting to try to summarize what has been experimentally established, and what remains partially or entirely unclear about turbulent premixed flames. The hope is that the results of this state of knowledge will be in agreement with the existing state of the art in the field of turbulent flames modeling, which can be found elsewhere [13,14]. However, it is necessary first to define precisely the framework of the present survey: we will discuss in Section 7.1.2 the methodology needed for the study of turbulent flames, considering three very basic questions about their nature, about the concept of "turbulent flame speed," and what are the main and sufficient quantities required to describe a turbulent premixed flame.

7.1.2 Three Basic Questions

7.1.2.1 What Is a "Turbulent Premixed Flame"?

Darrieus and Landau established that a planar laminar premixed flame is intrinsically unstable, and many studies have been devoted to this phenomenon, theoretically, numerically, and experimentally. The question is then whether a turbulent flame is the final state, saturated but continuously fluctuating, of an unstable laminar flame, similar to a turbulent inert flow, which is the product of loss of stability of a laminar flow. Indeed, should it exist, this kind of flame does constitute a clearly and simply well-posed problem, eventually free from any boundary conditions when the flame has been initiated in one point far from the walls.

However, in practice, the term of turbulent premixed flame is given to flames that are developing in a flow, or a gaseous medium, which is already turbulent before it sees the flame. When the turbulent flame is considered anchored in a flow, the oncoming flow is already turbulent, owing to some walls or obstacles upstream. When the turbulent flame propagates into a premixed gaseous medium, the latter medium is not exactly at rest but its velocity is supposed to fluctuate with classical characteristics of turbulence before the flame passes through. Of course, one should not assume that the presence of a flame will leave unchanged the velocity fluctuation levels; on the contrary, because of the heat release and the instability of laminar flames, it is expected that this modification does take place, and is important at least for some conditions. Indeed, the "flame-generated turbulence" has been envisaged very early by Karlovitz [15].

In fact, the clearly posed problem of the final state of an unstable laminar flame is a limiting case of turbulent flame for vanishing initial turbulence of the oncoming flow, but the general case, for any initial velocity fluctuations, is clearly of great interest in practical devices such as spark-ignited engines, turbojet, or gas turbine combustion chambers.

Though, in practice, the composition of the incoming flow is not always perfectly known and controlled, we have to restrict the present analysis to turbulent reactive

flows with spatially homogeneous compositions and negligible fluctuations of equivalence ratio or composition. If it is not the case, additional phenomena will complicate the picture.

7.1.2.2 Is the "Turbulent Flame Speed" an Intrinsic Well-Defined Quantity?

The question here is twofold: first, how to prescribe a precise experimental procedure for defining the "turbulent flame speed"? and second, is this quantity independent of the way used to initiate the flame? This is the case for a laminar flame, and the flame propagation velocity S_L as well as the characteristic laminar flame thickness δ_L is an intrinsic quantity.

It has to be first emphasized that similar to any turbulent phenomenon, a turbulent flame must be defined statistically: we have to consider an ensemble of individual flames, each one produced by apparently identical conditions, each one exhibiting strong fluctuations in time and space, and the quantities in which we will be interested must be defined as the statistical averages of the corresponding quantities for all the individual flames. Averaged quantities can be defined in this way locally and at a given instant. But when a turbulent flame is statistically steady, time averages can be used instead of statistical averages, and when statistical homogeneity in one direction or along one plane or surface holds, spatial averages along this direction or surface can be used. Because of this necessary averaging process, it is more convenient to consider "mean turbulent flame speed" S_T, or "mean turbulent flame thickness" δ_T. The term "mean" is usually omitted, and the mean turbulent flame thickness is often called "flame brush thickness." Let us consider the possible ways for defining these quantities in greater details.

We consider a turbulent premixed flame produced in an isotropic homogeneous turbulent velocity field with zero mean value and constant turbulence properties, i.e., nondecaying in time (then continuously stirred), not affected by gravity, ignited at an infinite plane; notice that we do not claim that such a flame is easy to obtain experimentally! The phenomenon is then statistically homogeneous on any plane parallel to the ignition plane and therefore we can use spatial averages. Then, there are two distinct ways for defining the turbulent flame speed, following the previously given principles, one based on a Eulerian description and the other relying on a Lagrangian point of view.

The Eulerian approach requires a measurement of the temperature or the progress variable at many sample points at a given normal distance from the ignition plane, at a given time elapsed since ignition. The progress variable introduced here can be for instance a normalized temperature or concentration that varies from

$c \equiv 0$ in the fresh reactants to $c \equiv 1$ in the fully burnt products. Then, the average of these values over each plane may be calculated, enabling us to plot the profiles of mean temperature or mean progress variable as a function of distance to the ignition plane. In this procedure, the number of sample points needed must be large enough in such a way that the calculated averages do not depend any more on this number and for the first-order statistical moments, 500 or 1000 points would normally suffice. Moreover, these sample points have to be sufficiently far away from each other so that the measurements are statistically independent; this implies that they must be separated by a distance larger than the turbulence length scale.

Then, the displacement velocity of the (mean) flame is defined as the displacement velocity, measured over a short period of time, of the position of these profiles. More exactly, we have to consider the displacement of a particular point on one of these profiles, where the mean temperature or the mean progress variable has a prescribed value. If the flame propagates in a steady regime, where the profiles move without any change in shape, the choice of point and profile does not matter. On the contrary, if the flame does not have a steady propagation, then *a priori* choice of the reference point and profile has to be done. The displacement velocity discussed above is the mean speed of the flame with respect to the initiation plane or the quiescent burnt gases. The classical turbulent flame speed S_T, with respect to the unburnt mixture, is not exactly this displacement velocity, because the unburned gases are pushed away by the expansion of burnt gases. For a steady propagation, it is simply the displacement speed divided by the unburnt-to-burnt gases density ratio. A natural definition for the flame brush thickness in a Eulerian frame of reference is that it is simply the distance between two values of mean temperature or mean progress variables, one close to the burnt gases temperature or progress variable, the other close to the unburnt gas temperature or progress variable, on the profiles found for a given time. If the flame propagation is steady, such a flame brush thickness is independent of time.

A Lagrangian framework is also easily conceivable. Let us consider, at any given time, one isothermal surface, or one surface where a progress variable, e.g., defined from a chemical species concentration, is constant. Owing to the turbulence, this surface is not a plane but is wrinkled and, moreover, may not be simply connected. We can calculate the averaged distance between the sample points on this surface and the ignition plane, and doing so at two close times, we can deduce the velocity of displacement of the mean flame surface. This velocity is the mean turbulent flame speed with respect to burnt gases and it must be corrected for the effect of gas expansion. Similar to the Eulerian description, the

choice of the value for the temperature (between the unburnt mixture temperature and the adiabatic combustion temperature), or equivalently the choice of the value of the progress variable between zero and unity, has no importance if the flame propagates with a constant mean structure and mean thickness. To measure this flame brush thickness, it is necessary to choose two particular values of the progress variable, one close to zero and one close to unity, and to calculate at any time the difference between the two mean positions corresponding to these two values.

In this simplified situation, can we really consider that the mean flame structure and thickness are steady, after certain delay and distance from initiation, and then the "turbulent flame speed" is a well-defined intrinsic quantity? Indeed, with the present state of knowledge, there is no certainty in any answer to this question. Of course, it is hardly possible to build an experiment with nondecaying turbulence without external stirring. In decaying turbulence, the independence of the turbulent flame speed on the choice of reference values of progress variable has been verified in neither experiment nor theory.

Many attempts are known to construct a burner where the flow would approach the theoretical ideal conditions of homogeneous and isotropic turbulence; in addition to the already mentioned fan-stirred bombs, very recently a cruciform burner has been developed by Shy et al. [16] where the flame speed was determined from the time elapsed between the passage of the flame at the locations of two fixed ionization sensors. However, the work by the same group later established that the flame speed varies by as much as 45% over a distance of 20 cm, see Figure 3 in Ref. [17]. This example is a very good illustration of many complex difficulties associated in an attempt to reproduce a simple theoretical picture in an experimental test rig. Nevertheless, from the very beginning and again now, the idea that there exists an intrinsically valid turbulent flame speed has been implicitly accepted in the scientific community.

7.1.2.3 *How to Describe a Turbulent Premixed Flame with a Few Well-Defined Quantities?*

If the turbulent flame is ever proven to have asymptotically a constant flame brush thickness and constant speed in constant, i.e., nondecaying, turbulence, then the aforementioned turbulent flame speed S_T and the flame brush thickness δ_T give a well-defined sufficient characterization of the flame in its asymptotic behavior. However, it is not proven up to now that the studied experimental devices have been large enough to ensure that this asymptotic state can be reached. Besides, the correct definitions for the turbulent flame speed or flame brush thickness, as given above, are far from being easy to apply in practice. As a consequence, many of the experiments have not used them, and the quantities that have been measured have been related to them only with additional hypothesis, the influence of which on results must be carefully considered.

In any circumstances, it can be expected that S_T and δ_T are algebraic functions of turbulence length scale and kinetic energy, as well as chemical and molecular quantities of the mixture. Of course, it is expedient to determine these in terms of relevant dimensionless quantities. The simplest possible formula, in the case of very fast chemistry, i.e., large Damköhler number $Da = (S_L\, l_T)/(\delta_L u')$ and large Reynolds $Re_T = (u'\, l_T)/(\delta_L\, S_L)$ and Péclet numbers, i.e., small Karlovitz number $Ka = \sqrt{Re_T}/Da$ will be $S_T/S_L = f(u'\,/\,S_L)$, but other ratios are also quite likely to play a role in the general case.

Anyway, it has been found without any doubt that many experiments, if not all, do show that the flame brush thickness indeed is not constant in the domain where it is practically interesting to study the flame. This does not necessarily imply that the flame has no asymptotic behavior and that S_T and δ_T do not exist, but this implies that an eventual asymptotic behavior is not the only one that deserves interest. Before the flame reaches an asymptotic state, its speed and thickness can be defined, though with more arbitrariness, but their values shall depend on time or position or both. In principle, these transient variables cannot be deduced from an algebraic formula in terms of the parameters describing turbulence, molecular processes, and chemistry, simply because the distance (duration or spatial distance) from the flame initiation has to play a role. Then, S_T and δ_T lose their interest. In this case, the mean structure of the turbulent flame must be calculated locally by using partial differential evolution equations in which a mathematical model of turbulent combustion replaces the simple knowledge of S_T and δ_T. The mathematical models already proposed are developed in terms of Eulerian averaged variables because this point of view is more suitable for experiments. The reader is referred to Refs. [13,14], or to the relevant chapters of this book, for a deeper insight into this methodology. Any such mathematical model must contain two sub-models: (1) one describing the turbulent transport, namely the scalar fluxes and Reynolds stresses, and (2) the other the so-called "mean reaction rate." To build these sub-models, necessary are the experiments that put clearly into evidence the governing physical factors. For the turbulent transport submodel, the phenomena of "flame-generated turbulence" and "counter-gradient diffusion" must be studied in adequately oriented experiments. For the "mean reaction rate" submodel, the small-scale structure of the turbulent flame, under different conditions, must be studied.

The numerous experimental studies that have been performed since 1940s do bring each their individual contribution to the task of studying each of these submodels; some of them brought a larger contribution, other ones a smaller contribution. We will present in the two following sections selected results obtained through such experimental studies. Only the clearest results, whose physical explanation appears well based, and which allow to understand the main and simplest features of turbulent premixed flames, are presented here. Many works are not mentioned, often with the reason that it will complicate unnecessarily the picture, for the purpose of the present short chapter and the selection has been guided, above all, by simplicity of arguments.

7.1.3 Propagating Turbulent Flames

7.1.3.1 Spherical Propagation in a Stirred Enclosure

Use of such kind of experimental device, i.e., a quasi spherical enclosure in which turbulence is continuously produced by fans has been pioneered by Sokolik, Karpov, and Semenov [5,18] and was followed by many other research groups; it is worth noting that a large amount of results have been gathered by the group of Bradley and his colleagues at Leeds University [19]. The rotational speed of the fans controls the turbulence kinetic energy, i.e., the velocity fluctuation levels, and the electrical power dissipated is directly related to the dissipation rate of the turbulence. The turbulent flame is ignited at the center of the bomb after filling the volume with a given mixture, and the pressure rise is measured, giving the mass burning rate and, with a few additional hypothesis, the "turbulent burning rate," which is supposed to be proportional to the turbulent flame speed previously defined.

The influence of turbulence has commonly been studied in terms of u', which is the square root of the turbulence kinetic energy, and the Taylor scale λ, and the so-called Karlovitz number $Ka = (u'\delta_L)/(\lambda S_L)$ characterizing the small-scale features of turbulent combustion. A typical set of results is given in Figure 7.1.1, showing clearly the tendency to "smoothed combustion" when the value of the Karlovitz number becomes large. In addition to this, it has also been found that the Lewis number $Le = D/\kappa$ of the deficient reactant has a strong influence on both flame speed and local extinctions of flamelets [12,18]. On the visualizations, such as the one presented in Figure 7.1.1, we can see that the structure of the flame brush seems to display "flamelets" structures when the Karlovitz number is not too large, in agreement with the structure of wrinkled flame. On the other hand, an increase in this parameter, usually obtained by an increase in turbulence, leads to a visual

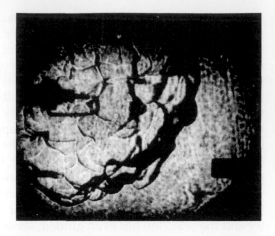
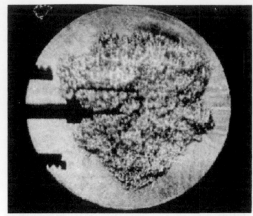

FIGURE 7.1.1
Experiments in closed vessels by Abdel-Gayed and Bradley [19], left: $Ka.Le = 0.003$ continuous laminar flame sheet, right: $Ka.Le = 0.238$ breakup of the continuous flame sheet. (Reprinted from Lewis, B. and Von Elbe, G., *Combustion, Flames and Explosions of Gases*, Academic Press, New York, 1961. With permission. Figure 204, p. 401, copyright New York Academic Press (Elsevier editions).)

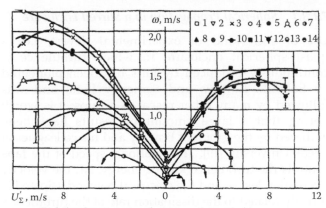

Turbulent burning rates of mixtures of air with: (a) propane: (b) methane in relation to fluctuation velocity for α of: 1) 0.6, 2) 0.65, 3) 0.8, 4) 0.9, 5) 1.0, 6) 1.2, 7) 1.4, 8) 0.7, 9) 0.8, 10) 1.0, 11) 1.1, 12) 1.2, 13) 1.4, 14) 1.6.

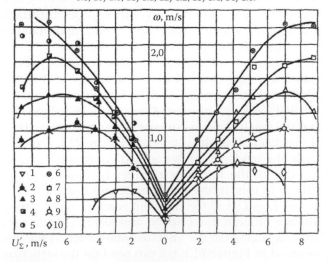

FIGURE 7.1.2
Turbulent mass burning rate versus the turbulent root-mean-square velocity by Karpov and Severin [18]. Here, α is the air excess coefficient that is the inverse of the equivalence ratio. (Reprinted from Abdel-Gayed, R., Bradley, D., and Lung, F.K.-K., *Combustion regimes and the straining of turbulent premixed flames*, *Combust. Flame*, 76, 213, 1989. With permission. Figure 2, p. 215, copyright Elsevier editions.)

disappearance, or a scrambling, of these fronts. This has been attributed to the flame stretching by turbulence, which can even lead to local flamelet extinctions; this response to flame-stretching effects is "naturally" sensitive to the Lewis number. However, this simple representation of turbulent flame as an ensemble of stretched laminar flamelets is also deficient in many respects.

The turbulent burning rate has been measured in these experiments for many different cases, as shown in Figure 7.1.2, displaying first an increase in the velocity fluctuations. Then, a saturation and even a decrease are observed for increasing values of the velocity fluctuations; this is attributed to flamelet extinctions. From these results, one could have thought that turbulent premixed flames are very likely to extinguish in the distributed combustion regime, but this question is not so simple: the recent experimental investigation of Shy et al. [16] seems to show that methane–air flames are very resistant to extinction even for remarkably large values of the strain rates (i.e., high values of the Karlovitz number Ka).

It is worth noticing that the "turbulent burning rates" reported in Figure 7.1.2 have been defined similarly but not exactly as the "turbulent flame speed" mentioned in Section 7.1.2. The mixture has been ignited at the center of the bomb and the dependence of the pressure on time has been recorded. This has enabled to determine the derivative of the burned mixture volume. This derivative is ascribed to a spherical surface whose volume is simply equal to the volume of fully burned products, thus leading to an estimate of the turbulent combustion rate.

7.1.3.2 Propagation in a Turbulent Box

The visualization of the turbulent flame is easier for the experiments carried out by Trinité et al. [20].

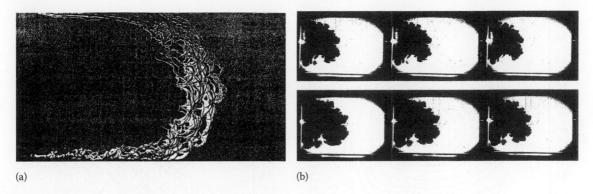

(a)   (b)

FIGURE 7.1.3
(a) Direct shadowgraphy of a turbulent flame propagating in a square cross-section combustion chamber. (b) Temporal evolution after ignition imaged thanks to Laser tomography, from top (right) to bottom (left). (Reprinted from Karpov, V.P. and Severin E.S., *Fizika Goreniya I Vzryva*, 16, 45, 1980. With permission. Figures 1 and 2, p. 42, copyright Plenum Publishing Corporation (Springer editions).)

In this instance, the flame is produced by a line of sparks located in a square cross-section combustion chamber equipped with quartz windows after it has been filled through a perforated plate. Figure 7.1.3a shows the image obtained by the direct shadowgraphy; the light crosses the entire 10 cm wide vessel and both the apparent thickness of the flame and the turbulent flame brush can thus be directly estimated.

It is perhaps worth noticing that a measurement of flame displacement speed is trivial in this configuration; however, a measurement of the flame burning rate, which is the flame speed relative to the fresh mixture, would require an additional set of measurements determining the flow velocity. This is quite different from the measurements performed in a "bomb," shown in Figure 7.1.2, where the burning rate readily deduced from the pressure rise is a nonlocal quantity averaged over the entire flame.

An individual isothermal surface can be traced with the help of laser tomography, also known as laser sheet imaging, where a laser sheet and oil droplets are combined to visualize the instantaneous flame surface in a plane. This technique is ideal when wrinkling of an isoline is of interest; besides, typically it shows the area occupied by the combustion products if the instantaneous flame thickness is small, such as a black area in

Figure 7.1.3b. Furthermore, Figure 7.1.4 clearly shows that the amount and size of wrinkles depend on the parameter u'/S_L. These photographs can also be used to investigate whether this isothermal surface that we can probably identify with a "flamelet" exhibits a fractal behavior and if so, what is its fractal dimension. The result obtained has been that the surface is not really a fractal because its fractal dimension varies with the scale of measurement, and this is directly related to the fact that the wrinkling factor depends on the ratio of turbulent kinetic energy to the flamelet propagation velocity u'/S_L, see Figure 7.1.4; also further details can be found in Ref. [21].

7.1.3.3 Spherical Propagation in a Grid Turbulence

An interesting alternative to closed volume apparatus is provided by an arrangement where a flame is ignited in a (large) homogeneous flow behind a turbulence-generating grid. The turbulence is slowly decaying downstream the grid, and the growing flame is carried by the mean flow velocity, but many detailed instantaneous images of the flamelets, as well as velocity measurements by particle image velocimetry (PIV), can be obtained.

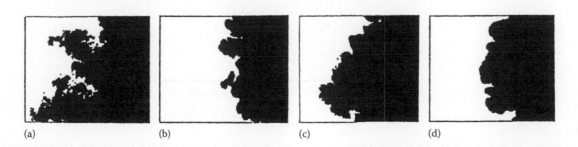

(a)   (b)   (c)   (d)

FIGURE 7.1.4
Tomographic (laser sheet) cuts of turbulent premixed flame fronts. Propane–air mixture is studied at the equivalence ratio 0.9. Case (a) $u'/S_L = 2.48$, case (b) $u'/S_L = 1.55$, case (c) $u'/S_L = 1.22$, case (d) $u'/S_L = 0.68$.

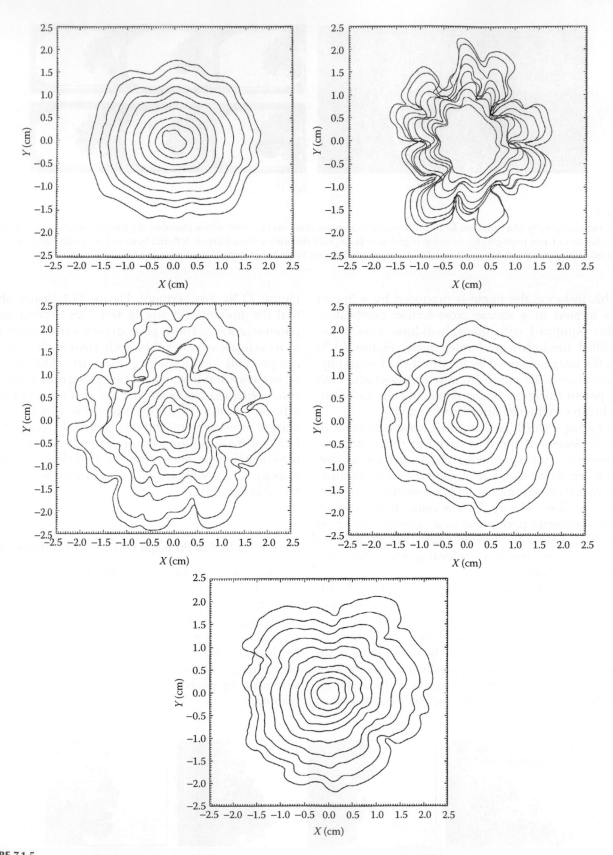

FIGURE 7.1.5

Spherical expanding flames. Left: CH_4–air stoichiometric ($\phi = 1$) flames in nearly homogeneous grid turbulence. The turbulence intensity $I_t = u'/U$ is 4% (top) and 12% (bottom). Right: $u'/S_L - 0.9$, influence of molecular diffusion through the Lewis number. Top: H_2–air ($\phi = 0.27$), middle: CH_4–air ($\phi = 1$), bottom: C_3H_8–air ($\phi = 1$). (Reprinted from Pocheau, A. and Queiros-Condé, D., *Phys. Rev. Lett.*, 76, 3352, 1996. With permission. Figure 2, p. 3353, copyright American Physical Society (APS).)

Figure 7.1.5 by Boukhalfa and his coworkers [12,22,23] do show the influence of both turbulence intensity (left) and molecular diffusion properties (right). The pictures on the left side agree with the findings of Figure 7.1.4. The pictures on the right side are obtained by using three distinct mixtures of hydrogen, methane, and propane from the top to the bottom. This corresponds to different values of the Lewis number and its influence is clearly delineated in the results: for Lewis number values lower than unity, the flame front displays successive lobes or cusps structures likely to give rise to the formation of hot products pockets in the fresh gases. It is also worth noticing that the wrinkling rate and consequently the available flame surface density appear to be significantly influenced by those molecular properties. However, this does not necessarily imply that the mean consumption rate increases since the local consumption rate can decrease owing to local curvature and strain rate effects.

The turbulent flame brush is easily measured in these experiments. Figure 7.1.6 clearly shows that the previously discussed steady state of propagation has not been reached with the present experiments. Nevertheless, it is clear that again the Lewis number (much lower for hydrogen flames) has a significant influence on this quantity. This influence is supposed to be related to the so-called thermal-diffusional instabilities of the flamelets, arising in these conditions. Many interesting features have been evidenced in those different studies: the main results are related to the topology of the flame front and its sensitivity to a large number of parameters related to both turbulence and molecular properties [12,22,23].

7.1.4 Stabilized Oblique Turbulent Flames

7.1.4.1 Turbulent Bunsen Flames

Turbulent Bunsen burners, although they have relatively low turbulence levels, are also good tools for studying the structure of turbulent premixed flames. Gülder et al. [11] and Gökalp et al. [24] have successfully employed this type of experiment to study wrinkled flames and their properties. Figure 7.1.7 displays the global structure of such a flame revealed by Laser tomography applied by Dumont, Durox, and Borghi [25]. This beautiful picture shows the large-scale wrinkles of the flame front with possible disruptions at the top of the flame. The green color is due to the 15 ns duration pulse at a 532 nm wavelength of the YAG laser used to obtain instantaneous tomographic images.

In this study, the flame can be classified as a wrinkled flame throughout most of the flow field. The main findings of [25] are related to both (1) the question of how the turbulent velocity field is affected by the chemical reaction and induced expansion phenomena and (2) the measurements of mean flame surface density and the

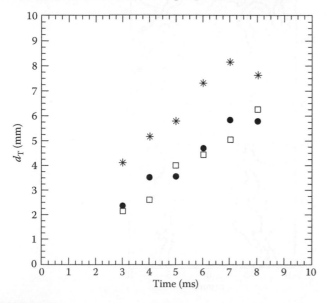

FIGURE 7.1.6
Temporal evolution of the flame brush thickness for the previously described mixtures of hydrogen, methane, and propane with air. (Reproduced from Renou, B. and Boukhalfa, M., *Combust. Sci. Technol.*, 162, 347, 2001. With permission. Figure 2, p. 353, copyright Gordon Breach Science Publishers (Taylor and Francis editions).)

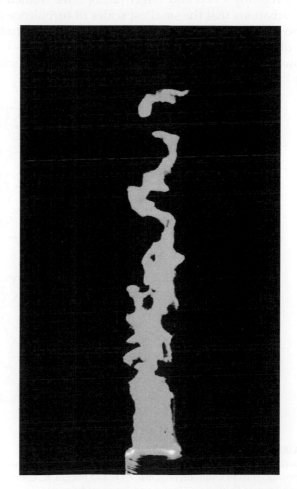

FIGURE 7.1.7
Methane–air Bunsen burner turbulent premixed flame.

evaluation of the mean chemical rate. Concerning the former point, it has been found that in this kind of open flow, the turbulence kinetic energy is only very slightly affected by the various heat release effects, which prove to be small and seem to compensate for each other in some way. A more interesting point is that a significant increase in turbulence length scale caused by the dilatation of the flow field has been evidenced in the flame brush. Another interesting result has been obtained from the measurements of the flame surface density. No substantial influence of stretching and curvature on flamelets has been found in this case, and this has allowed to evaluate the mean combustion rate $\overline{\rho\omega}$ as proportional to $\rho^u S_L \overline{\Sigma}$ with S_L the planar and unstretched velocity of propagation of the laminar flamelet, ρ^u and $\overline{\Sigma}$ being, respectively, the density of the fresh mixture and the available mean flame surface density. Finally, the eddy breakup (EBU) representation has shown a surprisingly good agreement with the obtained data.

More recently, experimental studies have been carried out using a similar device but with an annular external hot coflow of burned gases that allowed one to operate within a much larger velocity range. Chen et al. [26] and more recently Chen and Bilger [27,28] have studied the perturbations that the smallest scales of turbulence can impose to the local flamelet structures. Those studies are of paramount importance, first because they have allowed to get deeper insights into the local structure

of turbulent premixed flames and second because they evidenced turbulent combustion regimes significantly different from the oversimplified picture of a thin flame front with constant thickness just wrinkled by the turbulence flow field.

In the work of Chen et al. [26], premixed stoichiometric turbulent methane–air flames have been investigated with varying fuel–air mixture nozzle jet exit velocities: 30 (case F3), 50 (case F2), and 65 m s^{-1} (case F1). In terms of their location on the turbulent premixed combustion diagram, the three flames cover a large range of regimes in the turbulent combustion diagram from the flamelet regime to the borderline of the perfectly stirred reactor (PSR) regime. Thanks to advanced Laser diagnostics, the flow fields as well as the scalar fields have been characterized using two points and two components, laser Doppler anemometry (ADL) together with 2D Rayleigh thermometry and line Raman–Rayleigh laser-induced predissociation fluorescence (LIPF)-OH techniques, respectively.

From the 2D instantaneous Rayleigh temperature fields, such as shown in Figure 7.1.8, isotemperature contours can be obtained and they clearly show that the distance between the isothermal contours strongly varies at different locations, being deeply perturbed by turbulence, especially on the fresh reactants side.

The more recent work of Chen and Bilger [27,28] attempted to provide a criterion for the transition from

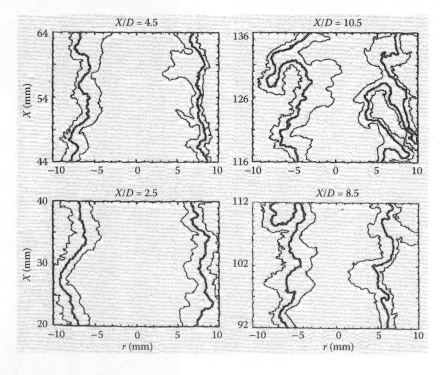

FIGURE 7.1.8
Perturbed flamelet structure as obtained from 2D instantaneous Rayleigh temperature fields for the case F1 (jet exit velocity is 65 m/s). (Reproduced from Dumont, J.P., Durox, D., and Borghi, R., *Combust. Sci. Tech.*, 89, 219, 1993. With permission. Figure 3.1, p. 233, copyright Gordon Breach Science Publishers (Taylor and Francis editions).)

flamelet to nonflamelet behavior. In this work, OH concentrations and three-dimensional gradients of the reaction progress variable have been measured in Bunsen flames with a combined two-sheet Rayleigh scattering and planar LIF-OH imaging technique. Instantaneous as well as conditional average of progress variable gradients have been evaluated; this revealed that noticeable flame thickening when compared with a premixed laminar flame of reference, both in preheating and in reaction zones, can be observed as turbulent intensity is increased. However, for sufficiently low levels of turbulence, strong correlations are found between the progress variable based on temperature c_T and its gradient $|\nabla c_T|$, which qualitatively follows the laminar flamelet behavior.

The results reproduced from Ref. [28] and presented in Figure 7.1.9 show that the internal structure of the "flamelets" within the studied flames displays strong departures from both unstretched laminar flamelet and stretched counterflow flamelets. Figure 7.1.9 supports the picture of the perturbed flamelet model recently introduced in Ref. [29]. In this model, depending on the local value of the ratio of laminar flame thickness and

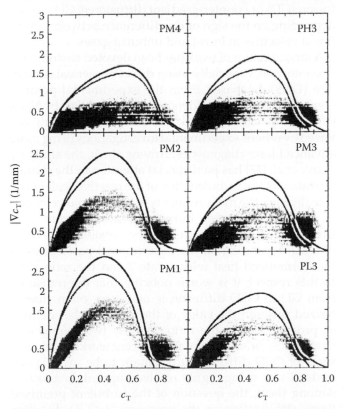

FIGURE 7.1.9
Instantaneous progress variable gradients versus progress variable value; the structure corresponding to a one-dimensional planar unstretched flame and to a Tsuji counterflow (unburnt to air) geometry is also depicted. (Reprinted from Chen, Y.C., Peters, N., Schneemann, G.A., Wruck, N., Renz, U., and Mansour, M.S., *Combust. Flame*, 107, 223, 1996. With permission. Figure 11, p. 234, copyright Elsevier editions.)

Kolmogorov length scale, the conditional probability density function (PDF) of the reactive scalar gradient is eventually decomposed into two distinct parts: (1) one following the laminar flamelet behavior within the reaction zone where the progress variable is greater than a given progress variable value c_T^* (approximately 0.7 for the results presented in Figure 7.1.9), (2) the other where reactive scalar gradients are determined by turbulent mixing representation, within the preheating zone for progress variable below c_T^*.

Such experiments have also allowed a better understanding concerning the turbulent combustion regimes, which extends the earlier foundation works mentioned in the introduction, and revisited by Barrère and Borghi [30,31] and others [32,33]. New intermediate combustion regimes have been delineated, thanks to the instantaneous flame front imaged through simultaneous 2D measurements.

This recent attempt differs from the previous classification where the wrinkled flamelet regime has been considered up to $\eta_K = \delta_L$. Chen and Bilger have proposed to tentatively classify the different turbulent premixed flame structures they observed among four different regimes:

1. Wrinkled laminar flamelet regime. The well-known ideal regime where the laminar flame structure is only wrinkled by turbulence without any modification of its internal structure.

2. Complex strain flame-front regime. Where the flame fronts are still lamella-like but thickened due to enhanced turbulent diffusivity. Scalar transport is expected to be counter-gradient in this regime.

3. Turbulent flame-front regime. Eddy-like contortions of the flame preheat and burned gases zones give rise to "out of front" islands and peninsula structures of intermediate progress variable values. Scalar transport becomes gradient-like.

4. Distributed flame-front regime. The instantaneous flame front and the average flame brush occupy approximately the same volume with energy containing eddies that can enter it. The fuel consumption zone may still be much thinner and lamella-like.

The authors introduced a new criterion to delineate the two intermediate regimes: $l_m/\delta_L = 1$, which is based on an interaction length scale: from Figure 7.1.10, it appears proportional to the Kolmogorov scale, but needs a Reynolds number higher than 10. This criterion defines the evolution from a lamella-like pattern (flame PM1 for instance) to a nonflamelet disrupted one (flame PM4).

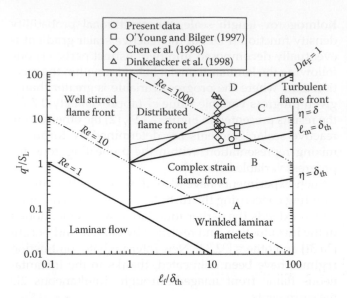

FIGURE 7.1.10

Premixed turbulent combustion regime diagram proposed by Chen and Bilger. Two intermediate regimes are delineated between distributed flame front and wrinkled laminar flamelets. (Reprinted from Chen, Y.C. and Bilger, R., *Combust. Flame*, 131, 400, 2002. With permission. Figure 9, p. 411, copyright Elsevier editions.)

In this respect, Figure 7.1.9 confirms that flames PM4 and PH3 lie in those highly perturbed regimes of turbulent premixed combustion. In this figure, we can also notice that intermediate states with progress variable values between 0.3 and 0.7 do appear (see case PM4 of Figure 7.1.9 for instance): finite rate chemistry effects are more and more clearly visible. For the same conditions PM4, one can also notice that instantaneous gradient values are diminished; it is now rather fixed by turbulence than by the laminar flamelet structure itself, an internal flamelet structure that is probably deeply modified by small-scale structures that perturb it.

In the search of identifying the small-scale properties of highly turbulent premixed flames, another interesting study has been recently carried out by Dinkelacker and coworkers [34]. In this work, highly resolved OH and temperature measurements have been compared to a compilation of recent experimental results including the one obtained by Chen and Bilger [28]. The respective influence of mean strain and of entrainment by small-scale eddies is discussed, and the fact that the flamelets are very much varying with time, in the same flame, is emphasized. The essential conclusion is that low (between 1 and 600) Reynolds number effects are not well taken into account in the proposed diagrams of [31–33], in qualitative agreement with the discussion of Chen and Bilger (who addressed only the range 1–10 in Reynolds number). Indeed, the folding process between flamelets, which is likely to give rise to flame thickening, is supposed to occur whatever the Reynolds number value in classical turbulent combustion diagrams

[30–33], whereas this process is observed only for experiments performed at sufficiently large Reynolds number. This emphasizes that further work is needed since several interrelated questions remain concerning, among others, the influence of mean strain rate on the flame structure and the Reynolds number dependency of the different limits used to delineate the different regimes of turbulent combustion.

Another question that has been largely discussed at least from the theoretical and modeling points of view is related to the occurrence of the so-called counter-gradient diffusion effects. Indeed, under the assumptions of infinitely fast chemistry and resulting infinitely thin flamelets, the early works of Libby and Bray [35] has shown that the turbulent transport of the reactive scalar progress variable follows:

$$\overline{\rho u_i'' c''} = \bar{\rho} \left(\bar{u}_i^b - \bar{u}_i^u \right) \tilde{c} (1 - \tilde{c})$$

where $\bar{u}_i^b$ and $\bar{u}_i^u$ denote the conditional velocities in burnt and unburnt mixtures, respectively. This relationship emphasizes the possibility that either gradient diffusion (GD) or counter-gradient diffusion (CGD) occurs depending on the sign of the difference between conditional velocities in burnt and unburnt gases.

A large amount of work has been devoted to study this phenomenon, especially using direct numerical simulation (DNS) databases. From the experimental point of view, CGD has been evidenced as early as 1980 by Moss [36]. This pioneering work has been more recently followed by others' experimental investigations using more advanced laser diagnostics. Among them, the study by Frank et al. [37] has paid special attention to the characterization of the turbulent flux of the reactive scalar variable through the measurements of conditional velocities in a turbulent Bunsen burner geometry. Figure 7.1.11 from Ref. [37] clearly evidences the possible occurrence of CGD depending on the experimental parameters as the normalized heat release rate τ and the ratio u'/S_L. In this respect, it is worth noticing that the transition from GD to CGD diffusion is now very often characterized through the value of the Bray number defined as proportional to the ratio of these two parameters, namely $\tau S_L / u'$. In spite of these recent works, many features of CGD diffusion still remain to be investigated in details by using highly resolved optical diagnostics. Among them, the question of the turbulent premixed flame stabilization in the presence of CGD diffusion requires special attention.

Studying the influence of increased operating pressure on Bunsen turbulent flames, Kobayashi and coworkers [38,39] have recently put into evidence possible effects of flamelets instability, including modification of length scales, in particular. Figure 7.1.12 shows this remarkable

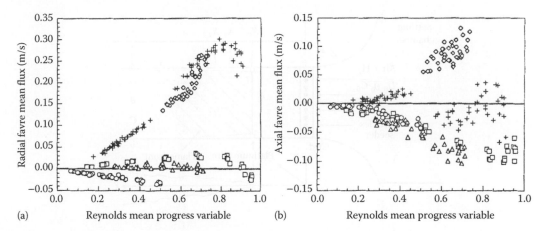

FIGURE 7.1.11

Radial and axial components of the Favre mean flux of progress variable obtained in a Bunsen burner geometry for different operating conditions. (Reproduced from Chen, Y.C. and Bilger, R., *Combust. Sci. Tech.*, 167, 187, 2001. With permission. Figure 19, p. 218, copyright Gordon Breach Science Publishers (Taylor and Francis editions).)

and quite surprising role of high operating pressure levels. Based on the classical works on hydrodynamical and thermodiffusive instabilities of laminar flames, the authors have shown that the region of wave number where the flame front is unstable extends towards larger wave numbers by increasing pressure. More specifically, with increase in pressure, the flame front becomes unstable to smaller-scale disturbances. Consequently, a significant influence on the turbulent burning velocity is observed at elevated pressures: S_T/S_L increases with u'/S_L up to 30 at 3.0 MPa. The observed increase is very fast at high pressures, particularly for weak turbulence intensity. This emphasizes how the effects associated to flamelet instability are able to increase the turbulence levels, leading to an increase in the turbulent burning velocity.

7.1.4.2 V-Shaped Flames in Grid Turbulence

The so-called turbulent "V-shaped flames" are the flames anchored behind a rod or a catalytic wire in a flow where turbulence is generated by an upstream grid. Trinité et al. [7,40] and Driscoll and Faeth [41] have studied such flames. Instantaneous images of rare beauty have been obtained from which it is very clearly seen that the turbulent flame brush width is continuously increasing downstream of the stabilizing rod, see Figure 7.1.13.

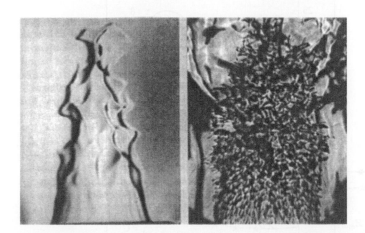

FIGURE 7.1.12

Instantaneous schlieren photographs of turbulent Bunsen burner flames at $P = 0.1$ MPa (left) and $P = 1.0$ MPa (right). The flow at $U = 2.0$ m/s is made turbulent, thanks to a perforated plate with hole diameter $d = 2.0$ mm. The burner exit diameter is 20 mm. (Reprinted from Frank, J.H., Kalt, P.A., and Bilger, R.W., *Combust. Flame*, 116, 220, 1999. With permission. Figure 9, p. 238, copyright Elsevier editions.)

FIGURE 7.1.13

Instantaneous images obtained in a turbulent premixed V-shaped flame configuration, methane and air in stoichiometric proportions. (Reproduced from Kobayashi, H., Tamura, T., Maruta, K., Niioka, T., and Williams, F.A., *Proc. Combust. Inst.*, 26, 389, 1996. With permission. Figure 2, p. 291, copyright Combustion Institute.)

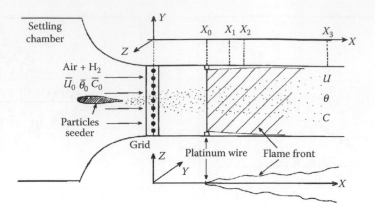

FIGURE 7.1.14
Experimental configuration used by Trinité and coworkers.

A peculiarity of this experiment is that, even if turbulence behind a grid is decaying and its scales are increasing, flame-generated turbulence has to be considered. It has been shown that turbulence can be directly modified by confining the flame between lateral walls, and by varying the longitudinal mean pressure gradient with this confinement. Such an experimental device has been studied among others by Trinité and his coworkers [7,40], see Figure 7.1.14.

Figure 7.1.15 shows a sample of turbulence fields, with and without combustion, measured in this configuration. The turbulence levels are significantly higher for the reactive than for the nonreactive flow. The authors have also tried to compare their measurements to the Bray–Moss–Libby (BML) representation, which is able to deal with counter-gradient phenomena and flame-generated turbulence, but further discussions are needed for this problem, and the reader is referred to [42]. This study was probably one among the first that evidenced flame-generated turbulence phenomena in turbulent premixed flames.

7.1.4.3 High Velocity Confined Oblique Flames

The interest of studying turbulent premixed flames for modern gas turbines and jet engine afterburners

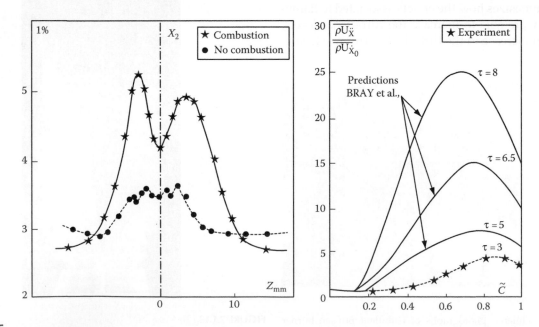

FIGURE 7.1.15
Turbulence measurements from reference [7]. Left: longitudinal turbulence intensity profiles with and without combustion in the X_2 section defined in Figure 7.1.14. Right: comparison with the BML prediction of the longitudinal turbulent energy. (From Escudié, D., Paranthoën, P., and Trinité, M., *Flames, Laser and Reactive Systems* (selected papers from the Eighth International Colloquium on Gasdynamics of Explosions and Reactive Systems), *Progress in Astronautics and Aeronautics Series*, AIAA Inc. publishers, pp. 147–163, 1983. With permission. Figure 1, p. 119, copyright AIAA.)

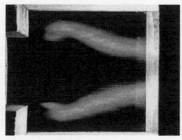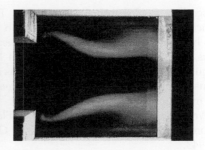

FIGURE 7.1.16

High velocity confined oblique flames: large-scale motion superimposed to turbulent combustion. Left: long-time exposure. Right: two short-time exposure photographs. (From Escudié, D., Paranthoën, P., and Trinité, M., *Flames, Laser and Reactive Systems* (selected papers from the Eighth International Colloquium on Gasdynamics of Explosions and Reactive Systems), *Progress in Astronautics and Aeronautics Series*, AIAA Inc. publishers, pp. 147–163, 1983. With permission. Figures 10 and 18, p. 155 and 161, copyright AIAA.)

stimulates experiments in high-velocity confined flows. The flame has then to be stabilized by either a sufficiently large bluff body or by a backward facing step in a side wall, or a pilot co-flowing flame. Such configurations have been studied among others by Moreau and Borghi [9], Magre et al. [43], and Deshaies, Bruel, and Champion [44]. These studies have revealed the same structure of the flame as found in Bunsen burners or V-shaped flames, but for much stronger turbulence, owing to the high velocity and large velocity gradients. Such experiments do allow one to build detailed databases for numerical modeling purposes. Fields of mean velocity, turbulence, temperature, and mean mass fraction of chemical species for different operating conditions have been measured. However, the particular feature of the turbulent flame in these devices is the occurrence of a large-scale fluctuation that appears to result from a particular coupling with the acoustical longitudinal modes of the duct.

Figure 7.1.16 clearly evidences these large-scale oscillations. In some cases, these oscillations can lead the flame to oscillate upstream and downstream of the bluff body, e.g., as found in the experimental study of Ganji and Sawyer [45,46]. This flame flapping complicates significantly the determination of statistical properties of combustion and the numerical calculation of such a flame. It may, however, be a good test case for large eddy simulation (LES) methods. Finally, it is worth noticing that such configurations have provided valuable experimental databases that are now widely used to validate turbulent combustion models [47–49].

7.1.5 Conclusions

From the present survey, the following conclusions can be drawn. First, the cited experiments have allowed a continuous and comprehensive study of turbulent premixed flames through the last decades. Many progresses have been made in their knowledge, thanks to the insight gained through experimental results, and it has led to the elaboration of experimental databases to test modeling proposals. Today, the extensive use of nonintrusive

optical diagnostics does allow not only a check of the model's capabilities by comparing, for instance, calculated and measured data but also an opportunity to evaluate directly the hypothesis on which a model would rely. This is a significant progress to propose realistic closures of turbulent premixed combustion.

Nevertheless, despite all these remarkable achievements, some open questions still remain. Among them is the influence of the molecular transport properties, in particular Lewis number effects, on the structure of turbulent premixed flames. Additional work is also needed to quantify the flame-generated turbulence phenomena and its relationship with the Darrieus–Landau instability. Another question is: what are exactly the conditions for turbulent scalar transport to occur in a counter-gradient mode? Finally, is it realistic to expect that a turbulent premixed flame reaches an asymptotic steady-state of propagation, and if so, is it possible, in the future, to devise an experiment demonstrating it?

References

1. G. Damköhler 1940, Der Einfluss der Turbulenz auf die Flammengeschwindigkeit in Gasgemischen, *Zs. Electrochemie* 6(11):601–626.
2. K.I. Shchelkin 1943, On the combustion in turbulent flows, *Zh.T.F.* 13(9–10):520–530.
3. B. Lewis and G. Von Elbe 1961, *Combustion, Flames and Explosions of Gases*, 2nd edn., Academic Press, New York.
4. M. Summerfield, S.H. Reiter, V. Kebely, and R.W. Mascolo 1955, The structure and propagation mechanisms of turbulent flames in high speed flows, *Jet Propulsion* 25(8):377–384.
5. V.P. Karpov, E.S. Semenov, and A.S. Sokolik 1959, Turbulent combustion in closed volume, *Doklady Akad. Nauka SSSR* 128:1220–1223.
6. A.V. Talantov, V.M. Ermolaev, V.K. Zotin, and E.A. Petrov 1969, Laws of combustion of an homogeneous mixture in a turbulent flow, *Fizika and Goreniya I Vzryva* 5(1):106–114.

7. D. Escudié, P. Paranthoën, and M. Trinité 1983, Modification of turbulent flow-field by an oblique premixed hydrogen-air flame, in: *Flames, Laser and Reactive Systems* (selected papers from the Eighth International Colloquium on Gasdynamics of Explosions and Reactive Systems), *Progress in Astronautics and Aeronautics Series*, AIAA Inc. publishers, pp. 147–163.

8. F.H. Wright and E.E. Zukowsky 1962, Flame spreading from bluff-body flame holders, *Proc. Combust. Inst.* 8:933–943.

9. R. Borghi and P. Moreau 1977, Turbulent combustion in a premixed flow, *Acta Astronautica* 4:321–341.

10. F.C. Gouldin 1987, An application of fractals to modeling premixed turbulent flames, *Combust. Flame* 68:249–266.

11. Ö.L. Gülder, G.J. Smallwood, R. Wong, D.R. Snelling, R. Smith, B.M. Deschamps, and J.C. Sautet 2000, Flame front surface characteristics in turbulent premixed propane–air combustion, *Combust. Flame* 120:476–416.

12. B. Renou, M. Boukhalfa, D. Puechberty, and M. Trinité 2000, Local flame structure of freely propagating premixed turbulent flames at various Lewis number, *Combust. Flame* 123:107–115.

13. R. Borghi 1988, Turbulent combustion modelling, *Prog. Energy Combust. Sci.* 14(4):245–292.

14. D. Veynante and L. Vervisch 2002, Turbulent combustion modelling, *Prog. Energy Combust. Sci.* 28(3):193–266.

15. B. Karlovitz, J.W. Denniston, D.H. Knapschaefer, and F.E. Wells 1953, Studies on turbulent flames, *Proc. Combust. Inst.* 4:613–620.

16. S.S. Shy, W.J. Lin, and J.C. Wei 2000, An experimental correlation of turbulent burning velocities for premixed turbulent methane–air combustion, *Proc. R. Soc. Lon. A* 456:1997–2019.

17. S.S. Shy, S.I. Yang, W.J. Lin, and R.C. Su 2005, Turbulent burning velocities of premixed CH_4/diluent/air flames in intense isotropic turbulence with consideration of radiation losses, *Combust. Flame* 143:106–118.

18. V.P. Karpov and E.S. Severin 1980, Effects of molecular transport coefficient on the rate of turbulent combustion, *Fizika Goreniya I Vzryva* 16(1):45–51, translated by Plenum Publishing Corporation.

19. R. Abdel-Gayed, D. Bradley, and F.K.-K. Lung 1989, Combustion regimes and the straining of turbulent premixed flames, *Combust. Flame* 76:213–218.

20. A. Floch, M. Trinité, F. Fisson, T. Kageyama, C.H. Kwon, and A. Pocheau 1989, *Proceedings of the Twelfth ICDERS*, University of Michigan, Ann Arbor, pp. 379–393.

21. A. Pocheau and D. Queiros-Condé 1996, Scale invariance of the wrinkling law of turbulent propagating interface, *Physical Review Letters* 76 (18):3352–3355.

22. B. Renou and M. Boukhalfa 2001, An experimental study of freely propagating premixed flames at various Lewis numbers, *Combust. Sci. Technol.* 162:347–371 (more informations through www.informaworld.com).

23. B. Renou, A. Mura, E. Samson, and M. Boukhalfa 2002, Characterization of the local flame structure and the flame surface density for freely propagating premixed flames at various Lewis number, *Combust. Sci. Technol.* 174:143–179.

24. I. Gokälp, I.G. Shepherd, and R.K. Cheng 1988, Spectral behavior of velocity fluctuations in premixed turbulent flames, *Combust. Flame* 71 (3):313–323.

25. J.P. Dumont, D. Durox, and R. Borghi 1993, Experimental study of the mean reaction rates in a turbulent premixed flame, *Combust. Sci. Technol.* 89:219–251 (more informations through www.informaworld.com).

26. Y.C. Chen, N. Peters, G.A. Schneemann, N. Wruck, U. Renz, and M.S. Mansour 1996, The detailed flame structure of highly stretched turbulent premixed methane–air flames, *Combust. Flame* 107:223–244.

27. Y.C. Chen and R. Bilger 2001, Simultaneous 2-D imaging measurements of reaction progress variable and OH radical concentration in turbulent premixed flames: Instantaneous flame front structure, *Combust. Sci. Tech.* 167:187–222 (more informations through www.informaworld.com).

28. Y.C. Chen and R. Bilger 2002, Experimental investigation of three dimensional flame-front structure in premixed turbulent combustion, Part I: Hydrocarbon–air Bunsen flames, *Combust. Flame* 131:400–435.

29. A. Mura, F. Galzin, and R. Borghi 2003, A unified PDF-flamelet model for turbulent premixed combustion, *Combust. Sci. Technol.* 175 (9):1573–1609.

30. M. Barrère 1974, Modèles de combustion turbulente, *Revue Générale de Thermique* 148:295–308.

31. R. Borghi 1985, On the structure and morphology of turbulent premixed flames, in: C. Bruno and S. Casci (Eds.), *Recent Advances in the Aerospace Sciences*, Plenum Press, New York, pp. 117–138.

32. J. Abraham, F.A. Williams, and F.V. Bracco 1985, A discussion of turbulent flame structure in premixed charge, SAE Paper 850343, in: *Engine Combustion Analysis:* New Approaches, p. 156.

33. N. Peters 1986, Laminar flamelet concepts in turbulent combustion, *Proc. Combust. Inst.* 21:1231–1250.

34. F. Dinkelacker 2003, Experimental validation of flame regimes for highly turbulent premixed flames, *Proceedings of the First European Combustion Meeting ECM2003*. See also the 27th *Symp.(Int.) on Combustion*, pp. 857–865, 1998.

35. P.A. Libby and K.N.C. Bray 1981, Countergradient diffusion in premixed turbulent flames, *AIAA J.* 19(2):205–213.

36. J.B. Moss 1980, Simultaneous measurements of concentration and velocity in an open premixed turbulent flame, *Combust. Sci. Tech.* 22:119–129.

37. J.H. Frank, P.A. Kalt, and R.W. Bilger 1999, Measurements of conditional velocities in turbulent premixed flames by simultaneous OH PLIF an PIV, *Combust. Flame* 116:220–232.

38. H. Kobayashi, T. Tamura, K. Maruta, T. Niioka, and F.A. Williams 1996, Burning velocity of turbulent premixed flames in a high pressure environment, *Proc. Combust. Inst.* 26:389–396.

39. H. Kobayashi, T. Nakashima, T. Tamura, K. Maruta, and T. Niioka. Turbulence measurements and observations of turbulent premixed flames at elevated pressures up to 3.0 MPa, *Combust. Flame* 108:104–117.

40. P. Goix, P. Paranthoën, and M. Trinité 1990, A tomographic study of measurements in a V-shaped H_2-air flame and a Lagrangian interpretation of the turbulent flame brush evolution, *Combust. Flame* 81:229–241.

41. M.S. Wu, S. Kwon, J. Driscoll, and G.M. Faeth 1990, Turbulent premixed hydrogen–air flames at high Reynolds numbers, *Combust. Sci. Technol.* 73(1–3):327–350.

42. K.N.C. Bray, P.A. Libby, G. Masuya, and J.B. Moss 1981, Turbulence production in premixed turbulent flames, *Combust. Sci. Technol.* 25:127–140.

43. P. Magre, P. Moreau, G. Collin, R. Borghi, and M. Péalat 1988, Further studies by CARS of premixed turbulent combustion in high velocity flow, *Combust. Flame* 71:147–168.

44. M. Besson, P. Bruel, J.L. Champion, and B. Deshaies 2000, Experimental analysis of combusting flows developing over a plane-symmetric expansion, *J. Thermophysics Heat Transfer* 14(1):59–67.

45. A.R. Ganji and R.F. Sawyer 1979, An experimental study of the flow field and pollutant formation in a two-dimensional pre-mixed turbulent flame, AIAA Paper 79-0017, *17th Aerospace Science Meeting*, New Orleans (Louisiana), January 15–17.

46. A.R. Ganji and R.F. Sawyer 1980, An experimental study of the flow field of a two-dimensional premixed turbulent flame, *AIAA J.* 18:817–824.

47. V. Robin, A. Mura, M. Champion, and P. Plion 2006, A multi Dirac presumed PDF model for turbulent reactive flows, *Combust. Sci. Technol.* 178:1843–1870.

48. A. Kurenkov and M. Oberlack 2005, Modelling turbulent premixed combustion using the level set approach for Reynolds averaged models, *Flow Turbulence Combust.* 74:387–407.

49. L. Duchamp de Lageneste and H. Pitsch 2001, Progress in the large eddy simulation of premixed and partially premixed turbulent combustion, in: Center for Turbulence Research (CTR, Stanford) Annual Research Brief.

7.2 Nonpremixed Turbulent Combustion

Jonathan H. Frank and Robert S. Barlow

7.2.1 Introduction

In nonpremixed combustion, the fuel and oxidizer streams are introduced separately, and combustion occurs after the fuel and oxidizer mix on the molecular scale. Many practical combustion devices, such as furnaces, steam boilers, diesel engines, liquid rocket motors, and gas turbine engines, involve turbulent nonpremixed combustion. In these devices, mixing occurs by a combination of turbulent stirring of the fuel and oxidizer streams and molecular diffusion. Turbulence greatly enhances the mixing process by increasing the surface area of the thin mixing layers where most of the molecular diffusion occurs. The interaction between turbulent mixing and combustion chemistry is extremely complex and remains an active research area. In this chapter, we provide an overview of some basic characteristics of turbulent nonpremixed combustion. The emphasis is on fundamental phenomena that have been experimentally studied in relatively simple burner configurations, which are also relevant to the understanding and predictive modeling of complex combustion systems. Detailed treatments of the theory, modeling, and applications of turbulent nonpremixed combustion are available elsewhere [1–5].

7.2.2 Basic Characteristics of Jet Flames

The structure of nonpremixed flames is governed by the coupling between mixing and chemical reaction. The relative importance of these processes is characterized by the Damköhler number, *Da*, which is the ratio of the rates of chemical reaction and fluid dynamic mixing. The extremes of the Damköhler number are designated as the "well-stirred" reactor ($Da \ll 1$) and the fast-chemistry ($Da \gg 1$) regimes, and at each extreme, it is the slower process that limits or controls the behavior of the system. In the "well-stirred" reactor regime, the reactants and products rapidly mix, and the chemical reactions proceed over an extended region of the reactor on a timescale that is much longer than the mixing time. In contrast, the fast-chemistry regime is characterized by thin reaction zones, in which reactions proceed to completion as soon as the reactants come in contact, such that the rate of conversion of reactants to products is limited by the rate of mixing. In their early theoretical work, Burke and Schumann modeled laminar nonpremixed flames as thin sheets using assumptions of an infinitely fast, irreversible one-step reaction ($Da = \infty$) [6]. The next improvement on this simplified model assumed infinitely fast, reversible combustion reactions, with the species and temperature at each location in the flame determined by local thermochemical equilibrium conditions. However, turbulent nonpremixed flames exhibit significant nonequilibrium behavior and involve a wide range of Damköhler numbers. The turbulent flow field produces temporal and spatial fluctuations in the mixing rates, which induce local fluctuations in the chemical reaction rates. Further advances in the modeling have sought to account for nonequilibrium and finite-rate chemistry effects that occur when the relevant Damköhler number is near unity [1–4,7].

Jet flames provide a simple canonical geometry for illustrating the essential features of turbulent nonpremixed flames. In Figure 7.2.1, chemiluminescence images, using different camera-exposure times, show the mean and fluctuating structure of a turbulent

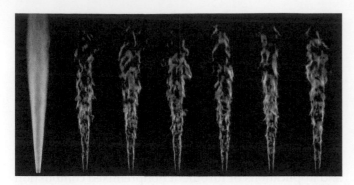

FIGURE 7.2.1
Chemiluminescence images of a turbulent $CH_4/H_2/N_2$ jet flame (Re_d = 15,200) measured with two different exposure times. The long-exposure image (far left) indicates the mean flame structure, and the six shorter exposures to the right illustrate the instantaneous turbulent structure.

nonpremixed jet flame. The fuel is an N_2-diluted mixture of CH_4 and H_2 that issues from the jet at an exit Reynolds number of $Re_d = Ud/\nu = 15,200$, where U is the

bulk exit velocity, $d = 8.0$ mm is the nozzle diameter, and ν is the kinematic viscosity. This particular flame has been the object of many experimental studies over the past 10 years, beginning with work by Bergmann et al. [8], and using a variety of measurement techniques in several laboratories around the world [9]. The long-exposure image on the left of the figure shows the mean envelope of the reaction zone, which is distributed across the mixing layer of the jet and the coflow. The six short-exposure images illustrate the complex instantaneous structure of the turbulent flames. The turbulent flow distorts the shape of the flame and produces a convoluted reaction zone with a wide range of length scales. These perturbations to the flame can result in significant variations in the local reaction rates. Because the reaction rates are highly nonlinear functions of the temperature, the measurements of the mean thermochemical properties of the flame are not adequate for predicting the production rates of the intermediate species and pollutants.

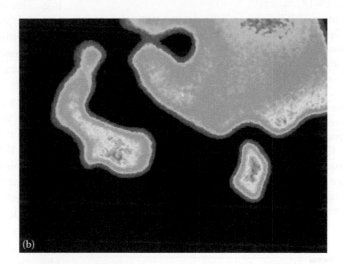

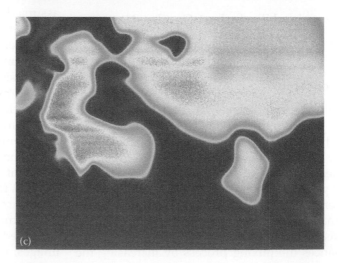

FIGURE 7.2.2
(a) Chemiluminescence image of a turbulent lifted $CH_4/H_2/N_2$ jet flame stabilized above the burner nozzle. The orange rectangle approximates the imaged area for (b) OH-LIF measurements and (c) temperature measurements by Rayleigh scattering.

As the jet exit-velocity increases, the flame becomes increasingly turbulent, but remains anchored to the rim of the burner nozzle. However, for sufficiently large jet velocities, the flame lifts off and stabilizes downstream of the nozzle, as illustrated in Figure 7.2.2a. The distance between the flame-stabilization location and the nozzle exit is referred to as the liftoff height. Partial premixing of the fuel and oxidizer occurs in the region upstream of the flame stabilization location, such that the stabilization region consists of a turbulent edge flame that propagates against the flow of a nonuniform mixture of fuel and air. An example of this complex flame structure is shown by the simultaneous OH-LIF and temperature measurements in Figure 7.2.2. This stabilization region has some characteristics of both nonpremixed and premixed flames, and this presents a challenge for combustion models. The lift-off height fluctuates as the local flow conditions vary in the turbulent jet. Detailed discussions of the stabilization mechanism of lifted flames are available elsewhere [2,10–12].

If the jet velocity increases further, after establishing a lifted flame, the flow reaches a condition for which a flame cannot be stabilized, and global extinction ensues. The velocity at which the flame extinguishes depends on the fuel composition and the degree of partial premixing. Global flame extinction should be avoided in both fundamental research and practical applications, and many approaches have been developed to stabilize flames. For the flames shown in Figures 7.2.1 and 7.2.2, the use of H_2 in the fuel mixture significantly increases the blowoff velocity relative to a CH_4/N_2 fuel mixture. Alternative approaches to increasing the robustness of CH_4 jet flames include partial premixing with an oxidizer and the use of pilot flames to help anchor the jet flame to the nozzle. Figure 7.2.3 shows an example of a partially premixed CH_4/air (1/3 by vol.) jet flame anchored by an annular pilot of lean premixed flames. At these flow conditions, the fuel-rich premixed chemistry is too slow to significantly affect the flame structure, and the flame behaves as a nonpremixed flame, with a single reaction zone. Such flames can be operated at higher exit velocities and higher Reynolds numbers than the corresponding simple jet flames, and they have been used extensively to investigate finite-rate chemistry effects and to develop models that account for these effects [4,9].

7.2.3 Mixture Fraction, Dissipation, and Finite-Rate Chemistry

The state of mixing between the fuel and oxidizer streams in nonpremixed flames is quantified by the mixture fraction, ξ. Conceptually, the mixture fraction is the fraction of mass that originates in the fuel stream, with 0.0 corresponding to the oxidizer stream and 1.0 corresponding to the pure fuel stream. The stoichiometric mixture fraction, ξ_{st}, indicates the condition for which the fuel

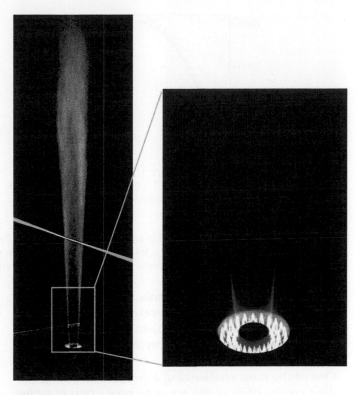

FIGURE 7.2.3
Chemiluminescence images of a turbulent partially premixed CH_4/air jet flame stabilized by premixed pilot flames.

and oxidizer are mixed in stoichiometric proportions. If a nonpremixed flame is modeled as a two-stream mixing problem with assumptions of fast chemistry, equal diffusivities of all species, and unit Lewis number (the ratio of thermal diffusivity to mass diffusivity), then the species mass fractions can be expressed solely as a function of the mixture fraction. The scalar dissipation rate, which is defined as $\chi = 2D_\xi(\nabla\xi \cdot \nabla\xi)$, where D_ξ is the corresponding diffusivity, quantifies the rate of molecular mixing, and is prominent in the theory and modeling of turbulent nonpremixed combustion. The reaction rates are proportional to the scalar dissipation rate via the following relationship: $w_i = -\rho(\chi/2)\,\partial^2 Y_i(\xi)/\partial\xi^2$ where w_i is the chemical production rate of species i, ρ is the density, and $Y_i(\xi)$ is the mass fraction of the species i as a function of mixture fraction [13].

The determination of mixture fraction in flames is challenging because it requires simultaneous measurements of all major species. Mixture-fraction measurement techniques use combinations of Raman scattering, Rayleigh scattering, and laser-induced fluorescence (LIF). Multidimensional mixture-fraction measurements are needed to determine the scalar dissipation. During the past two-and-a-half decades, the diagnostic capabilities for measuring mixture fraction in turbulent nonpremixed flames have evolved significantly, as described by Frank et al. [14]. The application of these techniques to a range of burner geometries has provided important

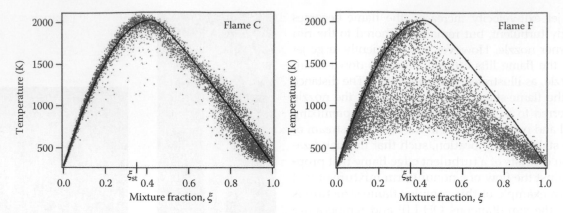

FIGURE 7.2.4

Scatter plots of temperature at $x/d = 15$ in turbulent CH_4/air jet flames with Reynolds numbers of 13,400 (Flame C) and 44,800 (Flame F). The stoichiometric mixture fraction is $\xi_{st} = 0.351$. The line shows the results of a laminar counterflow-flame calculation with a strain parameter of $a = 100\,s^{-1}$ and is included as a visual guide. (From Barlow, R.S. and Frank, J.H., *Proc. Combust. Inst.*, 27, 1087, 1998. With permission.)

insights into turbulent nonpremixed flames, and well-documented data sets are currently used for the development and validation of turbulent combustion models through the Turbulent Nonpremixed Flame (TNF) Workshop [15].

One of the most challenging aspects of modeling turbulent combustion is the accurate prediction of finite-rate chemistry effects. In highly turbulent flames, the local transport rates for the removal of combustion radicals and heat may be comparable to or larger than the production rates of radicals and heat from combustion reactions. As a result, the chemistry cannot keep up with the transport and the flame is quenched. To illustrate these finite-rate chemistry effects, we compare temperature measurements in two piloted, partially premixed CH_4/air (1/3 by vol.) jet flames with different turbulence levels. Figure 7.2.4 shows scatter plots of temperature as a function of mixture fraction for a fully burning flame (Flame C) and a flame with significant local extinction (Flame F) at a downstream location of $x/d = 15$ [16]. These scatter plots provide a qualitative indication of the probability of local extinction, which is characterized

by samples with strongly depressed temperatures. In Flame C, there is a very small probability of extinction, and the bulk of the data points are distributed along the curve that is obtained from a laminar flame calculation with a strain parameter of $a = 100\,s^{-1}$. In contrast, Flame F has a high probability of localized extinction with a significant fraction of samples exhibiting reduced temperatures. Accurate modeling of localized extinction and reignition is important for the development of practical combustion devices with low pollutant emissions and stable operating conditions.

Figure 7.2.5 provides a visualization of a localized extinction event in a turbulent jet flame, using a temporal sequence of OH planar LIF measurements. The OH-LIF measurements, combined with particle image velocimetry (PIV) reveal that a distinct vortex within the turbulent flow distorts and consequently breaks the OH front. These localized extinction events occur intermittently as the strength of the coupling between the turbulent flow and the flame chemistry fluctuates. The characteristics of the turbulent flame can be significantly altered as the frequency of these events increases.

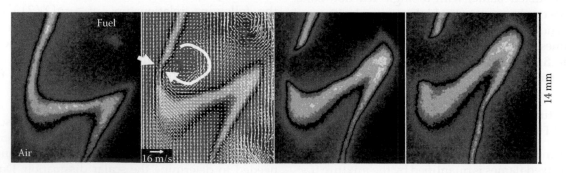

FIGURE 7.2.5

Temporal sequence of OH-LIF measurements captures a localized extinction event in a turbulent nonpremixed $CH_4/H_2/N_2$ jet flame ($Re \sim 20,000$) as a vortex perturbs the reaction zone. The time between frames is 125 μs. The velocity field from PIV measurements is superimposed on the second frame and has the mean vertical velocity of 9 m/s subtracted. (From Hult, J. et al., Paper No. 26-2, in *10th International Symposium on Applications of Laser Techniques to Fluid Mechanics*, Lisbon, 2000. With permission.)

Several experiments and large eddy simulations of turbulent jet flames have revealed thin sheet-like structures of high strain rate and high scalar dissipation rate that tend to be inclined to the flow, as shown in Figure 7.2.6. Two-dimensional (2D) imaging measurements of scalar dissipation in a piloted jet flame, obtained using the methods described by Frank et al. [14], are compared qualitatively with simulations of instantaneous scalar dissipation fields from two different LES models of a similar piloted jet flame, and the inclined structures of high scalar dissipation are evident in each frame. The importance of these structures in the overall combustion process is not fully understood and is the subject of ongoing research. However, there is evidence that local extinction may be caused by such structures and that a disproportionate amount of heat release may occur in these structures, relative to the volume they occupy. Therefore, combustion models may have to account for the effects of these structures to accurately predict some combustion phenomena.

7.2.4 Turbulence Structure and Length Scales

Turbulent nonpremixed flames contain a wide range of length scales. For a given flame geometry, the largest scales of turbulence are determined by the overall width of an unconfined jet flame or by the dimensions of the hardware that contain the flow. Therefore, the largest scales of turbulent motion are typically independent of Reynolds number. As the Reynolds number increases, the turbulent fluctuations in the velocity and mixture fraction cascade down to the progressively smaller eddies, increasing the dynamic range of the length scales. This extension to smaller scales is illustrated in Figure 7.2.7 by the OH-LIF measurements in turbulent H_2/Ar jet flames with Reynolds numbers ranging from 30,000 to 150,000. The largest length scales of the OH regions are comparable across the three sets of images. However, with increasing Reynolds number there is more fine-scale structure on the boundaries and within these large-scale structures.

The chemical reactions that drive combustion can occur only after the reactants are mixed at the molecular level by diffusion. While turbulent transport, or "stirring," takes place over a wide range of length scales, this final molecular mixing process is left to the smallest scales of turbulence, called the dissipation range. Based on the knowledge of nonreacting turbulent flows, we expect that the experimental resolution must approach the smallest scales of turbulence for the measurements of the mean scalar-dissipation rate to be accurate. The relevant length scale for determining the local resolution requirement is the Batchelor scale, λ_B. This scale represents, in an average sense, the smallest length over which turbulent fluctuations in a scalar quantity, such as mixture fraction or temperature, can occur. Scalar fluctuations at length scales near the Batchelor scale are rapidly dissipated by diffusion and must be continually fed by "energy" from turbulent fluctuations at larger scales (the corresponding scale for velocity fluctuations

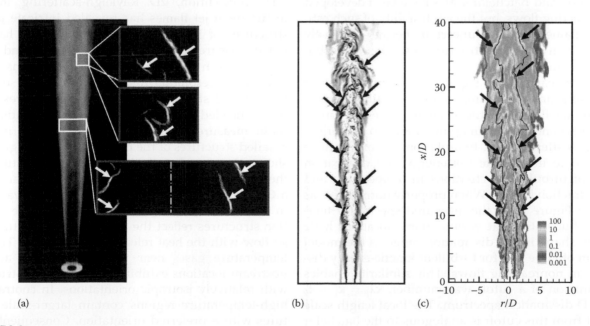

(a) (b) (c)

FIGURE 7.2.6
Qualitative comparison of the inclined structure of thin layers of high scalar dissipation in a piloted CH_4/air jet flame as revealed by (a) mixture fraction imaging, (b) LES with a steady flamelet library (a and b are adapted from Kempf, A. Flemming, F., and Janicka, J., *Proc. Combust. Inst.*, 30, 557, 2005. With permission.), and (c) LES with unsteady flamelet modeling. (Adapted from Pitsch, H. and Steiner, H., *Proc. Combust. Inst.*, 28, 41, 2000. With permission.)

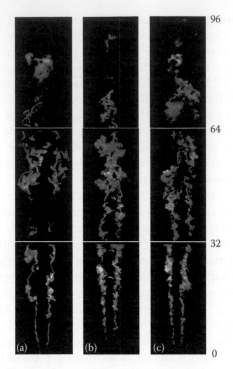

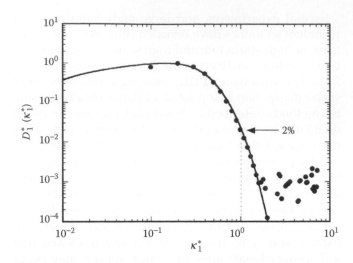

FIGURE 7.2.8
Model of 1D dissipation spectrum from Pope [19] (line) and measured, noise-corrected spectrum of the square of the radial gradient of fluctuating temperature in a $CH_4/H_2/N_2$ jet flame ($Re_d = 15,200$) (symbols). Each spectrum is normalized by its maximum value. The arrow indicates the 2% level, which corresponds to the normalized wavenumber $\kappa_1^* = 1$ according to the model spectrum. (From Barlow, R.S., *Proc. Combust. Inst.*, 31, 49, 2007. With permission.)

FIGURE 7.2.7
Composite OH-LIF images of turbulent nonpremixed H_2/Ar jet flames with Reynolds numbers of (a) $Re_d = 30,000$, (b) $Re_d = 75,000$, and (c) $Re_d = 150,000$. Numbers on right indicate the streamwise distance in the nozzle diameters ($d = 5$mm). (Adapted from Clemens, N.T., Paul, P.H., and Mungal, M.G., *Combust. Sci. Technol.*, 129, 165, 1997. With permission.)

is the Kolmogorov scale, η). Methods for estimating the Kolmogorov and Batchelor scales have been developed for nonreacting flows, but the applicability of such estimates to flames has been uncertain, because relatively little is known about the structure of small-scale turbulence in reacting flows.

Recent research has significantly improved our quantitative understanding of the structure of nonpremixed jet flames at the smallest scales of turbulence. Simultaneous line imaging of Raman scattering, Rayleigh scattering, and two-photon LIF of CO have been used to investigate the energy and dissipation spectra of turbulent fluctuations in temperature and mixture fraction [17,18]. When properly normalized, as shown in Figure 7.2.8, the measured spectra for temperature fluctuations at various flame locations have the same shape in the dissipation range as the model spectrum of Pope [19] for turbulent kinetic-energy dissipation in nonreacting flows. This similarity enables determination of a cutoff wavenumber, $\kappa\lambda_B = \kappa_1^* = 1$, in the 1D dissipation spectrum. The local length scale inferred from this cutoff is analogous to the Batchelor scale in the nonreacting flows. Furthermore, with Lewis number near unity in these flames, the 1D dissipation spectra for temperature and mixture fraction follow nearly the same roll off. These results represent a

breakthrough in the development of quantitative diagnostics for scalar-dissipation measurements in flames, because they suggest that local resolution requirements can be determined for complex flames using the relatively simple technique of Rayleigh scattering.

High-resolution, 2D Rayleigh-scattering imaging in turbulent jet flames has revealed intricate layered structures of high thermal dissipation and has provided measurements of dissipation spectra and length scales in both the radial and axial directions [20,21]. The spatial resolution that is required to resolve the thin-layered structures is greater than the resolution that is needed to measure the mean dissipation, and these measurements provide new insight into the detailed structures of the dissipation field. Figure 7.2.9 shows the samples of single-shot temperature and thermal dissipation measurements in the near field of a $CH_4/H_2/N_2$ jet flame ($Re_d = 15,200$). The variations in the thickness and spatial orientation of the dissipation structures reflect the interaction of the turbulent jet flow with the heat released by the flame. The low-temperature gases near the jet centerline at these upstream locations exhibit small turbulent structures with relatively isotropic orientations. In contrast, the high-temperature regions contain larger-scale structures with a preferred orientation. Consequently, the contributions of axial- and radial-temperature gradients to the dissipation field are similar near the jet centerline, but differ significantly in the high-temperature regions of the jet flame.

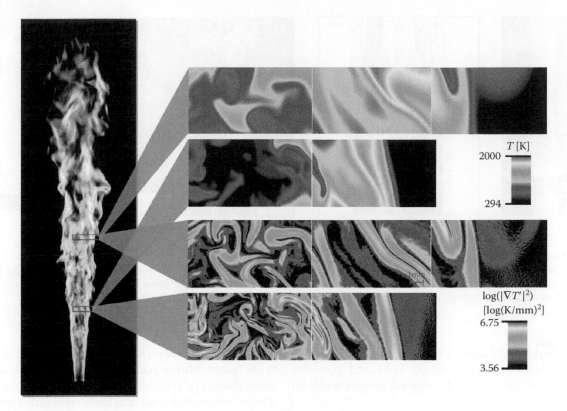

FIGURE 7.2.9
Instantaneous temperature and thermal dissipation measurements in a $CH_4/H_2/N_2$ jet flame ($Re_d = 15,200$) at $x/d = 10$ and 20. The thermal dissipation is displayed on a log scale to show the wide dynamic range.

The ability to resolve the dissipation structures allows a more detailed understanding of the interactions between turbulent flows and flame chemistry. This information on spectra, length scales, and the structure of small-scale turbulence in flames is also relevant to computational combustion models. For example, information on the locally measured values of the Batchelor scale and the dissipation-layer thickness can be used to design grids for large-eddy simulation (LES) or evaluate the relative resolution of LES results. There is also the potential to use high-resolution dissipation measurements to evaluate subgrid-scale models for LES.

7.2.4.1 Complex Geometries

Advancements in our fundamental understanding of turbulent nonpremixed combustion through studies of simple, canonical burner geometries are essential for developing and validating computational models that can predict the effects of interactions between turbulence and chemistry in flames. However, practical combustion devices often use complex burner geometries with swirling and recirculating flows that stabilize intense, highly turbulent flames with very high power densities. Consequently, the combustion research community has directed significant effort toward detailed studies

of flames and burners that have recirculating flows, swirling flows, and stabilization of detached flames by mixing with combustion products at high temperatures. Two examples are described here, and some additional examples are outlined by Barlow [9].

One method of stabilizing a flame in a high-velocity flow of air is to trap the combustion products in the recirculation zone, that forms downstream of a bluff body. The extended residence time of the recirculating flow allows time for combustion reactions to proceed, and the high-temperature products serve as a stable ignition source for the flame. Figure 7.2.10 shows a photograph of a bluff-body flame of CH_4/H_2 (equal parts by volume) and three computationally generated views of the structure of the recirculation zone. Fuel is injected through a 3.6 mm tube at the center of the bluff body, which is 50 mm in diameter and is surrounded by an air flow of up to 40 m/s [22]. The white rectangle in Figure 7.2.10a indicates the region represented in Figures 7.2.10b through d, which are obtained from LES [23]. The streamlines in Figure 7.2.10b represent the time-averaged structure of the recirculation zone, which has two annular vortices. Figure 7.2.10c is generated by integrating an instantaneous line-of-sight view of the simulated OH-radical concentration, and movies of such images can yield useful insights into the dynamics

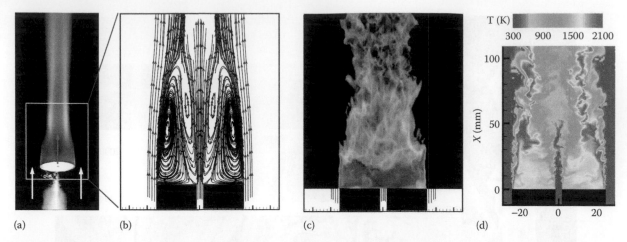

(a)   (b)   (c)   (d)

FIGURE 7.2.10

A bluff-body stabilized flame of CH_4/H_2 in air (designated HM1 by Dally et al. [22]): (a) time-averaged photograph of flame luminosity, (b) time-averaged streamlines from LES, (c) instantaneous visualization of OH "luminosity" from LES, and (d) instantaneous temperature field from LES. (b and d are adapted from Raman, V. and Pitch, H., *Combust. Flame*, 142, 329, 2005. With permission.)

of complex turbulent flames. Figure 7.2.10d shows a simulated, instantaneous temperature field and provides an indication of the range of resolved length scales in this flame.

Bluff-body flames can also exhibit local extinction, and the combination of recirculating flow, large-scale dynamics, and local extinction is a contemporary challenge for the advanced combustion models. However, these flames are still much simpler than those in a gas turbine combustor, for example. There is a strong motivation to perform detailed experiments on nonpremixed and partially premixed burners that include features of practical combustors. One such research target is the model gas turbine combustor, shown in Figure 7.2.11. This burner is designed to operate on gaseous fuels at atmospheric pressure. However, it is modeled after a liquid-fueled combustor used in small gas-turbine engines. In this combustor, two annular swirling flows of air surround a ring that injects fuel. The turbulent flame spreads out as a cone, and there are inner and outer recirculation zones. Detailed measurements of species and temperature revealed the detachment of the flame from the injector and the incidence of a significant degree of mixing between fuel and air before combustion [24]. Furthermore, combustion products from the inner- and outer-recirculation zones were also entrained into this mixing region just above the fuel injector.

Figure 7.2.12 shows scatter plots of instantaneous measurements of temperature and CH_4 mole fraction obtained at a height of 5 mm and at several radial locations, which are color-coded in the figure. The foremost observable characteristics are that there are no samples richer than 0.2 in the mixture fraction (1.0 being pure fuel) and that many samples remain at room temperature even within the limits of flammability. Many

samples also show an intermediate progress of reaction, with temperatures well below the calculated equilibrium (black) or strained laminar flame (orange) curves. These unreacted and partially reacted samples are from the highly strained mixing region above the injector jets. For measurement locations near the centerline ($r = 0$–2 mm) or outside the mixing layer ($r = 16$–30 mm), many samples are fully reacted and close to the equilibrium lines in the figures. These locations are in the inner

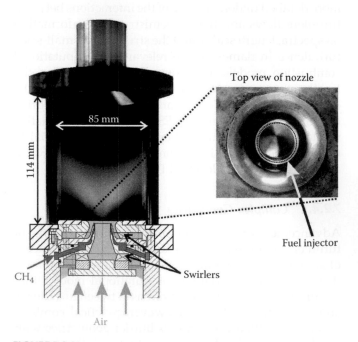

FIGURE 7.2.11

Diagram and photograph of a model gas turbine combustor operating on CH_4/air at atmospheric pressure. Fuel is injected from an annulus separating two swirling air streams. (From Meier, W., Duan, X.R., and Weigand, P., *Combust. Flame*, 144, 225, 2006. With permission.)

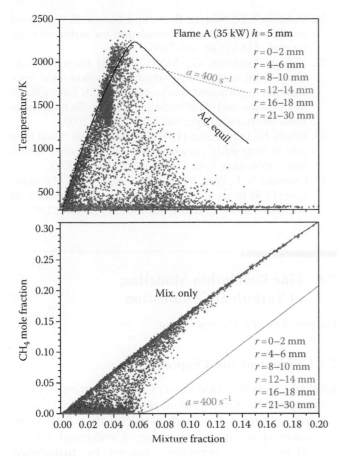

FIGURE 7.2.12
Scatter plots of temperature and CH_4 mole fraction versus mixture fraction in a model gas turbine combustor. (From Meier, W., Duan, X.R., and Weigand, P., *Combust. Flame*, 144, 225, 2006. With permission.)

and outer recirculation zones, respectively, where the mixing rates are slower than in the high shear regions of the burner.

In this burner configuration, fuel is injected directly into the combustion chamber and hence, one would initially categorize it as a nonpremixed burner. However, the overall combustion process is quite complex and involves features of nonpremixed, partially premixed, and stratified combustion, as well as the possibility that the autoignition of hot mixtures of fuel, air, and recirculated combustion products may play a role in stabilizing the flame. Thus, while one may start from simple concepts of nonpremixed turbulent flames, the inclusion of local extinction or flame lift-off quickly increases the physical and computational complexity of flames that begin with nonpremixed streams of fuel and oxidizer.

7.2.5 Summary

In this chapter, we described some of the basic characteristics of nonpremixed flames and provided a few examples of both simple and moderately complex flames and burner geometries for turbulent nonpremixed combustion. The central theme in nonpremixed combustion is that the structure and stability of a given flame depend on the coupling between turbulent mixing and chemical reactions. Mixture fraction (the state of mixing between fuel and oxidizer) and scalar dissipation (the rate of mixing at the molecular level) were identified as fundamental concepts and quantities. Local extinction, flame lift-off and stabilization, length scales of turbulent flames, and the structure of thin dissipation layers were discussed as examples of important interactions between fluid dynamics and chemistry. Piloted flames, bluff-body flames, and swirling flames were used to illustrate a range of methods for stabilizing flames in which the turbulent mixing rates were competitive with the critical rates of combustion reactions. These examples point towards the very complex nature of combustion in practical systems.

Nonpremixed combustion will continue to be important for many applications in power generation, transportation, and industrial processing. The need to develop advanced combustion systems with high efficiency and very low pollutant emissions places increasing demands on the computational design tools. Models for turbulent combustion systems will be predictive only if their underlying assumptions are soundly based on science and validated against well-documented test cases. Much of what is currently known about turbulent nonpremixed flames is based on experiments that use nonintrusive laser diagnostic techniques. However, the role of direct numerical simulations (DNS) and highly resolved LES in fundamental combustion research is increasing as a result of rapid advancements in computational hardware and methods for detailed simulations of flames (Westbrook et al. [25], Oefelein et al. [26]). The combination of closely coupled experiments and simulations is expected to significantly accelerate the development of predictive models for complex combustion systems over the next several years.

References

1. Bray, K. N. C., The challenge of turbulent combustion, *Proc. Combust. Inst.*, 26, 1, 1996.
2. Peters, N., *Turbulent Combustion*, Cambridge University Press, Cambridge, United Kingdom, 2000.
3. Vervisch, L., Using numerics to help the understanding of non-premixed turbulent flames, *Proc. Combust. Inst.*, 28, 11, 2000.
4. Bilger, R. W., Pope, S. B., Bray, K. N. C., and Driscoll, J. F., Paradigms in turbulent combustion research, *Proc. Combust. Inst.*, 30, 21, 2005.
5. Poinsot, T. and Veynante, D., *Theoretical and Numerical Combustion*, 2nd ed., Edwards, Philadelphia, 2005.

6. Burke, S. P. and Schumann, T. E. W., Diffusion flames, *Proc. Combust. Inst.*, 1, 2, 1928.

7. Pope, S. B., Computations of turbulent combustion: Progress and challenges, *Proc. Combust. Inst.*, 23, 591, 1990.

8. Bergmann, V., Meier, W., Wolff, D., and Stricker, W., Application of spontaneous Raman and Rayleigh scattering and 2D LIF for the characterization of a turbulent $CH_4/H_2/N_2$ jet diffusion flame, *Appl. Phys. B*, 66, 489, 1998.

9. Barlow, R. S., Laser diagnostics and their interplay with computations to understand turbulent combustion, *Proc. Combust. Inst.*, 31, 49, 2007.

10. Su, L. K., Sun, O. S., and Mungal, M. G., Experimental investigation of stabilization mechanisms in turbulent, lifted jet diffusion flames, *Combust. Flame*, 144, 494, 2006.

11. Muniz, L. and Mungal, M. G., Instantaneous flame-stabilization velocities in lifted-jet diffusion flames, *Combust. Flame*, 111, 16, 1997.

12. Pitts, W. M., Assessment of theories for the behavior and blowout of lifted turbulent jet diffusion flames, *Proc. Combust. Inst.*, 22, 809, 1988.

13. Bilger, R. W., The structure of diffusion flames, *Combust. Sci. Technol.*, 13, 155, 1976.

14. Frank, J. H., Kaiser, S. A., and Long, M. B., Multiscalar imaging in partially premixed jet flames with argon dilution, *Combust. Flame*, 143, 507, 2005.

15. Barlow, R. S., Editor: International workshop on measurement and computation of turbulent nonpremixed flames in http://www.ca.sandia.gov/TNF.

16. Barlow, R. S. and Frank, J. H., Effects of turbulence on species mass fractions in methane/air jet flames, *Proc. Combust. Inst.*, 27, 1087, 1998.

17. Wang, G. H., Barlow, R. S., and Clemens, N. T., Quantification of resolution and noise effects on thermal dissipation measurements in turbulent non-premixed jet flames, *Proc. Combust. Inst.*, 31, 1525, 2007.

18. Wang, G. H., Karpetis, A. N., and Barlow, R. S., Dissipation length scales in turbulent nonpremixed jet flames, *Combust. Flame*, 148, 62, 2007.

19. Pope, S. B., *Turbulent Flows*, Cambridge University Press, New York, 2000.

20. Frank, J. H. and Kaiser, S. A., High-resolution Rayleigh imaging of dissipative structures, *Exp. Fluids*, 44, 221, 2008.

21. Kaiser, S. A. and Frank, J. H., Imaging of dissipative structures in the near field of a turbulent non-premixed jet flame, *Proc. Combust. Inst.*, 31, 1515, 2007.

22. Dally, B. B., Masri, A. R., Barlow, R. S., and Fiechtner, G. J., Instantaneous and mean compositional structure of bluff-body stabilized nonpremixed flames, *Combust. Flame*, 114, 119, 1998.

23. Raman, V. and Pitsch, H., Large-eddy simulation of a bluff-body-stabilized non-premixed flame using a recursive filter-refinement procedure, *Combust. Flame*, 142, 329, 2005.

24. Meier, W., Duan, X. R., and Weigand, P., Investigations of swirl flames in a gas turbine model combustor—II. Turbulence-chemistry interactions, *Combust. Flame*, 144, 225, 2006.

25. Kempf, A., Flemming, F., and Janicka, J., Investigation of lengthscales, scalar dissipation, and flame orientation in a piloted diffusion flame by LES, *Proc. Combust. Inst.*, 30, 557, 2005.

26. Oefelein, J. C., Schefer, R. W., and Barlow, R. S., Toward validation of large eddy simulation for turbulent combustion, *AIAA J.*, 44, 418, 2006.

27. Hult, J., Josefsson, G., Aldén, M., and Kaminski, C.F., Flame front tracking and simultaneous flow field visualisation in turbulent combustion, in 10th International Symposium and Applications of Laser Techniques to Fluid Mechanics, Paper No. 26-2, Lisbon, 2000.

28. Pitsch, H. and Steiner, H., Scalar mixing and dissipation rate in large-eddy simulations of non-premixed turbulent combustion, *Proc. Combust. Inst.*, 28, 41, 2000.

29. Clemens, N.T., Paul, P.H., and Mungal, M.G., The structure of OH fields in high Reynolds number turbulent jet diffusion flames, *Combust. Sci, Technol.*, 129, 165, 1997.

7.3   Fine Resolution Modeling of Turbulent Combustion

Laurent Selle and Thierry Poinsot

7.3.1   Scope of This Chapter

Most of the combustion devices in engineering applications operate in the turbulent regime. Turbulent flames are more compact and more powerful than their laminar counterpart, mainly because of enhanced mixing and flame-surface wrinkling caused by turbulence. Turbulent combustion can be described as the study of the coupled interactions between turbulent fluid motion and chemical reactions. One way to classify numerical methods for turbulent combustion is to rank them according to the level of description for the turbulence: three categories can be roughly defined. The most accurate technique is the direct numerical simulation (DNS), in which all the structures of the flow are resolved in both space and time. The other extreme is the Reynolds averaged Navier–Stokes (RANS) methodology that solves the mean properties of all flow variables. A wide variety of methods offer an intermediate level of description, the most famous being the large-eddy simulation (LES), very-large-eddy simulation (VLES), detached-eddy simulation (DES), and unsteady RANS (U-RANS) [1]. Hybrid methods, coupling two or more of these models have also been explored [2]. Essentially, the driving idea of all these methods is to explicitly solve some of the nonstationary motion of the flow when or where the steady-state computations fail. Since the focus of this chapter is on "fine resolution" techniques, steady-state modeling (i.e., "classical" RANS) will not be discussed. Also, despite its undisputable practical usefulness for industrial computations, RANS modeling is not the state-of-the-art technique for quantitative predictions of turbulent reacting flows. Finally, stationary methods cannot address a number of crucial issues in turbulent combustion, such as ignition, extinction, and combustion instabilities.

Once the level of description for the turbulence is set, the same must be done for the chemistry. This leads to subclasses of numerical methods that are too numerous to discuss here [3], but rely on two major sets of assumptions. The first one is the number of species and reactions taken into account: *simple chemistry* models that use only a few species—oxidizer and fuel being the mandatory ones—opposed to the *detailed chemistry* models, taking into account hundreds of species and thousands of reactions. As it is appealing to believe that a more detailed model for chemistry will give "better" results, it must be pointed out that in many cases a very simple description of the flame preserving only global quantities, such as flame speed and final temperature can be strikingly predictive. The second set of assumptions is about the time and length scales of the reaction zone with respect to those of turbulence. The different combustion regimes control the type of model that must be developed. Predicting regimes in turbulent combustion is also a difficult task, as shown by the numerous diagrams found in the literature.

Last but not the least, on top of the fundamental considerations on the modeling of turbulent combustion, there is one technical aspect that should be pointed out: the advances in numerical simulations are tied in with those in computational power. Even though exponential, the increasing number of operations per second a single CPU can perform would have not allowed most of the computations presented in this chapter. It is the invention of parallel computing, with thousands of computers simultaneously working on the same problem, that truly unleashed the computational fluid dynamics for reacting flows.

This chapter is organized in the following manner: DNS techniques and their range of applicability are first described, then some successes of the U-RANS and LES methodologies are presented and finally, perspectives are offered about the future challenges of the computational turbulent combustion.

7.3.2 DNS

7.3.2.1 What Is the Use of DNS?

DNS results are usually considered as references providing the same level of accuracy as experimental data. The maximum attainable Reynolds number (*Re*) in a DNS is, however, too low to duplicate most practical turbulent reacting flows, and hence, the use of DNS is neither to replace experiments nor for direct comparisons—not yet at least. However, DNS results can be used to investigate three-dimensional (3D) features of the flow (coherent structures, Reynolds stresses, etc.) that are extremely difficult, and sometimes impossible, to measure. One example of such achievement for nonreacting

flows is the description of homogeneous isotropic turbulence (HIT) or of the boundary-layer structure [4–6]. Another valuable use of DNS is the realization of *a priori* studies, which consists of an evaluation of LES or RANS models from a DNS result [7–9]. Comparing DNS results with the LES of the exact same configuration enables further validation: this is called an *a posteriori* study [10]. A successful example of this strategy for nonreacting flows can be found in the DNS results from the Center for Turbulence Research (NASA Ames and Stanford University), which provided a breakthrough in the understanding and modeling of turbulent nonreacting flows in the early 1980s [11]. These successes were then reproduced in the 1990s in the field of turbulent combustion with the study of flame/vortex and flame/turbulence interactions, as well as in flame/wall interaction. A typical example of the use of DNS in reacting flows would be the investigation of the stabilization mechanism of diffusion flames through a triple-flame structure [12]. DNS was also useful in deriving turbulent combustion models used in RANS [13] and LES codes [14].

7.3.2.2 Examples of Modern DNS of Reacting Flows

Owing to the exponential increase in the computational power, today's DNS reaches fully turbulent Reynolds numbers and can use detailed chemical schemes, once affordable only for one-dimensional laminar flames computations.

7.3.2.2.1 Premixed Flame Front in HIT

The HIT—also referred to as grid turbulence—is the canonical configuration for the study of turbulence, its properties are well known from both experiments and theory, which makes it a good candidate for the study of flame/turbulence interactions. In this configuration (Figure 7.3.1), a turbulent flow of premixed gases is injected from the left and a V-shaped flame is stabilized on a "numerical wire,"—i.e., a hot spot at the tip of the flame [9]. In such a configuration, the injection of turbulence is a major challenge, because improper treatment of the boundary conditions can lead to spurious pressure oscillations and uncontrolled turbulence properties. In this example, the turbulent field is generated with a separate Navier–Stokes solver that generates grid turbulence and the boundary conditions are treated using the Navier Stokes Characteristic Boundary Conditions (NSCBC) technique [15].

7.3.2.2.2 Diffusion Flame in a Temporal Mixing-Layer

Together with HIT, the temporal mixing-layer (TML) is a useful configuration for the numerical study of turbulent flows. The TML configuration can be thought

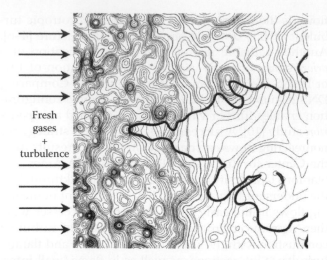

Fresh
gases
+
turbulence

FIGURE 7.3.1
DNS of a premixed flame in turbulence. (From Vervisch, L., Hauguel, R., Domingo, P., Rullaud, M., *J. Turbulence*, 5, 004, 2004.)

of as following the turbulent structures that develop on the edge of a jet at their convection speed. The computation presented in Figure 7.3.2 corresponds to a nonpremixed plane jet flame with CO/H_2 kinetics performed at Sandia [16]. The inner stream is a mixture of fuel and nitrogen while the outer streams contain air. A diffusion flame develops at the interfaces between the streams and interacts with the turbulence generated by the shear. The mesh contains 500 million grid points, which makes it one of the largest DNS with detailed chemistry (11 species and 21 reactions). At a jet Reynolds number of 9200, this computation (Figure 7.3.2) displays

FIGURE 7.3.2
Volume rendering of scalar dissipation rate in a DNS of a temporally evolving CO/H_2 jet flame, $Re = 9200$ [16]. The highest values of scalar dissipation rate (shown in red) exceed $30,000\,s^{-1}$.

the typical features of fully turbulent flows: intricate structures of very small size and high levels of scalar dissipation rate. This simulation was used to study finite-rate chemistry effects, such as extinction and reignition at an engineering-relevant Reynolds number. The size of this simulation makes the visualization and analysis of the results a challenge of its own: novel techniques using hardware-accelerated parallel visualization software had to be developed. Variations in the initial Reynolds number on this configuration revealed increasing levels of extinction and longer reignition times with increasing *Re*.

7.3.2.2.3 DNS of a Laboratory-Scale Flame

The detailed computation of a laboratory-scale flame poses several additional challenges when compared with the temporal simulations, such as HIT and TML. Of course, the ratio between the overall size of the flame and that of the finest turbulent and chemical structures requires a very large number of grid points. Consequently, the time for the whole flame to be statistically stationary often means that a great deal of computational time will be spent just to achieve a permanent regime. Finally, boundary conditions must be treated with great care for the flame not to be perturbed by numerical artifacts. The computation presented [17] is based on a setup used in experimental studies [18]. A preheated mixture of methane and air (800 K and 0.7 equivalence ratio) is injected from a slot burner (bottom center of Figure 7.3.3). On both sides of the burner, a low-speed coflow of burnt products is maintained to stabilize the flame. To mimic the experimental conditions, turbulence generated in a separate solver is introduced in the reactants' stream. Figure 7.3.3 presents a 3D view of the reaction zone. Typical features of premixed turbulent flames, such as intense flame-front wrinkling and fresh-gases pockets at the tip of the flame are offered on this snapshot. This particular study is conducted in a regime called "thin reaction zone" where small turbulent eddies can enter the preheat layer, but not the inner portion of the flame where chemical reactions occur. This particular regime poses several modeling issues that can be addressed with such DNS.

7.3.3 Unsteady RANS Methods

7.3.3.1 Concept and Use of U-RANS

The cornerstone of RANS methodology is to apply an operator of statistical averaging onto the DNS equations. Therefore, any flow for which the boundary conditions do not vary in time and does not keep the memory of its initial condition will become statistically stationary. This is why RANS results often consist of a single snapshot of the flow field. However, for flows that

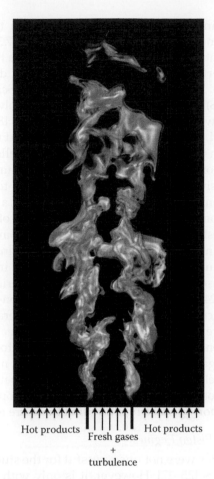

↑↑↑↑↑↑↑↑ ↑↑↑↑↑ ↑↑↑↑↑↑↑↑
Hot products   Hot products
               Fresh gases
                    +
                turbulence

FIGURE 7.3.3
DNS of a slot-burner premixed flame. (From Sankaran, R. et al., *Thirty-First International Symposium on Combustion*, 2007.)

are not statistically stationary—because of time-varying boundary conditions or other transient phenomenon—the RANS solution is time-dependent. Each solution then represents the averaged flow field, over all possible realizations, at a given time, hence, U-RANS. Two straightforward examples of such flows would be piston engine flows—because of moving boundaries—or the ignition sequence of inflammable gases—because of its transient nature.

7.3.3.2 Fire Safety: Pool Fires

Fire safety is a paradigm for which some, if not most, of the parameters that control combustion are unknown. For example, in a building fire, uncertainties in the exact composition of the fuel (carpets, furniture, concrete, etc.) do not enable precise predictions of heat transfer or flame temperatures. Another issue is the variety of time and length scales involved: radiations propagate at the speed of light while the buoyancy-driven advection is of the order of a few meters per second. Finally, on top of turbulence and combustion, these fires typically involve

the nonlinear coupling of many phenomena, such as soot formation, radiation, and heat transfer into solids. This accumulation of modeling issues is one reason why state-of-the-art computations of real-life-size fires are conducted within the U-RANS framework.

The example presented in Figure 7.3.4 simulates a large fire from a transportation accident in a cross-wind with objects, which are fully engulfed, and therefore, not seen in the image. The U-RANS methodology is adequate for this flow, because of its statistically transient nature—depletion of fuel with time. The simulation was conducted on 5000 processors on Red Storm supercomputer, which at the time of this computation was second in the world's fastest computer list. This computation solves 40 variables, which is very large, but keeping in mind the variety of phenomena represented in this computation, this number is still a great reduction from the actual number of independent variables. The reduction is the product of careful approximations based on the ratio of the timescales involved in this fire.

7.3.4 Large Eddy Simulation

7.3.4.1 LES Philosophy

The cornerstone of LES methodology is the self-similarity theory of Kolmogorov stating that even though the large structures of a turbulent flow depend on the boundary and initial conditions, the finer scales have a universal

FIGURE 7.3.4
Simulation of a fire resulting from a transport accident in a cross wind. The flame is visualized in yellow/red, while soot clouds are represented in black. (From Tieszen, S., private communication.)

behavior. Therefore, in an LES, while the evolution of the large eddies is resolved on the mesh, eddies smaller than the mesh are modeled using a so-called subgrid scale (SGS) contribution. In mathematical formalism, LES equations are obtained by applying a filter onto the DNS equations [19–21]. Because the filtering operation enables a drastic reduction in mesh size, LES is much cheaper than the DNS and the computation of a whole experimental setup becomes feasible.

There are, however, additional issues when one wants to perform the LES of an industrial device, and these are to be solved if the LES methodology is ever to be a useful tool. First, the geometric intricacy of most industrial combustion chambers cannot be represented with a Cartesian mesh: high-order methods on unstructured grids must be developed. Moreover, many issues about the boundary conditions are raised, such as boundary movement for blades or pistons, turbulence injection, and acoustic properties. Such challenges are not to be underestimated, as their impact on the structure of the flow might sometimes be greater than that of the turbulence model.

7.3.4.2 Examples of Practical Use of LES

This section presents a variety of reacting flows computed with the LES methodology. The cases presented in this study were chosen, because each features a different aspect of turbulent combustion and also addresses a specific technical difficulty.

7.3.4.2.1 Aeronautic Turbines

The major goals for aeronautic turbines design are a compact flame, stable combustion over a wide operating range, and low emission levels. Another crucial point

included in the certification procedure of a turbine is ignition and reignition procedures. Figure 7.3.5 shows the computation of an ignition sequence in a helicopter engine performed by the Centre Européen de recherche et de Formation Avancée en Calcul Scientifique (CERFACS) team, the unstructured LES solver AVBP [22]. The whole combustion chamber with 18 burners is resolved, along with the dilution holes and multiperforation in the chamber walls (20 million cells). As the computation of a single burner is already a cumbersome task, massively parallel software is needed to address the ignition of a whole engine [29]. Obviously, a code that handles complex geometries through the use of unstructured meshes is mandatory for such computation. On top of numerical issues raising from the intricacy of the configuration, for this engine the fuel is injected in liquid phase, described in the solver in a Eulerian framework [23,24]. In Figure 7.3.5, the combustion-chamber walls are colored with respect to the temperature; hot regions are yellow while cold gases are blue. The ignition sequence goes as follows: the top and bottom burner inject hot gases while the others inject a cold mixture of air and fuel. Ignition occurs at the interface between hot and fuel-loaded regions. The flame then propagates to the neighboring burners until all of them are ignited.

7.3.4.2.2 Piston Engines

RANS codes were not unsuccessful for the study of piston engines [25–27]. However, it is only with LES [30], for example, that the study of cycle-to-cycle variations becomes possible. For such studies, the solver must have moving-grid capabilities for the piston and the valves, while retaining all the required properties for LES, such as a high-order numerical method. From the point of view of modeling, the combustion model must handle

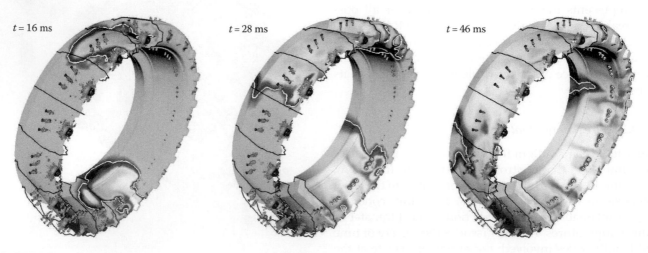

FIGURE 7.3.5
Ignition sequence of a helicopter engine. Hot gases (yellow) are injected through two burners and the flame propagates so that all 18 burners are eventually ignited. (From Boileau, M., Ignition of two-phase flow combustors. PhD, Institut National Polytechnique de Toulouse and CERFACS, 2007.)

ignition (by spark or by autoignition), flame/wall inter-actions, and extinction properly. Figure 7.3.6 is an example of recent LES in piston engines: this top view displays a slice through the combustion chamber of a single-piston engine that is colored with respect to the reaction rate. The four images correspond to different cycles at the same crank angle of 10°: the flame position and wrinkling depend on the cycle resulting in varia-tions in the instantaneous power of the engine. One can also point out regions where the combustion is almost quenched, as well as flame/wall interactions.

7.3.5 Perspectives

Over the past 15–20 years, the advances in numerical simulation of turbulent combustion have been more or less driven by the increase in computational power and the development of massively parallel computing, which have very good chances to continue.

The DNS in turbulent combustion has evolved from two-dimensional computations with simple chemistry to duplicate laboratory-scale experiments with detailed chemical schemes. Owing to its cost and the challenges in data processing and visualization, the DNS has remained as a tool for fundamental studies and labo-ratory-scale experiments of turbulent reacting flows. Obviously, larger Reynolds numbers will be attained and very complex chemical schemes will be imple-mented, but the computation of engineering devices with DNS should stay out of reach for some time. Also, DNS still has a key role to play as a validation tool for combustion models.

LES solvers, once devoted to academic configurations, can now handle the complex geometries and moving parts found in industrial applications. Thus, the LES is bound to replace RANS solvers in many industrial fields. Both the automotive and turbine industries are to switch to LES, because of its ability to predict transient phenomena as well as owing to its better performance globally in the prediction of the mean values. Massively parallel LES solvers will soon appear cost-competitive for these industries and the tremendous challenge of understanding and controlling combustion instabilities could be one of the tasks for LES in the future.

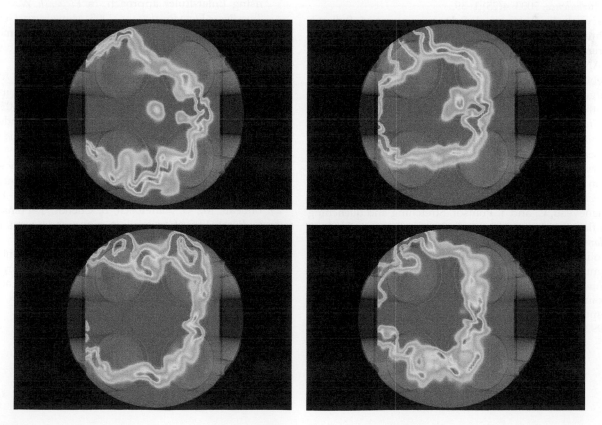

FIGURE 7.3.6
Top view of a piston engine at a crank angle of 10°. Cross-section colored by the reaction rate for four different cycles. (From Richard, S., Colin, O., Vermorel, O., Benkenida, A., Angelberger, C., and Veynante, D., *Proc. Comb. Inst.*, 31, 3059, 2007.)

References

1. Speziale, C.G., Turbulence modeling of time-dependent RANS and VLES—A review. *AIAA Journal*, 1998. 36(2): 173–184.
2. Rouson, D., S.R. Tieszen, and G. Evans, Modeling convection heat transfer and turbulence with fire applications: A high temperature vertical plate and a methane fire, in *Proceedings of the Summer Program*. 2002, Center for Turbulence Research, Stanford University. pp. 53–70.
3. Lindstedt, P., Modeling of the chemical complexities of flames. *Proc. Combust. Inst.*, 1998. 27:269–285.
4. Moin, P. and J. Kim, Numerical investigation of turbulent channel flow. *J. Fluid Mech.*, 1982. 118:341–377.
5. Moin, P. and J. Kim, The structure of the vorticity field in turbulent channel flow. Part 1. Analysis of instantaneous fields and statistical correlations. *J. Fluid Mech.*, 1985. 155: 441–464.
6. Kim, J. and P. Moin, The structure of the vorticity field in turbulent channel flow. Part 2. Study of ensemble-averaged fields. *J. Fluid Mech.*, 1986. 162:339–363.
7. Okong'o, N. and J. Bellan, Consistent large-eddy simulation of a temporal mixing layer laden with evaporating drops. Part 1. Direct numerical simulation, formulation and a priori analysis. *J. Fluid Mech.*, 2004. 499:1–47.
8. Vermorel, O., et al., Numerical study and modelling of turbulence modulation in a particle laden slab flow. *J. Turbulence*, 2003. 4(25):1–39.
9. Vervisch, L., R. Hauguel, P. Domingo, and M. Rullaud, Three facets of turbulent combustion modelling: DNS of premixed V-flame, LES of lifted nonpremixed flame and RANS of jet flame. *J. Turbulence*, 2004. 5(4):004.
10. Leboissetier, A., N. Okong'o, and J. Bellan, Consistent large-eddy simulation of a temporal mixing layer laden with evaporating drops. Part 2. A posteriori modelling. *J. Fluid Mech.*, 2005. 523:37–78.
11. Moin, P. and K. Mahesh, DNS: A tool in turbulence research. *Annu. Rev. Fluid Mech.*, 1998. 30:539–578.
12. Vervisch, L. and T. Poinsot, Direct numerical simulation of non-premixed turbulent flames. *Annu. Rev. Fluid Mech.*, 1998. 30:655–691.
13. Meneveau, C. and T. Poinsot, Stretching and quenching of flamelets in premixed turbulent combustion. *Combust. Flame*, 1991. 86:311–332.
14. Colin, O., et al., A thickened flame model for large-eddy simulations of turbulent premixed combustion. *Phys. Fluids*, 2000. 12(7):1843–1863.
15. Poinsot, T. and S.K. Lele, Boundary conditions for direct simulations of compressible viscous flows. *J. Comp. Phys.*, 1992. 101(1):104–129.
16. Hawkes, E.R., S. Sankaran, J.C. Sutherland, and J.H. Chen, Scalar mixing in direct numerical simulations of temporally-evolving plane jet flames with detailed CO/H_2 kinetics. *Proc. Combust. Inst.*, 2007. 31: 1633–1640.
17. Sankaran, R., E.R. Hawkes, J.H. Chen, T. Lu, C.K. Law, Structure of a spatially-developing lean methane-air turbulent bunsen flame. *Proc. Combust. Inst.*, 2007. 31: 1291–1298.
18. Filatyev, S.A., J.F. Driscoll, C.D. Carter, J.M. Donbar, Measured properties of turbulent premixed flames for model assessment, including burning velocities, stretch rates, and surface densities. *Combust Flame*, 2005. 141(1–2):1–21.
19. Peters, N., *Turbulent Combustion*. 2000, Cambridge University Press, Cambridge, USA.
20. Pope, S.B., *Turbulent Flows*. 2000, Cambridge University Press, Cambridge, USA.
21. Poinsot, T. and D. Veynante, *Theoretical and Numerical Combustion*. 2005, R.T. Edwards (Ed.), 2nd edn.
22. AVBP, AVBP Code: www.cerfacs.fr/cfd/avbp_code.php and www.cerfacs.fr/cfd/CFDPublications.html.
23. Moreau, M., B. Bedat, and O. Simonin, From Euler-Lagrange to Euler-Euler large eddy simulation approaches for gas-particle turbulent flows, in ASME Fluids Engineering Summer Conference, Houston. 2005, ASME FED.
24. Riber, E., et al., Towards large eddy simulation of non-homogeneous particle laden turbulent gas flows using Euler-Euler approach, in *Eleventh Workshop on Two-Phase Flow Predictions*. 2005, Merseburg, Germany.
25. Tahry, S.E., Application of a Reynolds stress model to engine-like flow calculations. *J. of Fluids Engineering*, 1985. 107(4):444–450.
26. Boudier, P., S. Henriot, T. Poinsot, T. Baritaud, A model for turbulent flame ignition and propagation in spark ignition engines. *Proc. Combust. Inst.*, 1992. 24:503–510.
27. Payri, F., et al., CFD modeling of the in-cylinder flow in direct-injection diesel engines. *Comput. Fluids*, 2004. 33(8):995–1021.
28. Celik, I., I. Yavuz, and A. Smirnov, Large eddy simulations of in-cylinder turbulence for internal combustion engines: A review. *Int. J. Eng Res.*, 2001. 2(2):119–148.
29. Boileau, M., G. Staffelbach, B. Cuenot, T. Poinsot, and C. Bérat, LES of an ignition sequence in a gas turbine engine. *Combust. Flame*, 2008. 154(1–2):2–22.
30. Richard, S., Colin, O., Vermorel, O., Benkenida, A., Angelberger, C., and Veynante, D., Towards large eddy simulation of combustion in spark ignition engines. *Proc. Comb. Inst.*, 2007. 31, 3059–3066.

8

Other Interesting Examples of Combustion and Flame Formation

CONTENTS

8.1   Candle and Jet Diffusion Flames: Mechanisms of Combustion under Gravity and Microgravity Conditions

Fumiaki Takahashi

8.1.1   Introduction

In flames formed by the condensed or gaseous fuel sources burning in the atmosphere, the combustion process is controlled by the mixing process rather than the rate of chemical reaction. The most familiar example is a candle flame, where the flow field is laminar and mixing occurs by molecular diffusion. Burke and Schumann [1] developed a theory of simplest laminar diffusion flames, where a gaseous fuel jet was issued into air flowing at the same velocity in a wider duct, with an infinitely thin flame-sheet model. In diffusion flames, however, the flame structure (i.e., spatial variations of velocity, temperature, and species concentrations) determines various characteristics, such as flame stabilization and pollutant formation. A detailed understanding of the structure, particularly the chemical aspect, of diffusion flames is only being achieved recently. Today, it is common to calculate the coupled time-varying fluid dynamics, heat transfer, and mass transfer with numerous elementary reactions throughout the flame. These results are then compared with the detailed, nonintrusive laser-based measurements of velocity, temperature, and species concentrations, even within the finite-thickness reaction zone itself.

Despite the differences, there are striking similarities in the physicochemical structure of the candle-like laminar diffusion flames of various hydrocarbons. Typically, the high-temperature reducing environment in the interior of the flame cracks the original fuel to the same smaller, unsaturated fragments that consequently diffuse to the oxidation region of the flame. Hence, the study of diffusion flames of much simpler fuels sheds light on many of the important phenomena present with flames of larger fuel molecules. Furthermore, experiments in the absence of gravity uncovered previously unknown phenomena and helped to validate analytical and numerical models.

This Chapter highlights the physical and chemical mechanisms of combustion of candle-like laminar diffusion flames: candles, gas jets, and liquid- or gas-fueled cup burners. First, a brief history on the understanding of candle-like flames is outlined, then general features of a burning candle are described, and finally, recent experimental and computational results of (methane, ethane, and propane) gas-jet and (*n*-heptane) cup-burner flames are presented. Particular emphasis is placed on the effects of the type of fuel and gravity on the structure of the flame-stabilizing region (base) and the trailing diffusion flame.

8.1.2   Historical Sketch

For thousands of Earth-years, candle-like flames were a leading lighting technology. Apart from providing light, the candle flame has been an object of fascination for centuries, inviting speculation on the nature of fire and the natural world. Several historical accounts of the study of flames, fire, or combustion can be seen in the literature [2–8]. Investigation of fires or flames, among other natural phenomena, was intensified during the scientific revolution in the seventeenth century. Francis Bacon [7] (1561–1626) postulated, with respect to the structure of a candle flame, that space is required for the fire to move, and if its motion is suppressed, for example, by a snuffer, it is instantly extinguished. Otto von Guericke (1602–1686) and Robert Boyle (1627–1691) independently demonstrated that a candle or charcoal did not burn in vessels exhausted of air, although it inflamed as soon as air was readmitted. By the early 1660s, Robert Hooke [8] (1635–1703) developed a concept of combustion, in which air possessed two quite separate components: reactive and inert parts. By inserting thin plates of glass and mica in a lamp or candle flame, he noticed that (1) the point of combustion appeared to be at the bottom part of the conical flame, where the oil rising up the wick became excited by the heat above it, and (2) the interior of the flame did not emit light. He also used powerful sunlight to project an image of a candle flame onto a whitewashed wall, whereby he could discern the dark interior and heat zones in the resulting shadow. John Mayow (1643–1679) observed that air is diminished in bulk by combustion and that the residual air is inactive, through an experiment with a burning candle in a bell-jar of air enclosed over water. Although these early investigators may have been on the verge of the discovery of the gas now called oxygen, they could not make it. Unfortunately, the phlogiston theory [2,5,6], which postulated that the essence of fire is a substance called phlogiston, dominated chemistry during the greater part of the eighteenth century. Finally, Antoine Lavoisier (1743–1794) overthrew the phlogiston theory and established his oxygen theory of combustion, after Scheele and Priestley independently discovered oxygen.

In the nineteenth century, Humphry Davy (1778–1829) speculated that the luminosity of flames is caused by the production and ignition of solid particles of carbon as a result of the decomposition of a part of the gas. Jöns Jakob Berzelius (1779–1848) is said to be the first to describe an ordinary candle flame as consisting of four distinct zones. Davy's protégé, Michael Faraday [9] (1791–1867) gave his Christmas lectures and accompanying demonstrations to a juvenile audience on "The Chemical History of a Candle" in 1848 and 1860. Around the turn of the century, modern combustion science was established based on the increased understanding of chemistry, physics, and thermodynamics.

In 1928, Burke and Schumann [1] gave a classic theory of circular and flat gaseous fuel, jet diffusion flames in an air duct with an infinitely thin flame-sheet model. Jost [10] pointed out that some of Burke and Schumann's results could be derived without solving the differential equation to predict the height of flames [11]. A new era began in combustion research after WWII; significant studies [12–16] were made of jet diffusion flames through improved measuring techniques. The picture of the Burke–Schumann flame was substantiated for a hydrogen flame by the gas composition data of Hottel and Hawthorne [12], obtained via gas chromatography. The detailed structure of a flat diffusion flame was investigated spectroscopically by Wolfhard and Parker [15], who showed that local states of chemical equilibrium may exist in a flame zone of a distinctly finite thickness. Parker and Wolfhard [16] made schlieren observations of candle flames. Gaydon and Wolfhard [11] used thermocouples to study the temperature field and revealed that the heat transfer to the wick is equal to the amount of heat required to heat up and vaporize the paraffin. Smith and Gordon [17] analyzed the gaseous products from a candle using a small quartz probe and a mass spectrometer. They found that the mechanism of precombustion reactions involved mainly the cracking of hydrocarbons predominantly to unsaturated compounds before the fuel-species had contact with oxygen. More recently, candles have been used to study various aspects of combustion phenomena: spontaneous near-extinction flame oscillations [18],

flame flickering [19], electric-field effects [20], elevated gravitational effects [21], fire safety [22], and smoke-point measurements [23].

The original Burke–Schumann theory was later refined [24–26]. Roper [26] developed a new theory, by relaxing the requirement of a single constant velocity, and estimated the reasonable flame lengths for both circular and noncircular nozzles. Numerical analyses of laminar diffusion flames have advanced over the last three decades, based on progresses in the reaction mechanisms and computer technologies [27–30]. It is now feasible to simulate, with reasonable accuracy, the transient flame phenomena with full chemistry in simple configurations (burner geometry, flow, and fuel).

In normal earth gravity ($1g$), buoyancy causes the hot gaseous products to rise, entraining air at the base of the flame. This air entrainment is of particular importance in the flame structure and stabilization [31–33]. In microgravity (μg), because of lack of buoyancy, the hot gaseous products tend to accumulate, and the flame enlarges and produces soot more readily [34–41]. The behavior of μg candle flames was investigated in the drop tower and aboard the spacecraft [36,39,41].

8.1.3 Candle Burning Processes

A well-balanced candle burns itself in a clean, steady-state, self-controlled flame. Figure 8.1.1 shows a photograph of a burning paraffin (typically 20–30 carbon atoms) candle in $1g$, exhibiting an elongated flame

FIGURE 8.1.1
Photograph of a 21 mm diameter paraffin candle flame in $1g$.

shape as a result of a rising flow. A (flat) wick curls itself sideways to burn out as it is exposed to the flame zone, and thereby, the wick length is kept constant to control the fuel vaporization rate and, in turn, the flame height. Allan et al. [23] found that soot emission was not possible for 13 different waxes tested when the wick size was below thresholds (diameter <1.8 mm or length <6 mm) and thus the flame height did not reach the smoke point [42].

The complex physical and chemical processes in a burning candle are reasonably well known. Figure 8.1.2 shows a conceptual sketch of a steady-state burning candle, showing various regions, major physical and chemical processes (right), and specific transport phenomena with arrows (left). The origin of the color of the dark-blue region at the base is chemiluminescence from the excited CH radicals in the reaction zone [11]. Heat transfer from the flame, by conduction and radiation, forms a pool of the melted wax at the top of the candle. The melting front moves steadily down the candle as the flame consumes the wax. The melted wax ascends through the wick by capillary action and vaporizes from the heat of the flame transferred primarily by conduction. Buoyancy induces an accelerating, ascending flow, thereby entraining the surrounding air into the lower part of the flame. The vaporized fuel ascends by convection and diffuses outwardly, while the fuel cracking (pyrolysis) reactions take place at high temperatures. The fuel fragments react with the oxygen from the surrounding air to form the diffusion flame. There are heat losses by radiation (primarily from CO_2 and H_2O vapor, and soot) from the flame zone to the surroundings. The convective flow immediately removes the combustion products and heat away from the flame zone.

Soot is formed on the fuel side of the flame zone owing to cracking and pyrolysis of the vaporized fuel at high temperatures. The soot particles are convected upward and eventually penetrate the flame zone, where they burn out by the reaction with oxygen that is diffused in from the surrounding air. Recent laser-diagnostic measurements in laminar diffusion flames revealed that soot is formed in a limited range of temperatures (approximately between 1300 and 1600 K) on the fuel side of the flame zone and that the soot volume fraction peaks in the middle heights of the flame [42]. The luminous part of the flame by soot incandescence may extend downstream beyond (i.e., higher than) the local stoichiometry contour for gaseous species. The heat and the major combustion products (H_2O vapor and CO_2) leave the flame primarily by convection at the tip.

In the absence of gravity, the properties of a candle flame change dramatically [36,39,41]. Figure 8.1.3 shows a candle flame on the Mir space station, in which the melt layer was hemispherical and much thicker than that in normal gravity, and the flame was smaller, spherical, and less sooty, uncovering the blue flame zone. There was significant circulation in the liquid phase (as a result of surface-tension-driven flow caused

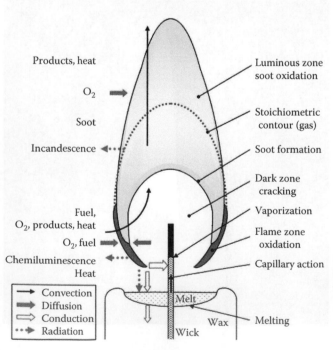

FIGURE 8.1.2
Physical and chemical processes in a candle in 1g.

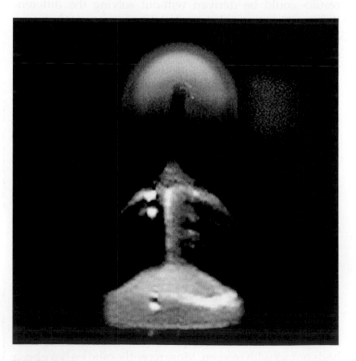

FIGURE 8.1.3
10 mm diameter candle flame in μg on the Mir. (From Ross, H.D., *Microgravity Combustion: Fire in Free Fall*, Ross, H.D., Ed., Academic Press, San Diego, 2001.)

by the temperature difference between the liquid wax near the flame and wax at the candleholder). After sometime, the molten ball of wax collapsed suddenly and the liquid mass moved back along the candleholder. The blue flame zone is much farther away from the wick compared with those in the normal gravity. In addition, it was observed that aerosol (condensed paraffin) was streaming out underneath the flame base and moving in the boundary layer produced by the liquid flow. The long life of the Mir candle flame (up to 45 min) burning in the oxygen-enriched atmosphere (mole fraction of 0.22–0.25) suggests that a steady microgravity flame can be achieved even in a quiescent environment. An effort to model candle flames in low gravity continues [43].

8.1.4 Jet Diffusion Flame Structure

Insight into the behavior of candle flames in normal and low gravity can be obtained from experiments and modeling of jet diffusion flames with simpler gaseous fuels. The NASA Glenn 2.2-Second Drop Tower has provided μg data [44,45] for gaseous hydrocarbon flames for comparison with $1g$ conditions. Figure 8.1.4 shows video images of methane (top row), ethane (middle), and propane (bottom) jet diffusion flames in $1g$ (left column) and μg (middle and right), formed on a circular fuel tube (2.87 mm i.d.) in still air in a vented combustion chamber (255 mm i.d. × 533 mm length). The fuel tube image is superimposed. The flames with "high" (left two columns) and "low" (right) fuel flow-rate levels are shown, in which the ratios of the flow rates of different fuels are maintained to achieve a constant oxygen requirement for each flow level, based on the molar stoichiometric expression. The luminous zone of soot in the $1g$ and "high" flow-rate propane flames (Figure 8.1.4d, g, and h) extended beyond and covered the tip of the blue flame zone. The base of the flame was <1 mm away from the tip of the fuel tube in $1g$, while in μg, it was 3–4 mm

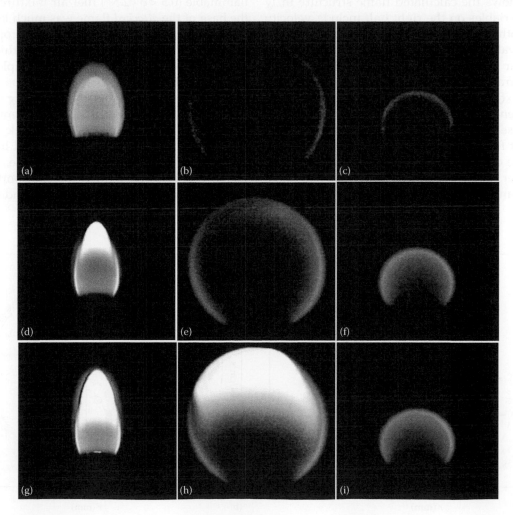

FIGURE 8.1.4
Video images of methane (top), ethane (middle), and propane (bottom) jet diffusion flames (2.87 mm i.d.) in still air in $1g$ (left) and μg (middle and right). The mean fuel jet velocity: (a, b) 13.5 cm/s, (c) 5.3 cm/s, (d, e) 7.7 cm/s, (f) 3.3 cm/s, (g, h) 5.6 cm/s, (i) 2.2 cm/s.

away. As the overall oxygen-consumption rate was kept constant for different fuels, the size (height and width) of the blue flame zone at each fuel flow level in $1g$ or μg was nearly identical. The flames in $1g$ were much thinner ($\approx 5.5\,mm$ width) than their counterparts in μg ($\approx 17\,mm$ width), because the buoyancy-induced convection moved the flame zone inward and resulted in higher oxygen transport rates per unit flame area. In the "low" flow-rate-level μg flames (right), the soot, formed initially at ignition transient, seemed to disappear even for propane, and their shape and size were comparable with those of the Mir flame (Figure 8.1.3).

Transient computations of methane, ethane, and propane gas-jet diffusion flames in $1g$ and $0g$ have been performed using the numerical code developed by Katta [30,46], with a detailed reaction mechanism [47,48] (33 species and 112 elementary steps) for these fuels and a simple radiation heat-loss model [49], for the high fuel-flow condition. The results for methane and ethane can be obtained from earlier studies [44,45]. For propane, Figure 8.1.5 shows the calculated flame structure in $1g$ and $0g$. The variables on the right half include, velocity vectors ($\mathbf{v}$), isotherms (T), total heat-release rate ($\dot{q}$), and the local equivalence ratio (ϕ_{local}); while on the left half: the total molar flux vectors of atomic hydrogen ($\mathbf{M}_H$), oxygen mole fraction (X_{O_2}), oxygen consumption rate ($-\hat{\omega}_{O_2}$), and mixture fraction (ξ), including stoichiometry ($\xi_{st} = 0.06$). General features of the propane flame are similar to those of methane and ethane flames [44,45]. Although soot formation is excluded in the model, the shape and size of the simulated flame match very well with the lower soot-free part of the observed flame (Figure 8.1.4g and h).

In $1g$ (Figure 8.1.5a), the velocity vectors show the longitudinal acceleration in the hot zone owing to buoyancy, and surrounding air is entrained into the lower part of the flame. The heat-release rate and the oxygen-consumption rate contours show a peak reactivity spot (called the reaction kernel) [44,45,50] at the flame base. The values at the reaction kernel were $\dot{q}_k = 196\,J/cm^3s$, $-\hat{\omega}_{O_2,k} = 0.000648\,mol/cm^3s$, $|v_k| = 0.250\,m/s$, $T_k = 1483\,K$, $X_{O_2,k} = 0.031$, $\phi_{local,k} = 0.99$, and $\xi_k = 0.060$. Hydrogen atoms and other chain radicals diffused onto both sides of the flame zone and in every downward direction around the flame base against the incoming flow with higher oxygen concentrations and gradients. Consequently, the chain-branching reaction, $H + O_2 \rightarrow OH + O$, and other chain reactions were enhanced to maximize the reactivity at the reaction kernel. Thus, the reaction kernel sustained stationary combustion processes within a residence time available in the flow, thereby holding the trailing diffusion flame (which had both lower reactivity and higher velocities) [44,45,50]. The thickness of the flammable ($0.5 < \phi < 2.5^{51}$) fuel/air mixture layer nearby the base of the *attached* flame was much lesser than the minimum quenching distance of the propane/air mixture ($2\,mm$) [51], and not sufficiently thick to form a combustion-wave reminiscent of the triple flame structure of lifted flames.

In $0g$ (Figure 8.1.5b), unlike in the $1g$ case, the fuel jet momentum dispersed and the centerline velocity decayed rapidly owing to the lack of buoyancy. As a result, the fuel molecules diffused in every direction and formed a quasi-spherical flame. The slow diffusion processes (1) limited the transport rates of the fuel and oxygen into the flame zone and (2) decreased

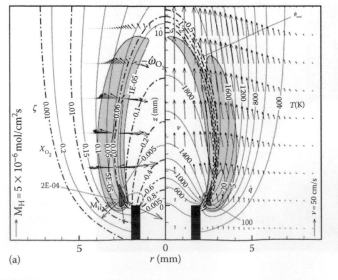
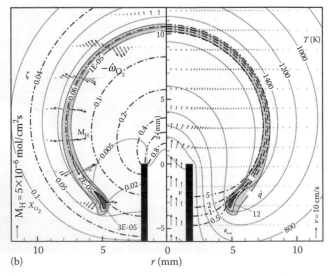

FIGURE 8.1.5

Calculated structure of propane jet diffusion flames (3 mm i.d., fuel velocity: 4.8 cm/s) in "still" air in (a) $1g$ and (b) $0g$.

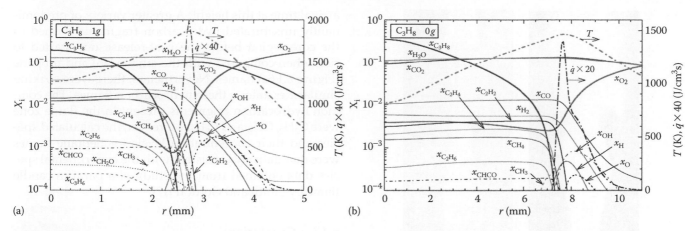

FIGURE 8.1.6
Calculated species mole fractions, temperature, and heat-release rate across propane jet diffusion flames in "still" air at a height of 3 mm in (a) 1g and (b) 0g.

the molar flux of hydrogen atoms by an order of magnitude as a result of lower concentration gradients, so that the reactivity in the reaction kernel decreased by an order of magnitude. The reaction kernel location is closely coupled with the local flow velocity and mixture reactivity [44]. The values of various variables at the reaction kernel were $\dot{q}_k = 14.8\,J/cm^3s$, $-\hat{\omega}_{O_2'k} = 0.000042\,mol/cm^3s$, $|v_k| = 0.0045\,m/s$, $T_k = 1299\,K$, $X_{O_2'k} = 0.017$, $\phi_{local,k} = 0.51$, and $\xi_k = 0.058$. The reaction kernel temperature in 0g was almost 300 K lower than in 1g.

Figure 8.1.6 shows the variations of the species mole fractions (X_i), temperature, and total heat-release rate across the propane flame in 1g and 0g at a height from the jet exit of 3 mm (i.e., in the trailing diffusion flame). The thickness of the flame zone, determined from the locus of positive values for the total heat-release rate, was approximately 2 mm in 1g and 3 mm in 0g. Hence, the flame zone was far from an infinitesimally thin flame-sheet model by Burke and Schumann [1]. In both 1g and 0g, general trends in the species mole fractions were typical of diffusion flames, i.e., chain radicals were formed on the air side at high temperatures, diffused onto the fuel side and dissociated the fuel into fragmental hydrocarbons, and finally CO and H_2 were oxidized to products. A distinct feature of both 0g and 1g flames, not accounted for any of the analytical models of diffusion flame shape, was that oxygen from the air penetrated into the fuel side of the flame zone through the quenched space between the flame base and the burner rim. Furthermore, for ethane [45] and propane, major hydrocarbon fragments and oxygenates burning in the reaction zone were C_2H_2, CH_4, C_2H_4, and CHCO, and thus the C_2 species oxidation reactions, CHCO $+O \rightarrow CO+CO+H$ and $C_2H_2+O \rightarrow CH_2+CO$ were major contributors to the total heat-release rate peak in

the inner region. For methane [44,50], the $CH_3+O \rightarrow CH_2O+H$ reaction was a dominant contributor to the heat-release rate peak. The exothermic final-product formation reactions, $CO+OH \rightarrow CO_2$ and $H_2+OH \rightarrow H_2O+H$, took place in a wide rage over the flame zone. On the air side, the $O_2+H+M \rightarrow HO_2+M$ and the subsequent $HO_2+OH \rightarrow H_2O+O_2$ reactions were very much exothermic, thus contributing to a secondary heat-release rate hump in the outer region. In 0g, the relative contributions of these HO_2 reactions with null activation energies became more significant, because of the lower flame temperature and total heat-release rate than those in 1g.

8.1.5 Cup-Burner Flame Structure

By examining the experimental and computational results obtained [52] for *n*-heptane flames in 1g, a further attempt was made to extract a general trend in the structure of higher aliphatic hydrocarbon flames. Figure 8.1.7 shows a photograph and the calculated fields of temperature (left half) and soot mass fraction (right half) of *n*-heptane diffusion flames in the coflowing air in the cup-burner apparatus ($\approx 30\,mm$ diameter). In the experiment (Figure 8.1.7a), in which liquid *n*-heptane was used, the blue flame base was anchored at the burner rim with an inward inclination as a result of the incoming buoyancy-induced flow. The color of the flame zone turned bright-yellow downstream owing to considerable soot formation. The computation (Figure 8.1.7b) of the gaseous *n*-heptane flame with a full-chemistry model [53] (197 species and 2757 elementary reaction steps), including global soot formation, captured well the transient flame flickering behavior, and the general trend in the soot field observed in the liquid-fuel flame (Figure 8.1.7a).

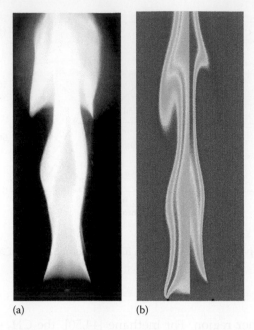

(a)   (b)

FIGURE 8.1.7
(a) Photograph of a liquid *n*-heptane cup-burner flame in 1*g*. (b) Calculated fields of temperature [left half] and soot mass fraction [right half] of a gaseous *n*-heptane diffusion flame in 1*g*. Burner diameter ≈ 30 mm.

Figure 8.1.8 shows the calculated structure of a gaseous *n*-heptane flame in 1*g* at a height of 10.8 mm (in the trailing diffusion flame). Because of the negligibly small gaseous fuel velocity (0.1 cm/s), which modeled the liquid pool flame, a near-stagnant recirculation zone was formed near the exit of the burner. As a result of high temperatures (>1000 K) and relatively long residence time, *n*-heptane pyrolyzed and disappeared by the flow

arrival time at this height. A greater variety of predominantly unsaturated hydrocarbon fragments formed in the core region before the heat-release rate started to rise, when compared with the ethane [45] and propane (Figure 8.1.6a) flames. Nevertheless, there were striking similarities among these hydrocarbon flames. The common hydrocarbon fragments present in the flame zone were C_2H_2, CH_4, and C_2H_4. Moreover, the calculated species and their concentration levels for heptane flames were qualitatively consistent with the experimental species data obtained from the study of a paraffin candle flame by Smith and Gordon [17].

8.1.6 Conclusions

The advancement in research tools has increased the understanding of candle and gas-jet diffusion flames. Notable insights were gleaned from the Burke–Schumann theory and refinements, a surge in the experimental work after WWII, nonintrusive laser diagnostics, progress in detailed computation, and low-gravity experiments. Similarities among diffusion flames of various hydrocarbons, including paraffin candles, arise from the cracking of the original fuel predominantly to unsaturated compounds on the fuel side of the flame zone. For higher (≥C_2) hydrocarbons, the common species burning in the flame zone, besides major intermediates (CO and H_2), are C_2H_2, CH_4, and C_2H_4 in the inner region, and HO_2 in the outer region. Although lack of buoyancy-induced flow acceleration in 0*g* enlarges the flame and makes the rates of flame processes an order-of-magnitude lower than in 1*g*, the general trend in the chemical kinetic structure typical of diffusion flames is maintained.

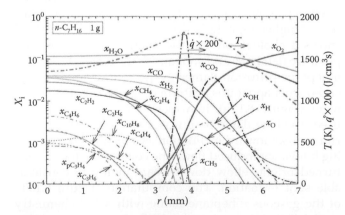

FIGURE 8.1.8
Calculated species mole fractions, temperature, and heat-release rate across a gaseous *n*-heptane diffusion flame (18 mm o.d., fuel velocity: 0.1 cm/s) in coflowing air (velocity: 10.7 cm/s) at a height of 10.8 mm in 1*g*.

Acknowledgments

This work was supported by the National Aeronautics and Space Agency, Washington, D.C. The author thanks Drs. V. R. Katta and G. T. Linteris for their assistance in refining the manuscript as well as for their long-term research partnership and contributions.

References

1. Burke, S.P. and Schumann, T.E.W., Diffusion flames, *Ind. Eng. Chem.*, 20, 998, 1928.
2. Bone, W.A. and Townend, D.T.A., *Flame and Combustion in Gases*, Longmans, Green and Co. Ltd., London, 1927, Chapter 1.

3. Fristrom, R.M., *Flame Structure and Processes*, Oxford University Press, New York, 1995, Chapter 1.

4. Weinberg, F.J., The first half-million years of combustion research and today's burning problems, *Proc. Combust. Inst.*, 15, 1 1975.

5. Williams, F.A., The role of theory in combustion science, *Proc. Combust. Inst.*, 24, 1, 1992.

6. Lyons, J.W., *Fire*, Scientific American Books, New York, 1985.

7. Bacon, F., *The New Organon*, Jardine, L. and Silverthorne, M., Eds., Cambridge University Press, Cambridge, UK, 2000, p. 124.

8. Chapman, A., England's Leonardo: Robert Hooke (1635–1703) and the art of experiment in restoration England, *Proc. R. Inst. Great Britain*, 67, 239, 1996.

9. Faraday, M., *A Course of Six Lectures on the Chemical History of a Candle*, Crookes, W., Ed., Harper & Brothers, New York, 1861.

10. Jost, W., *Explosion and Combustion Processes in Gases*, McGraw-Hill, New York, 1946.

11. Gaydon, A.G. and Wolfhard, H.G., *Flames, Their Structure, Radiation and Temperature*, 4th ed., Chapman and Hall, London, 1979, Chapter 6.

12. Hottel, H.C. and Hawthorne, W.R., Diffusion in laminar flame jets, *Proc. Combust. Inst.*, 3, 254, 1949.

13. Wohl, K., Gazley, C., and Kapp, N., Diffusion flames, *Proc. Combust. Inst.*, 3, 288, 1949.

14. Scholefield, D.A. and Garside, J.E., The structure and stability of diffusion flames, *Proc. Combust. Inst.*, 3, 102, 1949.

15. Wolfhard, H.G. and Parker, W.G., A spectroscopic investigation into the structure of diffusion flames, *Proc. Phys. Soc.*, A65, 2, 1952.

16. Parker, W.G. and Wolfhard, H.G., Carbon formation in flames, *J. Chem. Soc.*, 2038, 1950.

17. Smith, S.R. and Gordon, A.S., Precombustion reactions in hydrocarbon diffusion flames: The paraffin candle flame, *J. Chem. Phys.*, 22, 1150, 1954.

18. Chan, W.Y. and T'ien, J.S., An experiment on spontaneous oscillation prior to extinction, *Combust. Sci. Technol.*, 18, 139, 1978.

19. Buckmaster, J. and Peters, N., The infinite candle and its stability—a paradigm for flickering diffusion flames, *Proc. Combust. Inst.*, 21, 1829, 1988.

20. Carleton, F. and Weinberg, F., Electric field-induced flame convection in the absence of gravity, *Nature*, 330, 635, 1989.

21. Villermaux, E. and Durox, D., On the physics of jet diffusion flames, *Combust. Sci. Technol.*, 84, 279, 1992.

22. Hamins, A., Bundy, M., and Sillon, S.E., Characterization of candle flames, *J. Fire Prot. Eng.*, 15, 265, 2005.

23. Allan, K.M., Kaminski, J.R., Bertrand, J.C., Head, J., and Sunderland, P.B., Laminar smoke points of candle flames, presented at 5th US Combustion Meeting of the Combustion Institute, Paper No. D32, San Diego, CA, March 25–28, 2007.

24. Barr, J., Length of cylindrical laminar diffusion flames, *Fuel*, 33, 51, 1954.

25. Fay, J.A., The distributions of concentration and temperature in a laminar jet diffusion flame, *J. Aeronaut. Sci.*, 21, 681, 1954.

26. Roper, F.G., The prediction of laminar jet diffusion flame sizes: Part I. theoretical model, *Combust. Flame*, 29, 219, 1977.

27. Miller, J.A. and Kee, R.J., Chemical nonequilibrium effects in hydrogen-air laminar jet diffusion flames, *J. Phys. Chem.*, 81, 2534, 1977.

28. Mitchell, R.E., Sarofim, A.F., and Clomburg, L.A., Experimental and numerical investigation of confined laminar diffusion flames, *Combust. Flame*, 37, 227, 1980.

29. Smooke, M.D., Lin, P., Lam, J.K., and Long, M.B., Computational and experimental study of a laminar axisymmetric methane-air diffusion flame, *Proc. Combust. Inst.*, 23, 575, 1990.

30. Katta, V.R., Goss, L.P., and Roquemore, W.M., Numerical investigations of transitional H_2/N_2 jet diffusion flames, *AIAA J.*, 32, 84, 1994.

31. Robson, K. and Wilson, M.J.G., The stability of laminar diffusion flames of methane, *Combust. Flame*, 13, 626, 1969.

32. Kawamura, T., Asato, K., and Mazaki, T., Structure of the stabilizing region of plane, laminar fuel-jet flames, *Combust. Sci. Technol.*, 22, 211, 1980.

33. Takahashi, F., Mizomoto, M., and Ikai, S., Structure of the stabilizing region of a laminar jet diffusion flame, *J. Heat Transfer*, 110, 182, 1988.

34. Edelman, R.B., Fortune, O.F., Weilerstein, G., Cochran, T.H., and Haggard, J.B., Jr., An analytical and experimental investigation of gravity effects upon laminar gas jet-diffusion flames, *Proc. Combust. Inst.*, 14, 399, 1973.

35. Bahadori, M.Y., Edelman, R.B., Stocker, D.P., and Olson, S.L., Ignition and behavior of laminar gas-jet diffusion flames in microgravity, *AIAA J.*, 28, 236, 1990.

36. Ross, H.D., Sotos, R.G., and T'ien, J.S., Observations of candle flames under various atmospheres in microgravity, *Combust. Sci. Technol.*, 75, 155, 1991.

37. Walsh, K.T., Fielding, J., Smooke, M.D., and Long, M.B., Experimental and computational study of temperature, species, and soot in buoyant and non-buoyant coflow laminar diffusion flames, *Proc. Combust. Inst.*, 28, 1973, 2000.

38. Lin, K.-C., Faeth, G.M., Sunderland, P.B., Urban, D.L., and Yuan, Z.-G., Shapes of nonbuoyant round luminous hydrocarbon/air laminar jet diffusion flames, *Combust. Flame*, 116, 415, 1999.

39. Dietrich, D.L., Ross, H.D., Shu, Y., Chang, P., and T'ien, J.S., Candle flames in non-buoyant atmospheres, *Combust. Sci. Technol.*, 156, 1, 2000.

40. Urban, D.L., Yuan, Z.-G., Sunderland, P.B., Lin, K.-C., Dai, Z., and Faeth, G.M., Smoke-point properties of non-buoyant round laminar jet diffusion flames, *Proc. Combust. Inst.*, 28, 1965, 2000.

41. Ross, H.D., Basics of microgravity combustion, *Microgravity Combustion: Fire in Free Fall*, Ross, H.D. Ed., Academic Press, San Diego, CA, 2001, Chapter 1.

42. Glassman, I., *Combustion*, 3rd ed., Academic Press, San Diego, CA, 1996.

43. Alsairafi, A., Lee, S.T., and T'ien, J.S., Modeling gravity effects on diffusion flames stabilized around a cylindrical wick saturated with liquid fuel, *Combust. Sci. Technol.*, 176, 2165, 2004.

44. Takahashi, F. and Katta, V.R., Reaction kernel structure and stabilizing mechanisms of jet diffusion flames in microgravity, *Proc. Combust. Inst.*, 29, 2509, 2002.

45. Takahashi, F. and Katta, V.R., Further studies of the reaction kernel structure and stabilization of jet diffusion flames, *Proc. Combust. Inst.*, 30, 383, 2005.

46. Roquemore, W.M. and Katta, V.R., Role of flow visualization in the development of UNICORN, *J. Vis.*, 2, 257, 2000.

47. Peters, N., *Reduced Kinetic Mechanisms for Applications in Combustion Systems*, N. Peters and B. Rogg, Eds., Springer-Verlag, Berlin, Germany, 1993, p. 3.

48. Warnatz, J., *Combustion Chemistry*, W.C. Gardiner, Ed., Springer-Verlag, New York, 1984, p. 197.

49. Barlow, R.S., Karpetis, A.N., and Frank, J.H., Scalar profiles and NO formation in laminar opposed-flow partially premixed methane/air flames, *Combust. Flame*, 127, 2102, 2001.

50. Takahashi, F. and Katta, V.R., Chemical kinetic structure of the reaction kernel of methane jet diffusion flames, *Combust. Sci. Technol.*, 155, 243, 2000.

51. Lewis, B. and von Elbe, G., *Combustion, Flames and Explosions of Gases*, 3rd ed., Academic Press, New York, 1987.

52. Takahashi, F., Linteris, G., and Katta, V.R., Further studies of cup-burner flame extinguishment, 16th Annual Halon Options Technical Working Conference (HOTWC), Albuquerque, NM, May 2006.

53. Tsang, W. Progress in the development of combustion kinetics databases for liquid fuels, *Data Sci. J.*, 3, 1, 2004.

8.2 Combustion in Spark-Ignited Engines

James D. Smith and Volker Sick

8.2.1 Introduction

Spark-ignited internal-combustion engines have been in existence for well over 100 years, with the first example being presented by Nikolaus Otto in 1876. However, age should not be deceiving in this case, as significant research and development efforts are still being applied to this concept to achieve more power, better efficiency, and lower emissions. Fuel charge preparation and in-cylinder motion are the two main parameters affecting ignition and combustion, and therefore, have a large degree of influence on the aforementioned quantities. Depending on the method used for preparing the fuel/air mixture, the combustion regime ranges from near perfectly premixed to highly heterogeneous diffusion burning in a multiphase, liquid/vapor environment. In-cylinder motion is also crucial to engine operation at various speeds. As engine speed increases, so do in-cylinder turbulence levels, yielding faster burning velocities, without which engines cannot operate

on normal speed scales of the thousands of cycles per minute. Fuel charge preparation and in-cylinder flow are invariably intertwined, as certain preparation strategies rely on the directed flows to achieve reliable ignition, in addition to having a large degree of influence on the motion itself.

This chapter will treat the three modes of combustion in spark-ignited IC engines: homogeneous-charge spark-ignition (HCSI, premixed-turbulent combustion), stratified-charge spark-ignition (SCSI, partially premixed-turbulent combustion), and homogeneous-charge compression-ignition with spark assist (SACI, premixed-turbulent combustion). This range of fuel preparation and in-cylinder motion strategies covers the time span from the basic incarnation of the SI engine to future engine concepts that will probably play a major role in the future of ground transportation. The highlights and governing characteristics of each combustion mode will be discussed and visual examples of each case will be presented from a running engine with optical accesses that were obtained with a high-speed camera (12000 frames/s) that recorded combustion luminosity. The viewing area within the engine is presented in Figure 8.2.1, which shows prominent features, such as the fuel injector, spark plug, and intake/exhaust valves. Further details of the engine can be found in the work by Smith and Sick [1]. These visual examples will help to explain the observations made from pressure-based measurements, such as ignition delay, combustion duration, and heat release rates.

8.2.2 Homogeneous-Charge Spark-Ignition Engines

HCSI engines are the most common spark-ignition engines in existence. While methods for mixture preparation have varied from carburetors to port fuel injection, the overall goal is the same: to have a well-mixed fuel/air charge at all locations in the combustion chamber. With this concept, the air/fuel ratio is close to stoichiometric to promote reliable combustion and facilitate emissions after treatment. Since the ratio of air and fuel must be kept nearly constant, intake-air throttling is used to control engine power output. At low engine-load conditions, high throttling is necessary and has a negative consequence on the volumetric efficiency. Conversely, at high loads, near-ambient intake manifold pressures are used to draw in the maximum possible amount of air/fuel mixture to the combustion chamber.

Once the well-mixed air/fuel charge is inducted into the combustion chamber, the rising piston compresses and heats the mixture. At a point near the peak travel of the piston (top dead center; TDC), ignition is initiated through a spark plug. Normally, this occurs at a singular point; however, some instances of using two or more

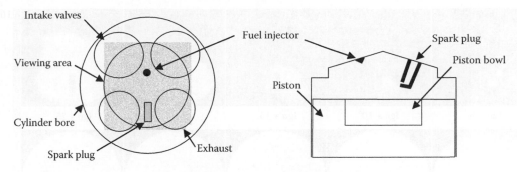

FIGURE 8.2.1
Schematic of viewing area for subsequent images. Bottom (left) and side (right) views of the combustion chamber are used in this chapter.

spark plugs per cylinder have been observed in production, most often to promote faster burning [2,3]. While the spark appears to be an instantaneous event, there are actually three distinct phases, each with unique timescales and energy-deposition characteristics [4,5]. The breakdown phase is marked by the establishment of a plasma channel bridging the electrodes of the spark plug. Voltages in the order of 10^5 V are common owing to the high impedance of the gap. This phase tends to last only for about 100 ns, as the voltage rapidly falls and current rises once the highly conductive plasma channel is established, which marks the beginning of the arc phase of the spark. Following this phase (usually 1–100 µs), the spark is characterized by a relatively long phase of energy discharge at moderate voltages, in the order of 500 V, known as the glow phase. Overall, the entire event tends to last for about 1.5 ms in SI engines, which spans tens of degrees of crankshaft rotation (CAD) at normal engine speeds.

The function of the spark is to provide the necessary activation energy to initiate a series of self-sustaining chemical reactions, collectively comprising the combustion process. In the HCSI engine, bulk flow in the region of the spark gap is generally low; however, steps are commonly taken to increase the turbulence levels, such as swirl and tumble flows. Owing to this low flow level, the initial reaction zone is nearly spherical between the spark-plug electrodes and expands radially outwards. Heat loss from the reaction zone to the electrodes has been found to be significant [6–8], and therefore, has to be taken into account when designing the spark plug. This point in the development of the flame is very crucial as a delicate balance is being struck between heat losses owing to radiation and conduction, and heat liberation through chemical reaction. For the flame to successfully grow, reactions must not only be fast enough to replenish energy lost through heat transfer, but also fast enough to expand the size of the flame. This process continues, relatively slowly, and is generally referred to as the ignition delay. This phase is marked by either 2% or 10% of the fuel being consumed or mass fraction burned (0%–2% mfb or 0%–10% mfb).

The ignition delay is followed by a period of rapid growth. As the flame grows in diameter, the surface begins to get stretched and distorted by turbulence, and becomes what is generally referred to as a wrinkled flame front.

As stated previously, turbulence plays a major role in the combustion process. While the Reynolds number (Re) is generally low in fluid mechanical terms (Re ~100–10,000) [9], declining large-scale vortices left over from the air/fuel induction event produce relatively high turbulence levels. This distortion is favorable in that it increases the surface area of the flame allowing more reactions to happen simultaneously, thus increasing the burning velocity and heat-release rate. In addition, turbulence levels tend to scale with engine speed [10], a phenomenon which allows engines to operate at rates of several thousand cycles per minute. The name "flame front" is used, as majority of the combustion-sustaining reactions occur in a thin region between the unburnt air/fuel mixture and the hot products of combustion. This hot zone in the middle of the flame is primarily responsible for thermal nitrogen-oxides (NO_x) formation. Figure 8.2.2 shows an image sequence of the flame growth in an HCSI engine. The high level of flame-front distortion and thin spatial dimension of the flame are easily observable from this image. This period of combustion is sometimes referred to as a "rapid burn phase" in which 80% of the mass of fuel is consumed (10%–90% mfb). It is during this phase that the piston will reach TDC and, therefore, begin the power stroke of the four-stroke cycle. It is highly desirable to phase combustion, such that the peak level of burning and peak cylinder pressure occurs just after the piston begins the power stroke. This minimizes the heat losses and allows maximum expansion of the hot combustion gases. To achieve this combustion timing, the spark is initiated to a certain degree before top dead center (BTDC) and is referred to as spark advance. The optimum level of spark advance is normally experimentally determined to be the minimum amount of spark advance yielding the maximum torque (MBT) for a given operating condition.

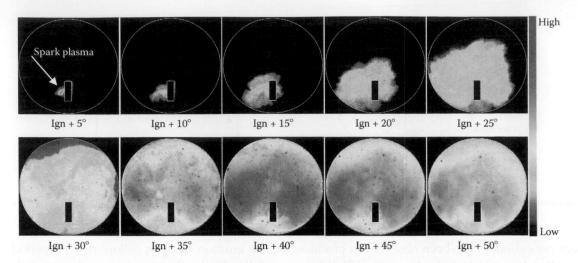

High

Low

Ign + 5°   Ign + 10°   Ign + 15°   Ign + 20°   Ign + 25°

Ign + 30°   Ign + 35°   Ign + 40°   Ign + 45°   Ign + 50°

FIGURE 8.2.2
Images of the flame propagation process in an HCSI engine. Color scale qualitatively represents burning intensity. The presence of the spark is highlighted in the first image.

Following the rapid-burn phase of engine combustion, reaction rates tend to slow rapidly. This is owing to the decreasing local temperatures as the piston travels downwards and also because the flame front begins to encounter the relatively cool walls and crevices of the combustion chamber and is, therefore, quenched. Any fuel not consumed owing to these effects contributes to combustion inefficiency and is potentially emitted as unburnt hydrocarbons (HC) in the exhaust. Fortunately, modern engine design has relatively dealt with these issues with combustion efficiencies, normally above 95% and near-zero HC emissions, successfully.

The development of models for HCSI combustion has been governed by the similarity of flame growth in HCSI engines and premixed turbulent flames. Thin laser-sheets of only 300 μm thickness were used to measure high-resolution cross sections of the temperature and OH radical distribution in flames of a propane-fueled engine. Figure 8.2.3 illustrates the structure where temperature and OH concentration are closely coupled with super equilibrium values for the OH radical close to the flame front [11].

The fact that the fuel/air ratio is spatially constant in HCSI engines, at least within a reasonably close approximation, allows substantial simplifications in combustion models. The burn rate or fuel consumption rate dm_b/dt is expressed as a function of flame surface area A_{fl}, the density of the unburnt fuel/air mixture ρ_u, the laminar burning velocity s_L, and the fluctuations of velocities, i.e., E as a measure of turbulence, u'.

$$\frac{dm_b}{dt} = A_{fl}\,\rho_u\,(u' + s_L)$$

Embedded in such models, in which variations were developed [12] are further detailed. The laminar burning velocity is expressed as a function of fuel type, fuel/air ratio, level of exhaust gas recirculation, pressure, temperature, etc. Furthermore, submodels have been developed to describe the impact of engine speed, port-flow control systems, in-cylinder gross-flow motion (i.e., swirl, tumble, squish), and turbulent fluctuations u'. Thus, with a wider knowledge base of the parametric impact of external variables, successful modeling of

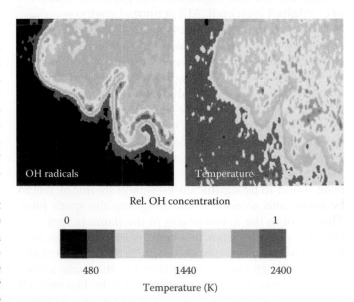

OH radicals   Temperature

Rel. OH concentration

0   1

480   1440   2400
Temperature (K)

FIGURE 8.2.3
Planar images obtained with laser-based imaging techniques in a propane-fueled research engine showing how closely coupled hydroxyl radical (OH) concentration is with temperature. (From Orth, A., Sick, V., Wolfrum, J., Maly, R.R., and Zahn, M., *Proc. Combust. Inst.*, 25, 143, 1994. With permission.)

HCSI combustion was possible. However, with newer combustion strategies that use strong fuel/air ratio stratification or homogenous charge compression ignition (HCCI)-like combustion, these simple relationships do no longer apply and modeling is far more complex [13,14].

8.2.2.1 Reprise

While the HCSI engine has proven itself to be a worthy powerplant, there are some plaguing issues that keep the overall thermal efficiency below 30% in most cases. Autoignition, commonly known as knock, limits the compression ratio that can be used. The use of higher, more efficient compression ratios causes the unburnt air/fuel mixture to autoignite before the flame front can consume the fuel. This creates damaging pressure spikes and unwanted driving characteristics and must therefore be avoided. Intake air-throttling also limits the efficiency of the engine at low- and part-load conditions where most driving is done. Finally, less precise fuel metering leads to increased wall-wetting, and consequently, the HC emissions, particularly during cold-starting. These are key concerns that must be addressed if the spark-ignition engine is going to remain competitive with the compression-ignition engine (Diesel) and other more modern combustion devices. The next two combustion modes will attempt to address these issues.

8.2.3 Stratified-Charge Spark-Ignition Engines

SCSI engines are most commonly known as direct-injected spark-ignition (DISI) engines. Since these engines can also be operated in a homogeneous charge regime, we will focus only on the stratified-charge subset of direct-injected engines. The name direct-injected (or direct-injection) refers to the fuel preparation strategy used, where fuel and air are injected and inducted, respectively, into the combustion chamber separately. Unlike the HCSI engine, the ratio of air to fuel can be infinitely varied, as long as it is within the flammability limits. During the stratified operation, a flammable mixture is maintained in key areas, such as near the spark plug, while maintaining overall fuel-lean conditions. This allows the removal of the throttle plate, as the power output can now be governed by the mass of fuel introduced into the combustion chamber and the excess air has little effect on combustion.

The key challenge to SCSI engine operation is successful fuel stratification in the combustion chamber. Three methods for this stratification are commonly referred to as wall guided, air guided, and spray guided [15]. Wall-guided systems utilize injector targeting at specific features (such as the piston) to achieve a reliable fuel cloud around the spark plug. This creates high levels of wall wetting (thus HC emissions) and is therefore less desirable. Air guiding uses complex combinations of in-cylinder bulk motion to achieve stratification. Unfortunately, the methods for achieving this motion can sometimes offset any gains in efficiency. The most promising candidate is the spray-guided system, which utilizes spray targeting directly at the spark plug. The spark is triggered close to the fuel-injection event and ignites a passing region of high fuel concentration. The key challenge in this concept is achieving reliable, stable operation. Cycle to cycle variations in fuel concentration can be large, and hence, significant research efforts have gone into the study of fuel preparation in these engines [16–22]. In some cases, complete misfire occurs, which impedes full-scale implementation in production. Even over the course of a single spark discharge (~15 CAD), the fuel concentration varies significantly in spatial distribution as highlighted in Figure 8.2.4, which is a sequence of high-speed laser-induced fluorescence imaging of the fuel distribution under motored conditions.

In contrast to ignition in the HCSI engine, strong bulk flow is introduced by the fuel-injection event. Since spark occurs within a short time-span of this event, residual flow through the gap is common and in the order of tens of meters per second [23]. This has the effect of stretching the spark plasma channel, sometimes to the point of detachment from the electrodes. It also makes the ignition process extremely sensitive to the relative orientation of spark plug and fuel injector, with some geometries yielding higher velocity conditions than the others. Figure 8.2.5 presents an image sequence of a single engine cycle under SCSI conditions. The spark can be seen to strike between the electrodes and immediately be stretched back by the residual flow left from the fuel-injection event. The geometry of the spark plug plays a major role in determining whether or not the spark would remain attached or stretch and re-strike throughout the discharge process [24]. This also has a key effect on the initial flame kernel. Cases in which the spark plug shields the flow lead to more spherical flame kernels, whereas less-protective spark plugs produce elongated flame kernels.

In addition to the spark occurring in a high-velocity environment, the conditions are also highly multiphase with both the liquid and vapor existing in the evaporating fuel cloud. It is still unclear as how this affects the physics of the spark process. However, the cooling effects of the evaporating liquid may be responsible for the increased ignition delays when compared with the HCSI engines. Figure 8.2.6 compares the mass fraction burnt (mfb) times for the HCSI, SCSI, and spark-assisted HCCI engines. As in the HCSI engine, the initial reaction zone is highly sensitive to both the heat transfer

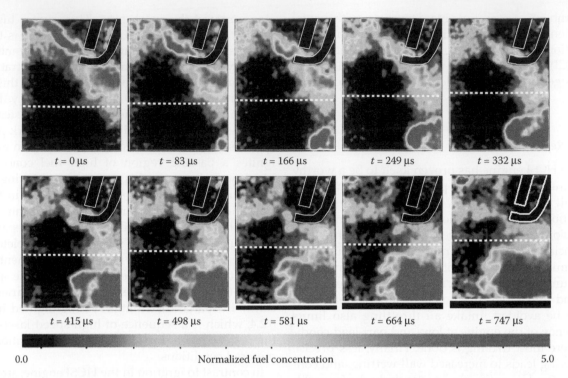

FIGURE 8.2.4
Images showing the fuel concentration near the time of ignition in a stratified charge direct-injected engine under motored conditions (1.0 represents stoichiometric). Time is relative to when spark would normally be initiated. (From Smith, J.D., Development and application of high-speed optical diagnostic techniques for conducting scalar measurements in internal combustion engines, *Mechanical Engineering*, University of Michigan, Ann Arbor, 2006.)

from the zone and the heat release through chemical reaction. Cooler temperatures, owing to spray evaporation, and lower heat release rates, owing to highly variable, non-optimal fuel distribution, may explain why SCSI engines have longer ignition delays.

The rapid-burn phase of SCSI engine combustion is a combination of both premixed and diffusion combustion. As in the HCSI engine, a typical premixed turbulent flame front propagates across the cylinder, originating from the spark plug. This phase tends to occur faster

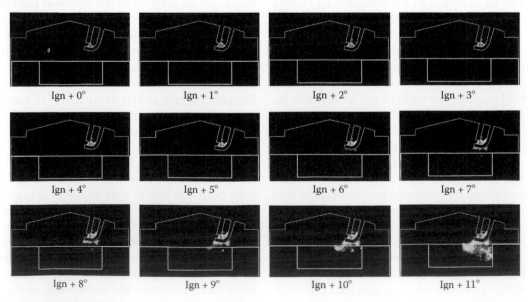

FIGURE 8.2.5
Side view of the ignition process in an SCSI engine showing the spark arc and early flame development.

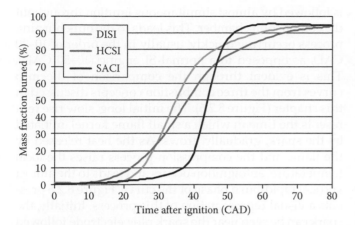

FIGURE 8.2.6
Mass fraction burnt (mfb) curves from each of the three engine concepts discussed in this chapter. Time is presented as crank-angle-degrees after ignition is initiated.

than that in the HCSI engine [25], which can be noted from the mfb curves shown in Figure 8.2.6. As the flame front progresses, the regions of varying fuel concentration are encountered leading to rich, diffusion-limited combustion in some regions, which leads to high soot production [26,27]. This is evident in the later images of Figure 8.2.7, where intense burning is visible in various regions of the combustion chamber, whereas the flame front is less visible. However, other imaging experiments again focused on the OH radical and verified the presence of a flame front. However, the signal was orders of magnitude lower than the soot luminosity and is therefore not visible in this string of images.

Combustion is generally confined to the inner regions of the combustion chamber during stratified combustion. This is mainly because the late fuel injection does

not allow large amounts of fuel to reach extremities and crevices. This helps lower heat transfer to the wall, in addition to the unburnt hydrocarbon emissions owing to crevice volumes. However, combustion efficiency tends to be lower with this concept, as not all fuels are present in flammable quantities and therefore, may not burn.

8.2.3.1 Reprise

SCSI engines offer significant improvements in the volumetric efficiency, owing to the removal of the throttle plate. Furthermore, wall-heat transfer is lowered and the better control over the injection event leads to better cold-starting and transient response. There are still concerns that need to be addressed before widespread usage can be realized. Reliable ignition and stable operation require highly repeatable and generally expensive fuel injectors. In addition, high fuel pressures are necessary to yield proper fuel atomization. This leads to higher parasitic losses from powering higher-performance fuel pumps. Perhaps the most concerning issue is the inability to use conventional exhaust, after-treatment devices (i.e., catalytic converters). These devices work most effectively when the engine is run near the stoichiometric air/fuel conditions. Since this concept operates on highly lean fuel, NO_x emissions cannot be mitigated by a traditional three-way catalyst and must therefore be dealt with separately, adding cost and complexity to the vehicle.

8.2.4 Spark-Assisted Compression Ignition

SACI engines, or spark-assisted HCCI engines, present unique benefits and challenges from the two previous operating modes. Using this mode, many of the benefits in performance and efficiency similar to SCSI can be

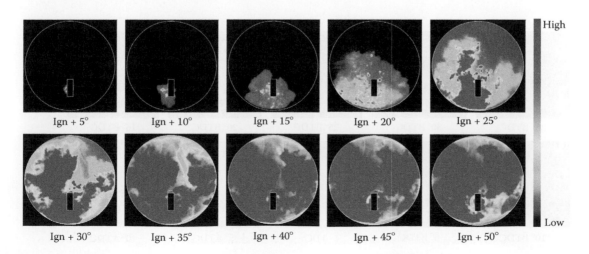

FIGURE 8.2.7
Images of the rapid-burn phase of combustion in a DISI engine. Stratification of the fuel distribution leads to areas of rich combustion and high soot production, as evidenced by areas of intense flame signal.

realized without the need for advanced exhaust after-treatment devices. Conversely, control of the combustion process is governed mainly by the thermal state of the combustion chamber and, therefore, presents unique challenges.

Under normal HCCI operating conditions, the fuel/air mixture is compression heated and autoignites at one or more locations, similar to a normal compression-ignition (i.e., diesel) engine. This removes the strict requirement of placing a flammable mixture in the vicinity of the spark plug, such as in a SCSI engine. Unlike the diesel engine, however, the fuel and air are well mixed and therefore combustion phasing cannot be controlled by the fuel injection timing. As a result of this, and the lack of a spark event, the only way to control combustion is to manipulate the thermal state of the combustion chamber. This is most commonly done through intake-air heating or inducting high amounts of hot exhaust gases. Both of these strategies work well in steady-state operation, but are less effective in transient operation, since changes in temperature tend to happen relatively slowly in comparison with an engine cycle. One method of addressing this shortcoming is to use a spark similar to that of a spark-ignition engine.

HCCI engines tend to operate at lean air/fuel ratios; therefore, the spark process is not enough to ensure reliable ignition. However, the combination of compression heating and the spark process allows for successful autoignition while still maintaining a method for controlling combustion timing. This concept should not be confused with a conventional HCSI engine, however, as the mode of combustion is strikingly different. Most obvious is the absence of a propagating flame, or flame front, as in both the previously discussed combustion modes. Conversely, a comparatively long ignition delay

is followed by almost simultaneous ignition throughout the combustion chamber. This leads to the bulk burning process occurring quickly, usually in the order of 10–20 CAD, as opposed to a normal SI engine (40–50 CAD). This is evident through the comparison of the mfb curves from the three combustion concepts discussed in this chapter (Figure 8.2.6). The initial long, slow rise represents the time in which a small flame kernel, initiated by the spark, gradually grows. As the heat released by the flame and the compression process raises the local temperature, autoignition occurs and leads to the almost vertical rapid-burn phase of the curve. Figure 8.2.8 presents a visual representation of this process. Initially, the spark can be seen near the spark plug electrode followed by the slow development of a flame kernel. In contrast to the HCSI and SCSI concepts, this flame does not grow or move significantly for the several crank angle degrees. In contrast, the bulk phase of combustion happens quickly and is most evident in images at 40 and 45 CAD after the start of ignition. To better illustrate this rapid-combustion phase, a sequence of images is presented in Figure 8.2.9, which focuses only on this phase.

8.2.5 Conclusion

Each of the three spark-ignition engine concepts presents its own unique advantages and disadvantages relating back to performance, emissions, and efficiency. It is therefore logical that the next generation of engines will not be classified as any of the aforementioned concepts, but rather as a multi-mode engine that combines all the three. Stratified charge can be employed at light loads, where pumping losses are generally greatest, but NO_x emissions are relatively low. The HCCI/SACI can be utilized at moderate

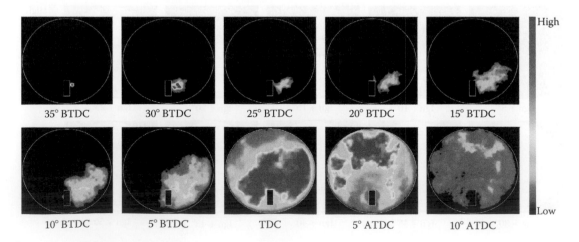

FIGURE 8.2.8
Images of combustion in a spark-assisted HCCI engine. Time is relative to the start of ignition. (Courtesy of Dr. Vinod Natarajan and Dr. Dave Reuss.)

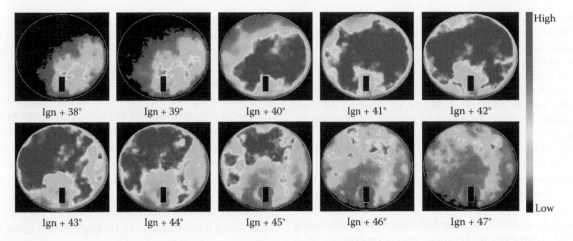

FIGURE 8.2.9
Images of the bulk/rapid-combustion phase in a spark-assisted HCCI engine. Time is relative to the start of ignition. Note that the majority of combustion occurs within 6–7 CAD. (Courtesy of Dr. Vinod Natarajan and Dr. Dave Reuss.)

loading conditions (i.e., freeway cruise conditions) where the ability to reduce NO_x using the three-way catalyst is desirable. Finally, homogeneous conditions can be realized by injecting fuel early in the intake stroke during high-load conditions. This provides the added benefit of improved volumetric efficiency through evaporative charge cooling.

The fruition of the multimode engine is relying on the advancement of engine technology. Better fuel injectors are needed for more repeatable fuel placement, particularly under stratified conditions. New intake- and exhaust-valve actuation schemes are needed to promote better thermal conditions for HCCI/SACI running, in addition to better combustion control. Assuming that technology does progress to this spark-ignition engine efficiency may rival that of a compression-ignition engine.

References

1. Smith, J.D. and V. Sick, A Multi-Variable High-Speed Optical Study of Ignition Instabilities in a Spray-Guided Direct-Injected Spark-Ignition Engine. SAE Paper 2006-01-1264, 2006.
2. Anderson, R.W., The Effect of Ignition System Power on Fast Burn Engine Combustion. SAE Paper 870549, 1987.
3. Anderson, R. and J.R. Asik, Ignitability Experiments in a Fast Burn, Lean Burn Engine. SAE, 830477, 1983.
4. Sher, E., J. Ben-Ya'Ish, and T. Kravchik, On the birth of spark channels. *Combustion and Flame*, 89: 186–194, 1992.
5. Lee, M.J., M. Hall, O.A. Ezekoye, and R. Matthews, Voltage, and Energy Deposition Characteristics of Spark Ignition Systems. SAE, 2005-01-0231, 2005.
6. Alger, T., B. Mangold, D. Mehta, and C. Roberts, The Effect of Sparkplug Design on Initial Flame Kernel Development and Sparkplug Performance. SAE, 2006-01-0224, 2006.
7. Hori, T., M. Shibata, S. Okabe, and K. Hashizume, Super Ignition Spark Plug with Fine Center & Ground Electrodes. SAE, 2003-01-0404, 2003.
8. Ko, Y. and R.W. Anderson, Electrode Heat Transfer During Spark Ignition. SAE, 892083, 1989.
9. Abraham, J., F.A. Williams, and F.V. Bracco, A Discussion of Turbulent Flame Structures in Premixed Charges. SAE, 850345, 1985.
10. Smith, J.R., The influence of turbulence on flame structure in and engine, in *Flows in Internal Combustion Engines*, T. Uzkan, Editor, ASME: New York, 1982, pp. 67–72.
11. Orth, A., V. Sick, J. Wolfrum, R.R. Maly, and M. Zahn, Simultaneous 2D single-shot imaging of OH concentrations and temperature fields in an SI engine simulator. *Proceedings of the Combustion Institute*, 25: 143–150, 1994.
12. Heywood, J.B., *Internal Combustion Engine Fundamentals*. 1st ed., 1988, New York: McGraw-Hill.
13. Zhao, F., D.L. Harrington, and M.C. Lai, *Automotive Gasoline Direct-Injection Engines*, Warrendale, PA: Society of Automotive Engineers, 2002.
14. Zhao, F., *Homogeneous Charge Compression Ignition (HCCI) Engines*, Warrendale, PA: Society of Automotive Engineers, 2003.
15. Zhao, F., M.C. Lai, and D.L. Harrington, Automotive spark-ignited direct injection gasoline engines. *Progress in Energy and Combustion Science*, 25: 437–562, 1999.
16. Fansler, T.D., M.C. Drake, B.D. Stojkovic, and M.E. Rosalik, Local fuel concentration, ignition and combustion in a stratified charge spark ignited direct injection engine: Spectroscopic, imaging and pressure-based measurements. *International Journal of Engine Research*, 4(2): 61–87, 2002.
17. Frieden, D. and V. Sick, Investigation of the Fuel Injection, Mixing and Combustion Processes in an SIDI Engine Using Quasi-3D LIF Imaging. SAE, 2003-01-0068, 2003.

18. Fujikawa, T., Y. Nomura, Y. Hattori, T. Kobayashi, and M. Kanda, Analysis of cycle-by-cycle variation in a direct injection gasoline engine using a laser induced flourescence technique. *International Journal of Engine Research*, 4(2): 143–154, 2003.

19. Kakuho, A., K. Yamaguchi, Y. Hashizume, T. Urushihara, T. Itoh, and E. Tomita, A Study of Air-Fuel Mixture Formation in Direct-Injection SI Engines. SAE, 2004-01-1946, 2004.

20. Smith, J.D., Development and application of high-speed optical diagnostic techniques for conducting scalar measurements in internal combustion engines, PhD dissertation in *Mechanical Engineering*. 2006, University of Michigan: Ann Arbor.

21. Smith, J.D. and V. Sick, Crank-Angle Resolved Imaging of Fuel Distribution, Ignition and Combustion in a Spark-Ignition Direct-Injection Engine. SAE Paper 2005-01-3753, 2005.

22. Smith, J.D. and V. Sick, Real-time imaging of fuel injection, evaporation and ignition in a direct-injected spark-ignition engine. In *ILASS-Americas*, Toronto, ON, 2006.

23. Fajardo, C.M. and V. Sick, Flow field assessment in a fired spray-guided spark-ignition direct-injection engine based on UV particle image velocimetry with sub crank angle resolution. *Proceedings of the Combustion Institute*, 31(2):3023–3031, 2007.

24. Smith, J.D. and V. Sick, Factors Influencing Spark Behavior in a Spray-Guided Direct-Injection Engine. SAE, 2006-01-3376, 2006.

25. Spicher, U., A. Kolmel, H. Kubach, and G. Topfer, Combustion in Spark Ignition Engines with Direct Injection. SAE, 2000-01-0649, 2000.

26. Aleifres, P.G., Y. Hardalupas, A.M.K.P. Taylor, K. Ishii, and Y. Urata, Flame chemiluminescence studies of cyclic combustion variations and air-to-fuel ratio of the reacting mixture in a lean-burn stratified-charge spark-ignition engine. *Combustion and Flame*, 136: 72–90, 2004.

27. Wyszynski, P., R. Aboagye, R. Stone, and G. Kalghatgi, Combustion imaging and analysis in a gasoline direct injection engine. SAE, 2004-01-0045, 2004.

8.3 Combustion in Compression-Ignition Engines

Zoran Filipi and Volker Sick

8.3.1 Introduction

Compression-ignition (CI) engines are often called Diesel engines after their inventor Rudolf Diesel (1858–1913). The invention was a result of Diesel's realization that high compression/expansion ratio is the key to increasing the engine efficiency. The Otto engine was already introduced at the time, but its efficiency was severely limited by the occurrence of knocking combustion, i.e., spontaneous, uncontrollable ignition. Diesel's solution was simple: inject fuel late in the compression process, just before the desired ignition time. A combination of very high compression ratios (CR), typically between 14 and 22, and unthrottled operation (the load is controlled by simply changing the amount of fuel injected), opened up the doors to the development of technology that to this day provides one of the most efficient fuel-conversion devices. Late injection results in very short mixture-preparation time and thus, the basic feature of a CI engine is operation with heterogeneous mixture. This enables operation with extremely lean overall fuel-to-air (F/A) ratio, since local values can be kept well within the flammability limits. However, the consequences are the two persistent problems with the diesel engine emissions: formation of nitric oxides (NO_x) and soot particles.

Mixture preparation and in-cylinder motion have a critical impact on autoignition, combustion, and formation of pollutants in a CI engine. Challenges arise from short timescales and transient nature of the processes, since combustion chamber volume, pressure, temperature, compositional and flow fields, all change rapidly during every cycle. Depending on the method used for preparing the fuel/air mixture, the combustion regimes range from near-perfectly premixed to highly heterogeneous diffusion burning in a multiphase liquid/vapor environment. Fuel/air mixture preparation and in-cylinder flow are invariably intertwined, and certain preparation strategies rely on directed flows for enhanced mixing and suppression of emission formation. Particularly, in-cylinder motion plays an important role in small, high-speed CI engines, where thousands of cycles occur every minute in a very small space. Significant increase in injection pressure up to 2000 bars and multiple-injections have recently gained popularity as a great complement or alternative to organized charge motion. Over a period of time, the direct injection concept has achieved absolute dominance over the divided chamber (prechamber or swirl-chamber) owing to the significant efficiency advantages.

This chapter will review the three main modes of combustion in a CI engine: (1) the *conventional* diesel process, (2) the *high-speed* (HSCI) light-duty engine process characterized by space and time constraints, and (3) the *low-temperature combustion* in a premixed diesel (PCI). The conventional process typically provides enough space for complete development of the diesel spray and does not rely on intense air motion in the cylinder. It currently dominates the heavy-duty truck engines, and is the main stay of large stationary and marine engines. Exhaust gas recirculation (EGR) can be used to reduce combustion temperatures and NO_x, but the conditions in the cylinder are heterogeneous and the soot–NO_x tradeoff limits the potential to reduce engine-out emissions. The cylinder bore of a typical HSCI engine is small (< 90 mm) and the spray impingement on the surface of the piston bowl is inevitable. Timescales are extremely small owing to high speeds, and the engine needs to perform well over a very wide range of operating conditions. Hence, a careful optimization of the piston-bowl shape, injection rates, and organized charge motion (swirl) is a prerequisite for achieving efficient operation and limiting the exhaust emission. The thermal efficiency is not as high as in the case of larger, conventional diesels, but is nevertheless superior to spark-ignition engines. Finally, the PCI engine offers the promise of preserving most of the high-efficiency potential, while avoiding the infamous soot–NO_x tradeoff and providing a "clean diesel" option. The idea is to enhance mixing to the point of achieving near-homogenous conditions, thus avoiding the soot-forming regions. Lean fuel/air mixture and dilution with recirculated residual keep the flame temperature low and allow simultaneous reduction of NO_x. The concept has been successfully demonstrated in both heavy CI engines and HSCI engines, but the speed/load range appears to be limited by the mixing phenomena and inability to avoid knocking combustion at high loads. The three modes of combustion will be discussed in individual sections and each case will be illustrated with visual examples. Lastly, the chapter ends with the summary and outlook.

8.3.2 Conventional Compression-Ignition Engine

Conventional CI or *diesel* engines are the prime-movers of choice for medium and heavy trucks, heavy construction and farming machinery, locomotives, and ships. The main reason for their widespread use is high efficiency of converting fuel energy to mechanical work, ranging from ~45% for smaller to >50% for large engines. Designs have been significantly refined over time, but the essence of the concept remains firmly linked to the original invention. Fresh air enters the cylinder during the intake process and mixes with whatever amount of exhaust residual might be present. The air often enters the cylinder at pressures higher than the ambient pressure owing to turbocharging. After the intake valve (or port) closes, the fresh charge is compressed by the piston to very high pressures and temperatures. The fuel is injected at high velocities through small holes on the injector nozzle just before the piston reaches the top dead center (TDC– the position providing minimum clearance volume). The piston top is shaped in a way that allows development of the spray, fuel atomization, and good mixing with air. The typical quiescent piston bowl is shown in Figure 8.3.1 [1]. Multiple sprays are used to ensure

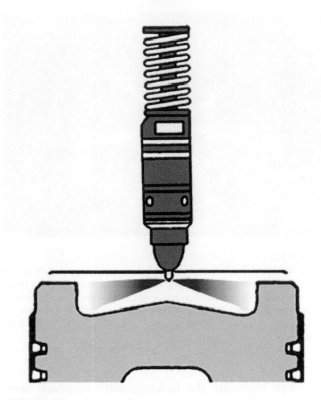

FIGURE 8.3.1
Cross section of the CI engine piston with a typical quiescent-bowl shape and a centrally located high-pressure fuel injector.

good utilization of air in the chamber, as shown in Figure 8.3.2 [2]. The symmetry of sprays depends on the particular nozzle type, and possible asymmetric penetration can be exaggerated at low needle lifts [3]. Fuel evaporates and mixes with air, and owing to very high gas temperatures, autoignites after a delay of only a few crank-angles. Fuel/air mixture prepared during the ignition-delay period burns rapidly and this is referred to as a premixed phase of burning. The injection continues after ignition, and the subsequent stage of the process controlled by mixing rates is called a diffusion phase. Rate of heat release (RHR) given in Figure 8.3.3a illustrates the differences between two stages of combustion. The distribution of the mass of fuel between the two phases changes with engine load, as illustrated in Figure 8.3.3b, showing a sequence of RHR profiles obtained during a step-change of fueling [4]. The premixed burning is much more dominant at low loads (relatively small amount of fuel injected), and diffusion burning is more dominant at high load (large amount of fuel injected). The mixture remains heterogeneous throughout the process; therefore, the overall F/A ratio has to be leaner than stoichiometric even at full load. The local F/A ratios vary significantly, with very rich hot pockets near the liquid core and extremely lean pockets near the periphery and away from the spray axis. Figures 8.3.4a

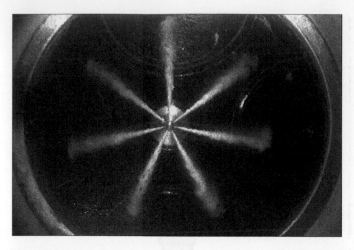

FIGURE 8.3.2
View of the typical conventional CI-engine spray targeting from the bottom. The images of the nonevaporative spray were obtained in an engine with a transparent piston top, using a fast camera and a pulsed laser for illuminating the sprays from the same side as the camera. (From Cronhjort, A. and Wåhlin, F., *Appl. Opt.*, 43(32), 5971, 2004. With permission.)

and 8.3.4b illustrate the heterogeneity typically seen in a CI combustion chamber. The flame image in Figure 8.3.4a is obtained in a production-style heavy duty CI engine with a high-speed camera and endoscope integrated in a Videoscope instrument [5]. The area shown represents a top view of the combustion chamber slice, between the two sprays. The nozzle is located at the top of the image, and the conditions shown correspond to part-load operation. The flame is clearly located close to the sides, on the edges of the

neighboring sprays. Analyzing the flame images using a two-color pyrometry technique [6–8] provides flame temperature maps as shown in Figure 8.3.4b. Processing of the radiation signal from hot particles is the foundation of the two-color pyrometry technique; hence, the existence of flame temperature maps indirectly confirms the presence of soot during the main phase of burning. Yellow and orange zones indicate hot pockets, and they are likely to be locally rich in fuel. The fact that these pockets are close to the edge of the flame emphasizes the emissions challenge for CI engines, since hot zones close to oxygen-rich regions stimulate formation of NO_x. In summary, the mixture preparation and combustion processes are extremely complex, and until recently, were not fully understood.

The advances of engine spray and combustion imaging techniques in the 1980s and 1990s provided a qualitatively new insight into the fundamentals of the diesel engine process, and indicated that analogies with steady flames in furnaces and gas turbines do not apply to the highly transient processes in a CI engine. A major breakthrough in understanding the diesel combustion phenomena came with the analysis performed by Dec [9] in 1997. He synthesized the results of combustion visualization at Sandia National Laboratories and elsewhere into a coherent conceptual model of diesel engine combustion. The idea about the temporal evolution of the fuel jet and a sequence of events occurring in a fully developed, reacting jet replaced the earlier notions of a diffusion flame that supposedly occurs around evaporating droplets or a pure-fuel spray core. The phenomenological model has been widely accepted and used

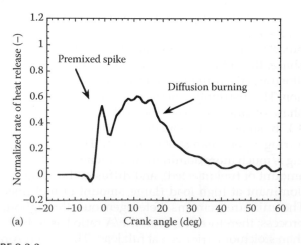

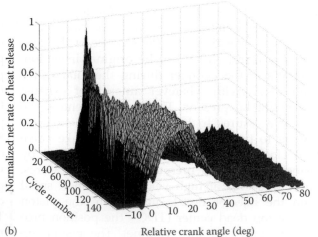

FIGURE 8.3.3
Rate of heat release obtained in a conventional CI engine: (a) the typical profile demonstrating a premixed spike followed by a diffusion burning phase, (b) sequence of rate of heat-release profiles obtained during a fueling change, from low to high. Lower loads display relatively more premixed burning (back), while at high loads diffusion part becomes more dominant (front).

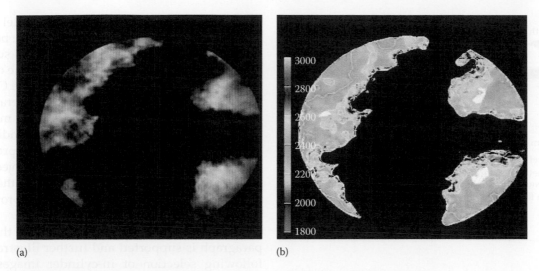

FIGURE 8.3.4

Images from a section of the heavy-duty CI engine's combustion chamber, capturing edges of two neighboring reacting sprays and a lean, cool zone in between (a) raw flame images and (b) flame temperature maps extracted from raw images using a two-color pyrometry technique. Note the heterogeneous nature of the process in the CI engine's cylinder. Obtained at 1200 rpm—30% load.

as guidance in the experimental and modeling studies aimed at improving diesel combustion and emissions, and hence, provides an excellent starting point for more detailed discussion of CI engine combustion.

Detailed engine-visualization studies require optical access to the combustion chamber, and one of the most advanced laboratory single-cylinder engine designed specifically for this purpose is shown in Figure 8.3.5a. Optical access is enabled through windows in the cylinder head and around the top of the cylinder liner, as well as through the quartz-glass piston top. A mirror positioned at a 45° angle in the lower part

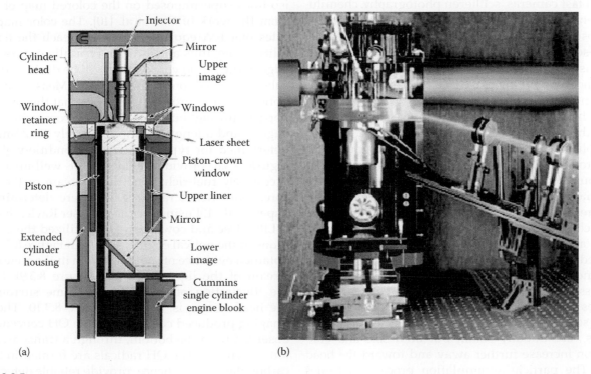

FIGURE 8.3.5

The single-cylinder engine with optical access: (a) cross section showing quartz windows and the extended piston construction [8] and (b) view of the engine during testing. (Courtesy of Dec, J.)

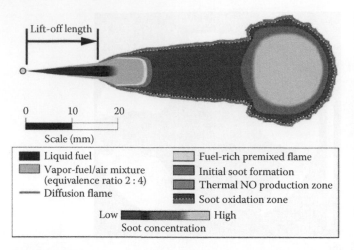

FIGURE 8.3.6

Conceptual model of conventional CI combustion characterized by a sequence of processes occurring in a fully developed reacting jet. (From Dec, J., A Conceptual Model of DI Diesel Combustion Based on Laser Sheet Imaging, SAE, 970873, 1997. With permission.)

of the extended piston allows one to collect the bottom-view images, as shown in a photograph of the operational setup in Figure 8.3.5b. Imaging in high-pressure vessels and rapid-compression machines can be a useful complement, as long as test parameters are representative of in-cylinder conditions. A range of visualization techniques used for the analysis of the CI engine includes, but is not limited to, direct imaging with fast cameras, schlieren photography, chemiluminescence imaging, laser-induced fluorescence (LIF), laser absorption scattering (LAS), and laser-induced incandescence (LII). Imaging in a plane, using a laser sheet has been particularly useful for understanding the phenomena taking place in a reacting diesel jet, as shown in Figure 8.3.6.

The schematic in Figure 8.3.6 captures the processes during the mixing-controlled burn following the premixed phase, as proposed by Dec [9]. The liquid core (dark brown in Figure 8.3.6) has been shown to persist throughout the injection process. The turbulent air entrainment facilitates the evaporation of fuel droplets downstream of the liquid core. A zone of relatively uniform, rich mixture with F/A equivalence ratio of 2–4, extends ahead and around the liquid core. A standing premixed flame (light-blue in Figure 8.3.6) forms at the boundary of the gaseous fuel/air zone, and owing to excessively rich conditions, the premixed flame produces polycyclic aromatic hydrocarbons (PAH—a known soot precursor) and solid particles. The soot particles are initially small, but both size and concentration increase further away and toward the head vortex. The particle accumulation process continues in the head-vortex zone surrounded by a thin diffusion flame. Consequently, the diffusion flame differs

from a traditional fuel/air burning model in a sense that it actually represents a reaction zone between the products of fuel-rich premixed flame and surrounding oxygen-rich charge. Particles that reach the outer edges of the diffusion flame are oxidized by the OH radicals and possibly by oxygen. The high temperature of the diffusion flame and proximity of oxygen molecules in the surrounding fresh charge create conditions very conducive for the production of NO_x. The production of NO_x will continue even after the end of injection, since temperatures remain high enough during the latter part of diffusion burning, and further mixing provides more oxygen for the reactions.

The conceptual model summarized in the previous paragraph is supported and further illustrated by the following selection of in-cylinder images obtained using advanced visualization techniques. Details of the events occurring prior to the full development of the reacting jet, such as initial spray development, autoignition, and premixed burning are discussed only briefly owing to space limitations. Figure 8.3.7 shows the images of the developing spray obtained in the high-pressure vessel under conditions typically seen in the CI engine cylinder. The visible-wavelength image determines the droplet optical thickness, while the UV image provides the joint optical thickness of the vapor and droplets. LAS analysis uses the two images to produce liquid and vapor concentrations, as shown in Figure 8.3.7c, with the F/A equivalence ratio iso-lines superimposed on the colored map obtained from the work of Gao et al. [10]. The color map indicates that F/A equivalence values reach the flammability range, thus autoignition typically occurs on the edge of the liquid core, which will be discussed in the subsequent section of this chapter. Measurements in optical engines provided evidence that the liquid core persists throughout the injection process in a firing engine, and a length of approximately 15–20 mm was determined for typical CI-engine conditions [9,11,12]. Figure 8.3.8 provides evidence of well-mixed, but excessively fuel-rich, zone downstream of the liquid core. Equivalence ratios of 3–4 were determined by Espey et al. [13] using the planar laser Rayleigh scatter (PLRS). Dec and coworkers [9,14] utilized the LII technique in the optical engine, shown in Figure 8.3.5a and obtained evidence of soot formation in the zone downstream of the liquid core (see Figure 8.3.9). Finally, the illustration of the diffusion flame surrounding the head vortex is given in Figure 8.3.10. The PLIF imaging produced contours of high OH concentration observed from the bottom, through a transparent piston crown [15]. The OH radicals are formed in hydrocarbon flames and, hence, provide reliable detection of combustion. The red color inside indicates high soot concentrations. Figure 8.3.11 provides a visual way of

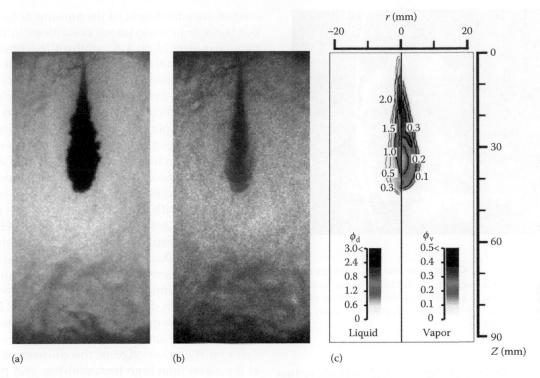

(a)   (b)   (c)

FIGURE 8.3.7
Diesel spray visualization in a high-pressure vessel: (a) raw image at UV wavelength, (b) raw image at visible wavelength, and (c) LAS-processed images indicating liquid and vapor concentrations along the spray axis. Ambient gas was nitrogen, pressure 4 MPa, and temperature 760 K. Nozzle-hole diameter of 0.125 mm, injection pressure of 90 MPa. The images were recorded at 0.6 ms after start of injection, and the total duration was 0.85 ms. (Courtesy of Nishida, K., University of Hiroshima, Higashi-Hiroshima, Japan.)

summarizing this discussion, with a sequence of two images illustrating sprays right after ignition (Figure 8.3.11a), and fully developed reacting sprays with high soot concentrations in the head vortex (Figure 8.3.11b). The images were obtained by Wang et al. [16] using

a fast camera in an optical engine with a transparent piston crown.

The high rates of NO_x formation in hot zones surrounding the head vortex can be offset by introducing a diluent in the form of recirculated exhaust gas.

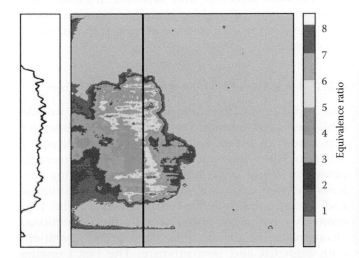

FIGURE 8.3.8
Vapor-phase fuel-distribution image converted to an equivalence-ratio field downstream of the maximum liquid-phase fuel penetration. Quantitative planar images are obtained in the optical engine using PLRS. (From Espey, C., Dec, J.E., Litzinger, T.A., and Santavicca, D.A., *Combust. Flame*, 109, 65, 1997.)

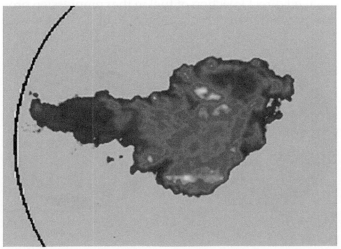

FIGURE 8.3.9
LII soot images indicating soot formation in the reacting jet and increased concentration downstream from the standing premixed reaction zone. (From Dec, J.E., *SAE Trans.*, 106, 1319, 1997. With permission.)

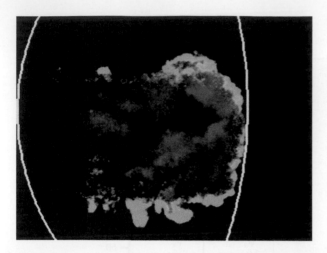

FIGURE 8.3.10
Combined PLIF images of OH (green) and PLII images of soot (red). The OH is an indication of the diffusion flame around the soot-rich zone. The flame is approaching the combustion chamber wall on the right. (From Dec, J.E. and Tree, D.R., *SAE Trans.*, 110(3), 1599, 2001. With permission.)

The flame maps shown in Figure 8.3.12 indicate reduction of peak temperatures with the increase in the EGR content, and this translates into tangible decrease in engine-out NO_x. While using EGR is a very effective way of reducing NO_x, its percentage and thus magnitude of NO_x reduction in a practical application is limited by the increased soot formation.

8.3.2.1 Reprise

The conventional CI-engine operates with a relatively high compression ratio and in an unthrottled manner. Delayed direct injection of fuel in the compression process prevents knocking combustion and enables load

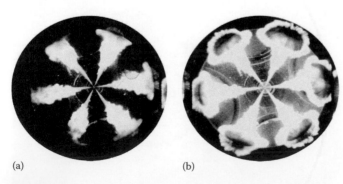

(a)   (b)

FIGURE 8.3.11
Images of the reacting jet in an optical CI engine: (a) evolving jet just after ignition and (b) fully developed reacting jet with dark zones indicating high-soot concentrations in the head vortex. (From Wang, T.-C., Han, J.S., Xie, S., Lai, M.-C., Henein, N., Schwartz, E., and Bryzik, W., Direct Visualization of High-Pressure Diesel Spray and Engine Combustion, SAE, 1999-01-3496, 1999. With permission.)

control via adjustment of the amount of fuel per cycle, but leads to heterogeneous conditions in the combustion chamber. High CR, unthrottled operation, and overall lean combustion enable very high brake thermal efficiency, above 45% for automotive engines and above 50% for larger stationary and marine engines. The mixture formation and combustion process is extremely complex and transient in nature, owing to reciprocating engine operation. Advances in engine visualization techniques have shown that analogies to burning in furnaces and gas turbines do not hold true in case of a diesel engine, and led to an understanding captured with a conceptual model of a reacting jet. The main features of the jet are a relatively short liquid core, fuel vapor-rich premixed zone ahead of it, a standing premixed flame creating soot particles, a hot zone downstream and within the head vortex characterized by accumulation and growth of soot particles, and a diffusion flame surrounding the vortex. The inevitable consequence of these conditions are soot and NO_x emissions, since only a portion of soot oxidizes on the outer edge of the diffusion flame, while at the same time high temperatures and proximity of oxygen molecules facilitate formation of NO_x. The NO_x emissions can be reduced by providing charge dilution with recirculated residual gas. The future of the CI engines, particularly heavy-duty, is bright because of their inherent ability to efficiently convert fuel energy to mechanical work, as well as their high power density. The emission problems can be partially mitigated with in-cylinder measures, but aftertreatment devices will be required for meeting future extremely stringent regulations. The in-cylinder measures include high-pressure injection and multiple injections for better mixing, EGR and/or premixed burning strategies. The latter will be covered in a separate section in this chapter.

8.3.3 High-Speed Compression-Ignition Engines

The high-speed, direct injection CI engines provide a high-efficiency option for propulsion of passenger cars. The high-speed, up to 5000 rpm, is required for achieving the target-specific power, i.e., a favorable power to weight ratio. Advanced turbocharging systems are used to increase the density of fresh charge and further improve the output power. The fuel economy advantage compared with the conventional SI engine is the result of the unthrottled operation, with high CR and lean mixture. The HSCI engine demonstrates the biggest advantage at part load, and this is where a typical passenger-car engine spends most of the time under realistic driving conditions. The biggest challenges are emissions, since passenger-car regulations set very stringent limits in most

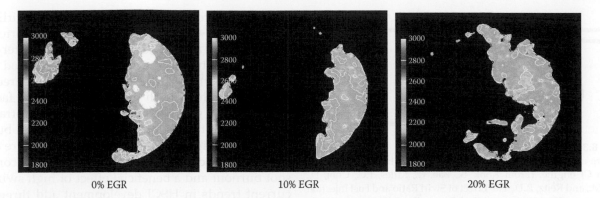

FIGURE 8.3.12
Variations of flame temperatures with exhaust-gas recirculation. Flame temperature maps are obtained by processing images obtained from a combustion chamber of a heavy-duty diesel engine using two-color pyrometry. Images are taken at 2° after the TDC at 1200 rpm low-load condition.

countries. In addition, the HSCI engine needs to perform favorably over a much wider range of speeds than a heavy-duty engine. The sturdier structure of the HSCI engine compared with the typical SI, as well as a very sophisticated and expensive high-pressure injection system, lead to increased cost. The fuel economy benefits outweigh the challenges in the markets with high fuel prices; hence, HSCI engines have already captured more than 50% of the European passenger car market.

The process in the HSCI engine differs significantly from what was described in the previous section in one important way. The dimensions of the cylinder are much smaller (Bore of 75–90 mm); therefore, the spray impingement on the combustion chamber wall is inevitable [17,18]. Figure 8.3.13 shows a sequence of spray and combustion images obtained in the rapid compression machine (RCM) using a fast camera [18]. The RCM combustion chamber and test parameters are designed to provide conditions representative of engine conditions. Observing from left to right, the first image shows initial spray penetration, and the second illustrates the ignition occurring on the side of the liquid core. Experimentation in engines with high-swirl indicates that the ignition typically occurs in the recirculation zones on the lee side of the spray [17,19]. In the third image, the flames completely engulf the sprays even though the amount of energy release is still small. The last image illustrates a typical impinging reactive spray, with darker sooty regions near the wall owing to limited air entrainment. Careful combustion chamber design and intensified charge motion are needed to counteract the adverse effects. The re-entrant bowl in the piston (see Figure 8.3.14) is designed in a way that directs the impinging spray back toward the center, to better utilize the available air [20]. In addition, the intense swirl is generated to speed up mixing and enhance evaporation of the fuel film on the wall. Near TDC, the squish flow is combined with swirl, thus creating a very complex flow field [17]. Figure 8.3.15 provides the highlights of the effect of swirl intensity on main phases of HSCI combustion. Every horizontal section consists of

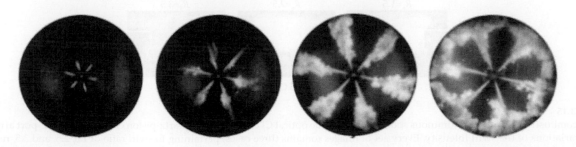

FIGURE 8.3.13
Images of spray and combustion in the rapid compression machine obtained for conditions representative of typical HSCI-engine operation. The sequence of four images covers the period immediately after injection—far left, and until the full development of a reacting jet—far right. (From Lu, P.-H., Han, J.-S., Lai, M.-C., Henein, N., and Bryzik, W., Combustion Visualization of DI Diesel Spray Combustion inside a Small-Bore Cylinder under Different EGR and Swirl Ratios, SAE, 2001-01-2005, 2001. With permission.)

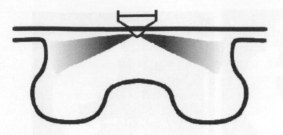

FIGURE 8.3.14
Typical re-entrant piston-bowl design for a small, high-speed direct-injection CI engine. (From Kook, S., Bae, C., Miles, P.C., Choi, D., Bergin, M., and Reitz, R.D., The Effect of Swirl Ratio and Fuel Injection Parameters on CO emission and Fuel Conversion Efficiency for High-Dilution, Low-Temperature Combustion in an Automotive Diesel Engine, SAE, 2006-01-0197, 2006. With permission.)

three sets of combustion lumino-sity images, obtained by Miles [19] in an optical single-cylinder engine for different levels of swirl ratios, i.e., 1.5, 2.5, and 3.5. The piston-bowl outline is shown on both the bottom- and the side-view images. The shape is adjusted to compensate for spatial distortions. Figure 8.3.15a illustrates the early part of premixed burning, with flames being

swept away from the spray axis owing to swirling-air motion. High swirl enhances mixing and shortens the ignition delay, as indicated by the much brighter image on the far right. During the mixing-controlled phase (Figure 8.3.15b), the differences between the three cases diminish, perhaps with the exception of the fact that hot luminous gases seem to be more concentrated in the center for the high-swirl case. In the final burnout phase, the images on the far right of the Figure 8.3.16c show increased patchiness, which is a sign of complete soot burnout and a beneficial effect of high swirl. The current trends in HSCI development add three more measures to the arsenal of tools for improving combustion and reducing emissions, namely the ultra-high pressure fuel injection, multiple injections, and exhaust gas recirculation.

8.3.4 Premixed CI Engine: An Ultra-Low Emission Concept

The conventional CI combustion mode with highly stratified in-cylinder conditions creates a perennial soot–NO_x trade-off that severely limits the

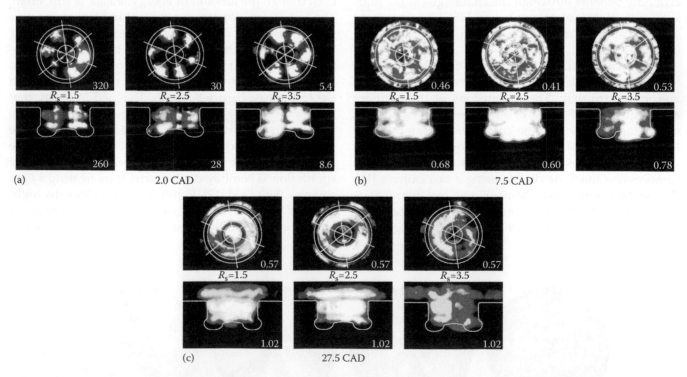

FIGURE 8.3.15
Images of combustion luminosity (luminous soot) obtained in the optical CI engine with a quartz-piston crown and intake-port arrangement, allowing variations of the swirl intensity. Every set of images contains three cases, pertaining to swirl ratio of 1.5, 2.5, and 3.5, respectively, from left to right. The images illustrate (a) early part of premixed burning at 4° CA after TDC, 8° CA after the start of injection, (b) early mixing controlled burn—7.5° CA after TDC, and (c) final burnout—27.5° CA after TDC. (From Miles, P., The Influence of Swirl on HSDI Diesel Combustion at Moderate Speed and Load, SAE, 2000-01-1829, 2000. With permission.)

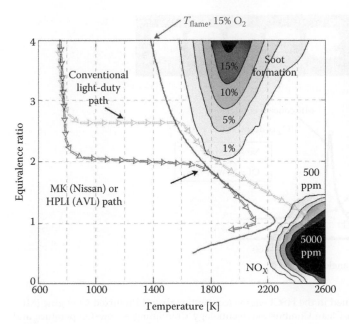

FIGURE 8.3.16
An equivalence-temperature plot for diesel-engine combustion indicating soot and NO_x forming regions. The blue line indicates typical progress of mixing-combustion in conventional engine, while the red line illustrates one way of achieving low-emission premixed combustion. (From Miles, P.C., *Proceedings of the THIESEL 2006 Conference: Thermo- and Fluid-Dynamic Processes in Diesel Engines*, Valencia, Spain, 12–15 September, 2006.)

take over and initiate the bulk burning in the cylinder. Mixing can be enhanced by prolonged ignition delay owing to very retarded injection (MK system), or using split injection with a very early first squirt (UNIBUS system). Massive amounts of cooled EGR are essential for achieving the desired equivalence vs. temperature trajectory for MK system, described by Miles [27] and shown in Figure 8.3.16. Direct comparison of the conventional HSCI engine process and the PCI–MK process in Figure 8.3.17 illustrates the major points [28]. The conventional engine displays a typical rate of heat-release profile with a premixed spike and the diffusion phase, and combustion images indicate very stratified conditions throughout the cycle. In contrast, the injection in the MK-type engine is late and ignition delay is extended owing to the relatively low gas-temperature and dilution. At the time of autoignition, the charge is well mixed and MK combustion images are very plain–there are no structured flame fronts, as combustion seems to be driven entirely by chemical kinetics. The NO_x and soot emissions are reduced to almost negligible amounts. However, the low-temperature conditions lead to incomplete combustion and increased amounts of unburnt hydrocarbons (HC) and carbon monoxide (CO) in the exhaust. The HC and CO are much easier to remove using aftertreatment than the NO_x and soot and hence, the PCI concept is attractive.

8.3.5 Conclusion

The CI internal combustion engine converts fuel energy to mechanical work with very high efficiency, thanks to its ability to operate with a high compression ratio and in an unthrottled manner. The power density is high and in most cases enhanced through the use of turbocharging. The load is adjusted by directly changing the amount of fuel injected in the cylinder. Direct injection just before ignition and small timescales for mixture formation lead to heterogeneous conditions and a very complex combustion process. Local composition/temperature conditions cause formation of soot and NO_x—a major challenge of CI engines, particularly for automotive applications. Advances of engine visualization techniques in the recent years provided better understanding of the key phenomena and enabled development of the conceptual model of diesel combustion. The pace of research accelerated beyond anything seen previously and led to the impressive achievements of emission reduction technologies.

possibilities for cleaning up the diesel exhaust. A fundamental understanding about the underlying phenomena was provided by Kamimoto and Bae [21] in the form of the F/A equivalence vs. flame temperature plot, shown in Figure 8.3.16. Locally high F/A ratios cause soot formation, while high-temperature combustion at near-stoichiometric conditions increases NO_x production. The essence of the low-temperature combustion concept is the desire to avoid these regions and realize the idea of the homogenous charge compression-ignition (HCCI) engine using diesel-engine hardware. Providing enough time for mixing prior to ignition should reduce the local F/A ratio below the threshold critical for soot formation. Keeping the mixture lean and using dilution to reduce availability of oxygen will reduce the flame temperature and avoid NO_x formation. The practical implementation is often called a premixed compression-ignition engine (PCI), and some of the best-known examples are already described ([22]—Nissan MK; [23]—Toyota UNIBUS; [24]—AVL HCLI; [25,26]). Regardless of the actual implementation, the idea is to extend the physical delay sufficiently and allow chemical kinetics to

The future of CI engines will be significantly influenced by the efforts to develop clean diesel concepts. In the context of heavy-duty diesel engines, increasing

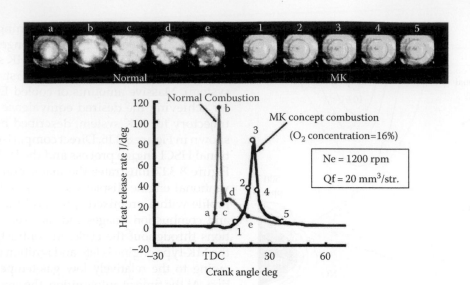

FIGURE 8.3.17
Comparisons of heat-release rates and combustion photographs obtained in the HSCI engine (conventional) and Premixed CI engine (MK). (From Kimura, S., Aoki, O., Kitahara, Y., and Aiyoshizawa, E., Ultra-Clean Combustion Technology Combining a Low-Temperature and Premixed Combustion Concept for Meeting Future Emissions Standards, SAE, 2001-01-0200, 2001. With permission.)

the fuel injection pressure for better mixing, and addition of EGR already demonstrated significant benefits. Organized air motion in the cylinder, such as swirl, can be added too. Swirl has proved to be very critical for small, high-speed engines (HSCI), owing to more severe space and timescale constraints. However, a novel mode of combustion brings a truly fundamental breakthrough and provides an ultra-low emission option for both the CI and HSCI engines. The PCI engine is a realization of the low-temperature combustion idea: enhanced mixing and delayed ignition homogenizes the charge in the cylinder to the level that allows avoiding compositional/temperature regions responsible for the formation of NO_x and soot. Careful optimization can sustain the LTC operation over a wide range of conditions, but there is typically an upper load limit imposed by the ringing combustion.

Advances of air, EGR, and fuel injection systems will support continuous improvements and development of clean and efficient concepts. It is quite likely that future CI engine will operate with a mix of modes, clean PCI in the large part of the range, and advanced conventional under extreme conditions. The aftertreatment devices will be necessary for achieving near-zero tailpipe emissions of pollutants, but their size and cost will decrease with further advances of combustion strategies. In summary, the high efficiency and low CO_2-emission potential, combined with advanced measures for mitigating emissions of pollutants, will continue to make the CI engine a fuel-converter of choice for many automotive, industrial, and marine applications.

References

1. Merrion, D. F., Diesel Engine Design for the 1990s, SAE special publication SP-1011, SAE, Warrendale, 1994.
2. Cronhjort, A., Wåhlin, F., Segmentation Algorithm for Diesel Spray Image Analysis, *Applied Optics*, 43(32), 5971–5980, 2004.
3. Han, J. -S., Wang, T. C., Xie, X. B., Lai, M. -C., Henein, N., Harrington, D. L., Pinson, J., and Miles, P., Dynamics of Multiple-Injection Fuel Sprays in a Small-bore HSDI Diesel Engine, SAE, 2000-01-1256, 2000.
4. Assanis, D. N., Filipi, Z. S., Fiveland, S. B., and Syrimis, M., A methodology for cycle-by-cycle transient heat release analysis in a turbocharged direct injection diesel engine, SAE, 2000, 2000-01-1185—*SAE Transactions, Journal of Engines*, 109(3), 1327–1339.
5. Jacobs, T., Filipi. Z., and Assanis, D., The Impact of Exhaust Gas Recirculation on Performance and Emissions of a Heavy-Duty Diesel Engine, SAE, 2003-01-1068, 2003.
6. Schmidradler D. and Werlberger P., Engine videoscope thermovision: Vision based temperature measurement for diesel engines, Unigraphics solutions: Integrated solutions for complex processes in automotive product development, 32nd ISATA Conference, ISATA, Vienna, Austria, 1999-25-0212, 1999.

7. Matsui, Y., Kamimoto, T., and Matsuoka, S., A Study on the Time and Space Resolved Measurement of Flame Temperature and Soot Concentration in a D.I. Diesel Engine by the Two-Color Method, SAE, 790491, 1974.

8. Ladommatos, N. and Zhao, H., A Guide to Measurement of Flame Temperature and Soot Concentration in Diesel Engines Using Two-Color Method—Part 1: Principles, SAE, 941956, 1994.

9. Dec, J. E., A Conceptual Model of DI Diesel Combustion Based on Laser-Sheet Imaging, *SAE Transactions*, 106, Sec. 3, 1319–1348, 970873, 1997.

10. Gao, J., Matsumoto, Y., and Nishida, K., Effect of Injection Pressure and Nozzle Hole Diameter on Mixture Properties of D.I. Diesel Spray, JSAE, 20065442, 2000.

11. Espey, C. and Dec, J. E., The Effect of TDC Temperature and Density on the Liquid Phase Fuel Penetration in a DI Diesel Engine, *SAE Transactions*, 104(4), 1400–1414, 1994, SAE, 952456.

12. Siebers, D. and Higgins, B., Flame Lift-Off on Direct Injection Diesel Sprays under Quiescent Conditions, SAE, 2001-01-0530, 2001.

13. Espey, C., Dec, J. E., Litzinger, T. A., and Santavicca, D. A., Planar Laser Rayleigh Scattering for Quantitative Vapor-Fuel Imaging in a Diesel Jet, *Combustion and Flame*, 109: 65–86, 1997.

14. Dec, J. and Kelly-Zion, P., The Effect if Injection Timing and Diluent Addition on Late-Combustion Soot Burnout in a DI Diesel Engine Based on Simultaneous 2-D Imaging of OH and Soot, SAE, 2000-01-0238, 2000.

15. Dec, J. E., and Tree, D. R., Diffusion-Flame/Wall Interactions in a Heavy-Duty DI Diesel Engine, *SAE Transactions*, 110, Sec. 3, 1599–1617, 2001-01-1295, 2001.

16. Wang, T. -C., Han, J. S., Xie, S., Lai, M. -C., Henein, N., Schwartz, E., and Bryzik, W., Direct Visualization of High-Pressure Diesel Spray and Engine Combustion, SAE, 1999-01-3496, 1999.

17. Hentschel, W., Schindler, K. -P., and Haahtele, O., European Diesel Research Idea—Experimental Results from DI Diesel Investigations, SAE, 941954, 1994.

18. Lu, P. -H., Han, J. -S., Lai, M. -C., Henein, N., and Bryzik, W., Combustion Visualization of DI Diesel Spray Combustion inside a Small-Bore Cylinder under Different EGR and Swirl Ratios, SAE, 2001-01-2005, 2001.

19. Miles, P., The Influence of Swirl on HSDI Diesel Combustion at Moderate Speed and Load, SAE, 2000-01-1829, 2000.

20. Kook, S., Bae, C., Miles, P. C., Choi, D., Bergin, M, and Reitz, R. D., The Effect of Swirl Ratio and Fuel Injection Parameters on CO emission and Fuel Conversion Efficiency for High-Dilution, Low-Temperature Combustion in an Automotive Diesel Engine, SAE, 2006-01-0197, 2006.

21. Kamimoto, T. and Bae, M., High Combustion Temperature for the Reduction of Particulate in Diesel Engines, SAE, 880423, 1988.

22. Kimura, S., Aoki, O., Ogawa, H., Muranaka, S., and Enomoto, Y., New Combustion Concept for Ultra-Clean High-Efficiency Small DI Diesel Engines, SAE, 1999-01-3681, 1999.

23. Hasegawa, R. and Yanagihara, H., HCCI Combustion in a DI Diesel Engine, SAE, 2003-01-0745, 2003.

24. Weißbäck, M., Csató, J., Glensvig, M., Sams, T., and Herzog, P., Alternative brennverfahren—ein ansatz für den zukünftigen pkw-dieselmotor, *Motortechnische Zeitschrift*, 64: 718–727, 2003.

25. Gatellier, B., Ranini, A., and Castagné, M., New developments of the nadi concept to improve operating range, exhaust emissions and noise, *Oil & Gas Science Technology—Revuede IFP*, 61(1), 7–23, 2006.

26. Jacobs, T. J., Bohac, S. V., Assanis, D. N., and Szymkowicz, Lean and Rich Premixed Compression Ignition Combustion in a Light-Duty Diesel Engine, SAE, 2005-01-0166, 2005.

27. Miles, P. C., In-cylinder Flow and Mixing Processes in Low-Temperature Diesel Combustion Systems, *Proceedings of the THIESEL 2006 Conference: Thermo- and Fluid-Dynamic Processes in Diesel Engines*, Valencia, Spain, September 12–15, 2006.

28. Kimura, S., Aoki, O., Kitahara, Y., and Aiyoshizawa, E., Ultra-Clean Combustion Technology Combining a Low-Temperature and Premixed Combustion Concept for Meeting Future Emissions Standards, SAE, 2001-01-0200, 2001.

8.4 Deflagration to Detonation Transition

Andrzej Teodorczyk

8.4.1 Introduction

A detonation wave in a gaseous combustible mixture can be initiated directly if sufficient energy is released in a small volume. A strong shock wave of some required duration must be generated to initiate the chemical reactions, which rapidly couple with the wave to form detonation front. If energy lower than critical is used for ignition, the shock wave will progressively decouple from the chemical reaction front. Such deflagration wave may under some favorable conditions accelerate and undergo transition to a detonation wave. This mode of detonation initiation in tubes or channels is referred to as deflagration to detonation transition (DDT).

The DDT can be observed in a variety of situations, including flame propagation in smooth tubes or channels, flame acceleration caused by repeated obstacles, and jet ignition. The processes leading to detonation can be classified into two categories:

- Detonation initiation resulting from shock reflection or shock focusing

- Transition to detonation caused by instabilities near the flame front, the flame interactions with a shock wave, another flame or a wall, or the explosion of a previously quenched pocket of combustible gas

The first category is essentially a direct initiation process where the shock strength is sufficient to autoignite the gas with the reaction front rapidly coupled with the shock front forming detonation wave. For accidental explosions, where the shock is produced by an accelerating flame, this process becomes much more probable when the shock interacts with a corner or a concave wall that can produce shock focusing. Shock initiation is an important mechanism in maintaining the propagation of quasi-detonations in a channel or a tube filled with obstacles. It has also been observed to promote detonation for relatively slow flames propagating toward an orifice, a corner, or a concave wall.

The second category of DDT processes, which occur in smooth tubes, is considerably more complex, because it involves gas-dynamic coupling of turbulent flow with chemical reactions and a variety of instability processes.

It has been first suggested theoretically by Zel'dovich et al. [1,2] and then experimentally observed by Lee et al. [3] that reactivity gradients (chemical induction-time gradients), associated with temperature and concentration nonuniformities in the combustible mixture, may cause flame acceleration and DDT. The formation of an induction-time gradient can produce a spatial time sequence of energy release. This sequence can then produce a compression wave that is gradually amplified into a strong shock wave that can autoignite the mixture and produce DDT. This mechanism was named as SWACER (shock wave amplification by coherent energy release) by Lee at al. [3]. The SWACER mechanism may also lead to DDT in a flame jet, because of a flame–vortex interaction that promotes a suitable temperature and concentration gradient.

Few review papers have been published in the past, in which the different fundamental aspects of the DDT problem are discussed in detail [4–7].

8.4.2 DDT in Smooth Tubes

8.4.2.1 Introduction

Depending on the fuel concentration, initial and geometrical conditions, flame propagation in smooth tube leading to DDT progresses through a series of regimes, as schematically shown in Figures 8.4.1 and 8.4.2.

The DDT process can be divided into four phases [5]:

- Deflagration initiation. A relatively weak energy source, such as an electric spark, ignites the mixture and a laminar flame is first formed. The mechanism of laminar flame propagation is via molecular transport of energy and free radicals from the reaction zone to the unburnt mixture ahead of it.

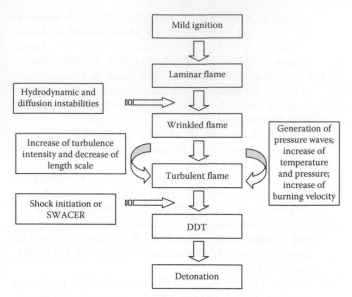

FIGURE 8.4.1
Regimes of flame propagation leading to DDT.

- Flame acceleration. The laminar flame expands and generates the unsteady flow upstream. This flow becomes turbulent owing to the interaction with tube wall and causes wrinkling and acceleration of the flame. At initial stages, the flow and flame acceleration is caused by thermal expansion of hot combustion products. The accelerating flame generates acoustic waves, which coalesce in pressure waves and then shock waves. These waves interact with turbulent vortices in the flow ahead of the flame and further increase turbulence intensity. They also increase the mixture temperature and pressure, thus increasing its chemical reaction rate. Pressure waves also create further vorticity via multiple reflections from the walls and between themselves. These pressure waves have a feedback to the flame front, because the burning rate depends on temperature, pressure, and turbulence intensity of the unburnt mixture. At later stages, various processes, such as flame–vortex interactions, shock–flame interactions, and microexplosions of vortices, as well as hydrodynamic instability mechanisms, such as Rayleigh–Taylor (RT), Richtmyer–Meshkov (RM), and Kelvin–Helmholtz (KH) become responsible for the increase in the flame surface area, energy release rate and flame speed, and, eventually, the strength of leading shock wave. Finally, a feedback mechanism is established in which an increase in the chemical reaction rate results in a greater effective flame-propagation velocity and velocity of unburnt mixture ahead

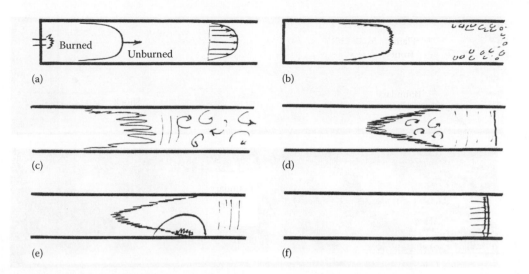

FIGURE 8.4.2
Progress of a DDT event in a smooth tube with a closed ignition end: (a) the initial configuration showing a smooth laminar flame front and the laminar flow ahead; (b) first wrinkling of the flame and vortices in the boundary layer generated by the upstream flow; (c) breakdown into turbulent flow and a corrugated flame; (d) production of pressure waves ahead of the turbulent tulip flame; (e) local explosion within the flame; and (f) transition to detonation. (From Shepherd, J.E. and Lee, J.H., *Major Research Topics in Combustion*, Springer, New York, p. 439, 1992. With permission)

of the flame. Larger velocity in the upstream gas increases the turbulence intensity and vorticity, which in turn increases the reaction rate. This feedback process accelerates the flame to high velocities, up to approximately 1000 m/s. Figure 8.4.3 shows the shadow photographs of the initial stages of flame propagation in smooth channel with rectangular cross-section of 50 mm^2 × 50 mm^2 in stoichiometric hydrogen/oxygen mixture at initial pressure of 0.075 MPa [8].

- Formation of explosion centers. A local explosion center is formed as a pocket of combustible mixture within the flame brush or ahead of it, but behind the leading shock wave. This pocket reaches the critical ignition conditions and explodes (explosion within the explosion as first named by Oppenheim [9]). The laser-light schlieren photographs of Urtiew and Oppenheim [10] revealed details of DDT resulting from the volume explosion at random location between the shock wave and the flame, depending on local temperature and concentration.

- Formation of a detonation wave. The explosion of local pocket creates a strong shock wave, which rapidly merges with the reaction front into a supersonic detonation front, which is self-sustaining. Figure 8.4.4 presents a sequence of photographs obtained with a stoichiometric hydrogen/oxygen mixture initially at 0.073 MPa, demonstrating the transition with an explosion in the vicinity of the flame front at the upper wall. Its kernel is distinct in the frame corresponding

to 55 μs. As a result of this explosion, a spherical wave is formed, which propagates transversely across the channel, burning all the mixtures between the highly turbulent flame front and the precursor shock. Subsequently, the spherical front of the "explosion in the explosion" penetrates through the shock wave, producing a self-sustained detonation wave.

Figure 8.4.5 presents the streak, direct photograph illustrating the stages of transition to detonation after a weak ignition and flame acceleration phase. Four main regions may be identified:

- Initial shock-flame complex. A leading shock and turbulent flames (4 and 5) propagate together. Energy release leads to the acceleration of the leading shock and the flame.
- Local explosion leading to transition to detonation (8).
- An overdriven detonation (6) after transition and retonation wave (9).
- Steady-state detonation (7) after decay of overdriven detonation wave.

The bottom part of Figure 8.4.5 illustrates the pressure histories associated with the transition to detonation events in an unobstructed channel:

a. Slow deflagration—after rapid increase in pressure associated with a pressure wave, a slow pressure increase is seen.

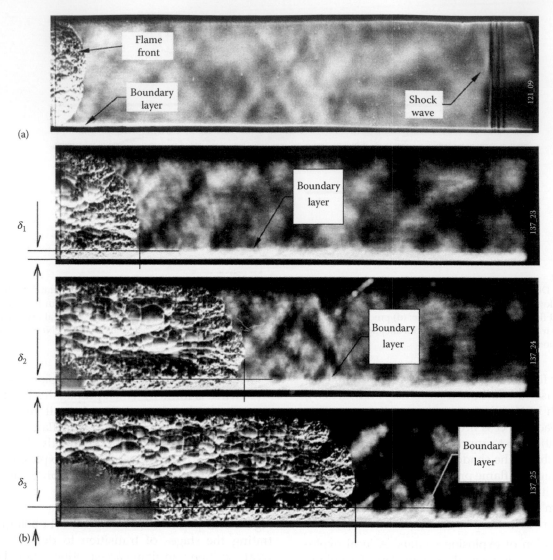

FIGURE 8.4.3
(a) Shadow photograph of early stage of flame propagation ($P_a = 0.075\,\text{MPa}$, window at 210–440 mm from ignition point). The mixture was ignited by means of a weak electric spark with the energy of 20 mJ. (b) Shadow photograph of the later stage of turbulent flame propagation from (a). (From Kuznetsov, M., Maksukov, I., Alekseev, V., Breitung, W., and Dorofeev, S., *Proceedings of the 20th International Colloquium on the Dynamics of Explosions and Reactive Systems*, Montreal, 2005.)

b. Fast deflagration—the flame position is much closer to the precursor shock wave.

c. Overdriven detonation—a transition to detonation that has just occurred and the detonation is significantly overdriven with the peak pressure, well in excess (2–3 times) of the value usually associated with a steady Chapman–Jouget (CJ) detonation. This peak pressure generated during the transition process is a particular point of concern in the industry.

d. Stable detonation—a steady detonation wave with velocity and pressure close to CJ values.

Transition to detonation in channels without obstacles was recently successfully simulated numerically [11,12]. In these simulations, it was shown that shock compression of the unreacted mixture forms the hot spots resulting from shock–shock, shock–wall, and shock–vortex interactions. The hot spots contain temperature gradients that produce spontaneous reaction waves and detonations.

8.4.2.2 Historical Review of DDT Studies in Smooth Tubes

The process of DDT is of intense interest ever since the discovery of the detonation wave in the 1880s [13,14].

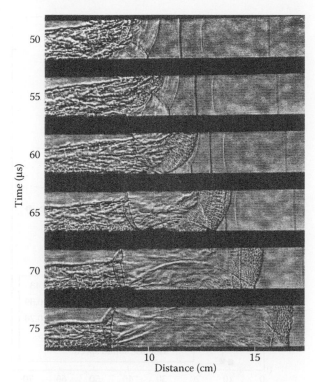

FIGURE 8.4.4
Stroboscopic schlieren record of the DDT process with onset at flame front in $2H_2 + O_2$ initially at a pressure of 0.073 MPa. (From Urtiew, P.A. and Oppenheim, A.K., *Proc. R. Soc. A*, 295, 13, 1966. With permission)

In the 1930s, Bone et al. [15] using rotating mirror camera observed the action of shock waves propagating into the unburnt mixture ahead of the accelerating flame, and postulated that the detonation wave was initiated as a result of preignition of the shock-compressed mixture.

In the 1950s, the more descriptive schlieren records of the interactions between pressure waves and deflagration fronts were obtained [16–18], and Oppenheim [9] introduced the hypothesis of the "explosion in the explosion" (of the detonating mixture) occurring in the regime of accelerating flame to explain the sudden change in the velocity of the combustion wave observed in the experiments.

In the 1960s, Oppenheim et al. [10,19,20] succeeded in obtaining photographs with better resolution by means of schlieren technique with microsecond flash and then with the very short (less than 10^{-8} s) laser light pulses. This facilitated the attainment of a stroboscopic set of essentially still photographs that revealed many details of DDT. At the same time, Soloukhin [21] published a series of streak photographs taken with schlieren system and Denisov and Troshin [22] discovered that detonation leaves a record of its passage in the form of imprint on a wall coated with the thin layer of soot.

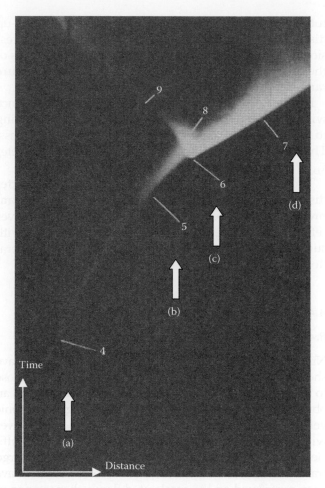

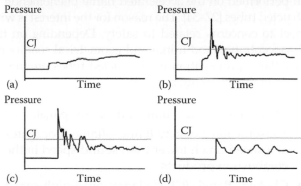

FIGURE 8.4.5
Streak direct image showing the general phases observed during a transition to detonation event following turbulent flame acceleration; 4—slow flame, 5—fast accelerating flame, 8—explosion in front of the flame, 9—retonation wave, 6—overdriven detonation, 7—steady detonation wave (From Lee, J.H., *Advances in Chemical Reaction Dynamics*, Rentzepis, P.M. and Capellos, C., Eds., 246, 1986.); Below the image: the sketch showing the typical pressure histories expected at locations (a)–(d). (Courtesy of G. Thomas.)

8.4.2.3 Experimental DDT Distances

There are some experimental data available on the effects of tube diameter, initial pressure, and temperature on the run-up distance to detonation for smooth tubes [23–26]. These data show a decrease in run-up distance with initial pressure according to the expression $x_{DDT} = f(p^{-m})$ where m depends on the properties of the mixture and lies in the range 0.4–0.8 for the pressure range from 0.01 to 0.65 MPa.

In some studies, an increase in the run-up distance with tube diameter was reported, but this may be owing to the hidden influence of such factor as tube roughness. The ratio of the run-up distance to the tube diameter x_{DDT}/D was found to be in the range of 15–40.

The DDT process in short tubes may occur at shorter distances than in long tubes owing to mixture precompression and flame interaction with the pressure waves reflected from the far end. This effect, together with surface roughness, plays a key role in the flame acceleration process.

8.4.3 DDT in Obstructed Channels

8.4.3.1 Introduction

Numerous experimental studies and accidents have shown that if a combustible gas mixture is not too close to the flammability limits, then a flame propagating in an obstacle field can accelerate very rapidly to high supersonic velocities. Such high-speed flames can drive shock waves with substantial overpressures. If the mixture is sufficiently sensitive, the highly accelerated flame may undergo transition to detonation. Numerous research studies have been performed on the accelerated flame phenomenon in obstructed tubes [27–34]. The reason for the interest is with respect to concerns related to safety. Depending on the fuel concentration and initial and geometrical conditions, steady flame propagation in obstructed tubes progresses in a one of the following regimes:

- Flame quenching—flame fails to propagate
- Subsonic low-velocity flame—flame propagates at a speed much lower than sound speed in the combustion products
- Choked flame (CJ deflagration)—high-speed flame propagating with the velocity close to sound speed in the combustion products (600–1200 m/s)
- Quasi-detonation—flame propagates with the velocity between the sound speed in the combustion products and CJ value
- Detonation—flame velocity is close to CJ value

From the practical point of view, the most important aspects of the accelerated flame phenomenon are with respect to the steady-state propagation of very high-speed flames, transition to detonation, and propagation of sub-CJ detonations (quasi-detonations).

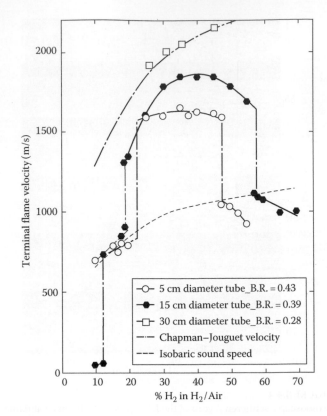

FIGURE 8.4.6
Flame velocity versus fuel concentration for H_2/air mixtures in the 10 m long tubes of 5, 15, and 30 cm internal diameter with obstacles (orifice plates); BR = $1 - d^2/D^2$ – blockage ratio, where d is the orifice diameter and D is the tube diameter. (From Lee, J.H., *Advances in Chemical Reaction Dynamics*, Rentzepis, P.M. and Capellos, C., Eds., 246, 1986.)

Figure 8.4.6 shows the plot of terminal flame velocity versus fuel concentration for hydrogen/air mixture [7]. After ignition, the flame accelerates rapidly and after propagating past a number of obstacles over a distance of about half to one meter, it approaches a steady-state velocity. The self-quenching regime of the flame was not observed in these experiments owing to the blockage ratios that were not sufficiently high. The low-velocity deflagration regime was observed and transition to the choking regime was distinct to occur at about 12.5% H_2. Subsequently, the quasi-detonation regime was observed for the mixtures 20%–50% H_2. For the 30 cm tube and low obstacle-blockage ratio, the normal CJ detonation was also observed.

8.4.3.2 Fast Deflagrations

The high-speed flames propagate in a tube with repeatable obstacles with the steady-state velocity, which is maintained for the rest of their passage over the obstacles. In some cases, the steady-state flame propagation velocity of the combustion products may approach the speed of sound. This level of flame velocity appears to be the maximum achievable by a turbulent flame in the non-detonative mode of combustion. It has been suggested that such maximum flame speed is prescribed, which limited the gas dynamically by the process of frictional and thermal choking [33].

Figure 8.4.7 illustrates the two time sequences of schlieren photographs of fast deflagration regime in the channel with obstacles. The pictures clearly show the decoupled structure of the leading shock wave followed by a reaction front, and both propagate with an averaged steady velocity of about 700 m/s. The leading shock wave is formed by the coalescence of the pressure waves generated continuously in the violent turbulent-flame brush.

No ignition was observed behind the leading shock wave, since for this shock velocity, the temperature behind the shock was only around 500 K. The leading shock wave, when reflected from the bottom wall, first appeared as a regular reflection and later on underwent the transition to a Mach reflection. Ignition did not occur behind either the regular reflected wave or in the Mach stem (in contrast to the quasi-detonation regime, as will be shown later). When the leading shock wave reached an obstacle and was partly reflected from it,

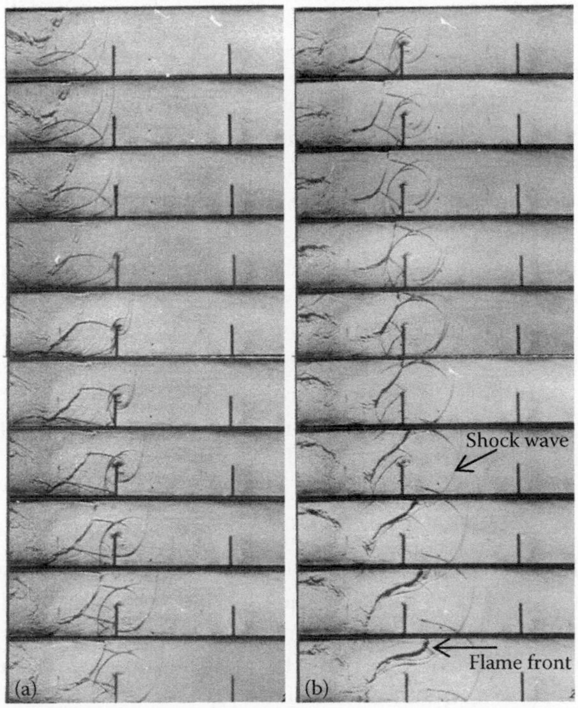

again ignition was not observed in the region close to the obstacle. The reflection of the incident shock wave at the obstacle generated a cylindrical reflected shock wave, which propagated transversely (upward toward the top wall as well as backward toward the flame front and interacted with it).

As the shock wave reflected from the bottom wall passed through the flame, the turbulent flame structure became smoother after the shock interaction process. This is owing to the Markstein instability effect, which in this case is in the stabilizing direction (i.e., shock moves from high to low density fluid), so that the flame perturbations were smoothened out. The process of flame stabilization is further continued by flame interaction with the curved cylindrical shock reflected from the obstacle, which is clearly illustrated in Figure 8.4.7. The sudden change in the energy release rate associated with the shock wave–flame interaction also results in the generation of pressure waves, as was demonstrated first by Markstein [35].

The pressure waves when reflected from the top wall interact with the flame again and cause a destabilizing effect on the flame front. The flame is accelerated toward a denser medium and the growth of the perturbations thus turbulizes the flame front via Rayleigh–Markstein instability mechanism.

The last five frames of Figure 8.4.7b show the flame propagation over the obstacle. Rapid acceleration of the flame and its turbulization are again clearly visible as the flame is convected along the accelerating converging flow, past the obstacle. The similar character of flame propagation over the obstacles has been observed experimentally in a single obstacle configuration by Wolanski and Wójcicki [36] and by Tsuruda and Hirano [37].

The structure of a turbulent high-speed deflagration in stoichiometric hydrogen/oxygen mixture in the channel is illustrated in Figure 8.4.8. The large roughness of the top and bottom walls is modeled by small cylindrical obstacles 2.5 mm in diameter. The structure consists of a series of compression waves in the front, followed by a highly turbulent reaction zone. The leading

FIGURE 8.4.7
Propagation of a high-speed deflagration in obstacle-filled channel illustrating the stabilizing effect of the reflected shock interaction with the flame front (a), and the accelerating effect and the turbulization of the flame as it passes over the obstacle (b). (From Teodorczyk, A., Lee, J.H.S., and Knystautas, R., *Prog. Astr. Aeron.*, 138, 223, 1990. With permission.)

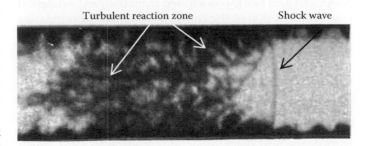

FIGURE 8.4.8
Structure of the turbulent high-speed deflagration propagating in a very rough channel; stoichiometric H_2/O_2 mixture at 150 torr.

compression waves are not strong enough to cause auto-ignition, so that the trailing reaction zone propagates with the characteristic V-shape with the leading edges at the wall, where intense turbulence is generated by the wall roughness as well as shock reflections on the obstacles take place. The shock-flame complex, which propagates with the velocity of about 1000 m/s is only 40 cm apart the weak (~1 mJ) electric spark ignitor.

8.4.3.3 Transition from Deflagration to Detonation

A distinction should be made between DDT in smooth and rough tubes, since wall roughness plays a very strong part in both the propagation of deflagration and detonation. In smooth tubes, the onset of detonation is marked by an abrupt change in the propagation speeds. Typically, predetonation flame velocity is less than 1000 m/s and the CJ detonation speed is over 2000 m/s. A very strong local explosion always occurs at the onset of detonation, so that the detonation wave formed is highly overdriven initially and decays subsequently to its CJ value. The shock wave from this local explosion that propagates back into the combustion product gases is always observed.

For very rough tubes, the flame acceleration is much more rapid as shown in the previous section. Transition to detonation is also clearly marked by a local explosion and abrupt change in the wave speed. The wall roughness controls the propagation of the wave by providing [5]:

1. Means for generating strong large-scale turbulence, thus a larger fraction of the average flow kinetic energy can be randomized
2. Means for generating strong shock reflections and diffractions and thus, an additional mechanism for the randomization of the average flow energy via these complex wave-interaction processes
3. Means for generating high local temperatures for autoignition via shock reflections (normal and Mach), which otherwise is not possible by the incident shock themselves (without reflections)

Shepherd and Lee [5] concluded from their experimental observations that in the complete absence of boundaries for shear and wave generation, as in a pure spherical geometry, the flame through its own self-turbulization mechanism of instability cannot provide sufficient randomization of the mean flow kinetic energy to cause DDT, except in the extremely sensitive mixtures. With very rough-walled tubes, the obstacles provide an efficient mode of flow randomization through large-scale turbulence and wave reflections leading to DDT, much sooner than in the smooth tubes.

Figure 8.4.9 shows the time sequence of schlieren photographs of DDT in very rough channel. It is clearly seen

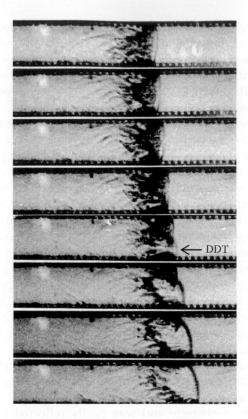

FIGURE 8.4.9
Time sequences of schlieren photographs showing DDT in very rough tube; stoichiometric H_2/O_2 mixture at 100 torr; 2 μs between frames.

that, as in the smooth tubes, the transition to detonation is associated with the abrupt change in the propagation velocity. The fast deflagration before DDT propagates with the speed of 1400 m/s; however, after transition, the detonation velocity becomes 3000 m/s. In contrast to the smooth tubes, in this case, the turbulent flame fully overcomes the leading shock at the moment of transition to detonation. The onset of detonation occurs at thick flame brush and no retonation wave is observed. This suggests that detonation is triggered by the gradual amplification of pressure disturbances, rather than through a local hot spot. Similar conclusions were formulated by Yatsufusa et al. [38].

8.4.3.4 Quasi-Detonations

In the early studies [22,24,39] on propagation of detonation in very rough tubes, the steady propagation velocities as low as 50% of the normal CJ value have been observed. Such low-velocity detonations have been referred to as quasi-detonations [4].

The studies by Teodorczyk et al. [40–42] and Chan et al. [30] conclusively demonstrated that the mechanism of detonation initiation in quasi-detonation regime is owing to autoignition via shock reflections. These studies show that normal shock reflections from the obstacle, Mach reflection of the diffracted shock on the bottom wall, and Mach reflection of the reflected shock from the top wall can lead to autoignition. The role of the obstacle is to promote strong shock reflections leading to high local temperatures for autoignition. Detonations are initiated from these local "hot spots," but are later destroyed by diffraction quenching around the obstacles themselves. Hence, for quasi-detonations, the diffraction around the obstacles destroys the initiated detonation while shock reflections, resulting from the interactions of the decoupled shock with the obstacle and the tube, will give rise to local hot spots and reinitiate the detonation.

In the quasi-detonation regime, the continuous periodic detonation failure owing to diffraction by the obstacles and reinitiation by shock reflections constitutes the principal mechanism of propagation.

Figure 8.4.10 shows two time sequences of schlieren photographs of quasi-detonation. In Figure 8.4.10a, detonation reinitiation occurs at the Mach stem on the bottom wall. However, prior to complete reinitiation of the decoupled wave by the upward-growing detonation, reflection and, subsequently, diffraction of the detonation occur again by encountering another obstacle. In the sixth frame of Figure 8.4.10a, the curved, diffracted, and reflected shock with a reaction zone close behind is clearly evident. However, as this cylindrical

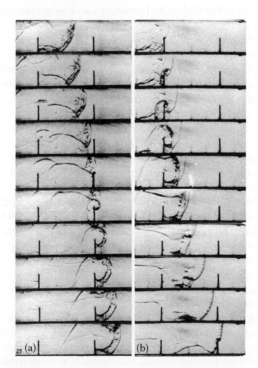

FIGURE 8.4.10

Propagation of quasi-detonation in obstacle array in stoichiometric H_2/O_2 mixture; (a) initial pressure 140 torr, detonation reinitiation via Mach reflection at the bottom wall; (b) initial pressure 120 torr, detonation reinitiation by normal Mach stem reflection from the obstacle with subsequent enhancement via reflection from the top wall; 6 µs between frames. (From Teodorczyk, A., Lee, J.H.S., and Knystautas, R., *Prog. Astr. Aeron.*, 138, 223, 1990. With permission.)

deflagration expands, progressive decoupling of the reaction zone occurs. In the last frame, this entire shock and reaction-zone complex is decoupled. With higher obstacle density (i.e., more obstacles per unit length), more frequent attenuations by the diffraction of the reinitiated detonation occur. This explains the decrease in the "averaged" velocity of the quasi-detonations with increasing obstacle density.

Figure 8.4.10b shows the reinitiation process owing to reflection of the Mach stem from the obstacle. The rapid expansion of the reflected shock owing to auto-explosion is clearly evident from the comparison of frames 2 and 3. In the present case, the diffraction causes the reinitiated detonation to fail, and it becomes a cylindrical deflagration with progressive decoupling of the reaction zone from the shock front. However, the normal reflection of the leading shock from the top wall causes a reinitiation process. The cellular detonation subsequently sweeps down to engulf the entire decoupled front.

Depending on the obstacle height and spacing, as well as on the vertical height of the channel, one or more of the above-described mechanisms can occur. However, the propagation mechanism comprises continuous reinitiation and attenuation by diffraction around the obstacles. This mechanism essentially is identical to that of a normal detonation, where reinitiation occurs when the transverse waves collide and the reinitiated wave fails between collisions. In quasi-detonations, the reinitiation is controlled by obstacles. In general, the obstacles and walls provide surfaces for the reflection and diffraction of shock and detonation waves.

The main regimes of flame propagation observed in experiments with obstructed channels, choking flames, quasi-detonations, and detonation, were reproduced in the recent numerical simulations by Gamezo et al. [43]. The simulations have shown in detail that at initial stages, the flame and flow acceleration is caused by thermal expansion of hot combustion products. At later stages, shock–flame interactions, RT, RH, and KH instabilities, and flame–vortex interactions in the obstacle wakes become responsible for the increase in the flame surface area, the energy release rate, and, eventually, the shock strength. Transition to detonation occurs at hot spots created by shock reflections at the corners between obstacles and the wall. The first DDT starting the quasi-detonation regime occurred when the Mach stem, created by the reflection of the leading shock from the bottom wall, collided with an obstacle. The same reignition mechanism was observed in experiments previously described.

8.4.4 Criteria for DDT

Results of numerous DDT experiments in tubes with orifice plates and channels with obstacles have demonstrated that the necessary criteria for DDT are

1. Minimum passage diameter $d \geq \lambda$, where d is the size of the unobstructed passage in a tube or channel with obstacles
2. Minimum scale, $L \geq 7\lambda$, where L is a general characteristic size defined as $L = \frac{(H+S)/2}{1 - d/H}$, where H is the channel height and S is the distance between the obstacles
3. Minimum tube diameter for smooth tubes, $D \geq \lambda/\pi$, where D is the internal tube diameter

---

References

1. Zel'dovich, Ya.B. et al., On the development of detonation in a non-uniformly preheated gas, *Astronaut. Acta*, 15, 313, 1970.
2. Zel'dovich Ya.B., Regime classification of an exothermic reaction with non-uniform initial conditions, *Combust. Flame*, 39, 211, 1990.
3. Lee, J.H.S., Knystautas, R., and Yoshikawa, N., Photochemical initiation and gaseous detonations, *Acta Astronaut.*, 5, 971, 1978.
4. Lee, J.H. and Moen, I., The mechanism of transition from deflagration to detonation in vapour cloud explosion, *Prog. Energy Combust. Sci.*, 6, 359, 1980.
5. Shepherd, J.E. and Lee, J.H., On the transition from deflagration to detonation, in *Major Research Topics in Combustion*, Hussaini, M.Y., Kumar, A., and Voigt R.G., Eds., Springer, New York, p. 439, 1992.
6. Lee, J.H., Fast flames and detonations, in *The Chemistry of Combustion Processes*, American Chemical Society, Sloane, T.M., Ed., *ACS Symposium Series*, No. 249, 1984.
7. Lee, J.H., The propagation of turbulent flames and detonations in tubes, in *Advances in Chemical Reaction Dynamics*, Rentzepis, P.M. and Capellos, C., Eds., D. Keidex Publ. Co. London, p. 246, 1986.
8. Kuznetsov, M. et al., Effect of boundary layer on flame acceleration and DDT, *Proceedings of the 20th International Colloquium on the Dynamics of Explosions and Reactive Systems* on CD, Montreal, 2005.
9. Oppenheim, A.K. and Stern, R.A., On the development of gaseous detonation - analysis of wave phenomena. *Proc. Combust. Inst.*, 7, 837, 1959.
10. Urtiew, P.A. and Oppenheim, A.K., Experimental observations of the transition to detonation in an explosive gas, *Proc. R. Soc.*, A 295, 13, 1966.
11. Khokhlov, A.M. and Oran, E.S., Numerical simulation of detonation initiation in a flame brush: The role of hot spots, *Combust. Flame*, 119, 400, 1999.
12. Gamezo, V.N., Khokhlov, A.M., and Oran, E.S., The influence of shock bifurcations on shock-flame interactions and DDT, *Combust. Flame*, 126, 1810, 2001.
13. Berthelot, M. and Vieille, P., Sur la vitesse de propagation des phenomenes explosifs dans les gaz *C.R. Acad. Sci.*, Paris 94, 101, seance du 16 Janvier; 822, séance du 27 Mars; 95, 151, seance de 24 Juillet, 1822.
14. Mallard, E. and Le Chatelier, H., Recherches experimentales et theoriques sur la combustion des melanges gaseux explosifs, *Ann. Mines*, 8, 4, 274, 1883.
15. Bone, W.A., Fraser, R.P., and Wheeler, W.H., A photographic investigation of flame movements in gaseous explosions. VII. The phenomenon of spin in detonation. *Philos. Trans.*, *R. Soc.*, London, A 29, 1935.
16. Schmidt, E., Steinicke, H., and Neubert, U., Flame and schlieren photographs of the combustion of gas-air mixtures in tubes. *TDI-Forschungsheft*, 431, Ausgabe B., Band 17, Deutscher Ingenieur-Yerlag, Dusseldorf, 1951.
17. Greifer, B. et al., Combustion and detonation in gases, *J. Appl. Phys.*, 28, 3, 289, 1957.
18. Salamandra, G.D., Bazhenova, T.Y., and Naboko, I.M., Formation of detonation ware during combustion of gas in combustion tube. *Proc. Combust. Inst.*, 7, 851, 1959.
19. Oppenheim, A.K., Laderman, A.J., and Urtiew, P.A., Onset of retonation, *Combust. Flame*, 6, 3, 193, 1962.
20. Oppenheim, A.K., Urtiew, P.A., and Weinberg, P.J., On the use of laser sources in schlieren-interferometer systems, *Proc. R. Soc.*, A 291, 279, 1966.
21. Soloukhin, R.I., Perekhod goreniya v detonatsiru v gazakh (Transition from combustion to detonation in gases), *P.M.T.F.*, 4, 128, 1961.
22. Denisov, Yu. and Troshin, Ya.K., Struktura gazovoi detonatsii v trubakh (Structure of gaseous detonation in tubes), *Zim. Tekhn.*, *Fiz.*, 30, 4, 450, 1960.
23. Egerton, A. and Gates, S.F., On detonation in gaseous mixtures at high initial pressures and temperatures, *Proc. R. Soc.*, *Lond. A*, 114, 152, 1927.
24. Schelkin, K.I. and Sokolik, A.S., Detonation in gaseous mixtures, *Soviet. Zhurn. Phys. Chem.*, 10, 479, 1937.
25. Bollinger, L.E., Fong, M.C., and Edse, R., Experimental measurements and theoretical analysis of detonation induction distances, *J. Am. Rocket Soc.*, 31, 5, 588, 1961.
26. Kuznetsov, M. et al., DDT in a smooth tube filled with a hydrogen–oxygen mixture, *Shock Waves*, 14, 3, 2005.
27. Dorge, K.J., Pangritz, D., and Wagner, H.G., On the influence of several orifices on the propagation of flames: Continuation of the experiments of Wheeler. *Z. Fur Phys. Chemie*, 127, 61, 1981.
28. Moen, I.O. et al., Flame acceleration due to turbulence produced by obstacles. *Combust. Flame*, 39, 21, 1980.
29. Moen, I.O. et al., Pressure development due to turbulent flame propagation in large-scale methane-air explosions. *Combust. Flame*, 47, 31, 1982.
30. Chan, C., Moen, I.O., and Lee, J.H., Influence of confinement on flame acceleration due to repeated obstacles. *Combust. Flame.*, 49, 27, 1983.
31. Hjertager, B.H. et al., Flame acceleration of propane-air in a large-scale obstructed tube, *Prog. Astr. Aeron.*, 94, 504, 1984.
32. Lee, J.H., Knystautas, R., and Freiman, A., High speed turbulent deflagrations and transition to detonation in H_2-air mixtures, *Combust. Flame*, 56, 227, 1984.
33. Lee, J.H., Knystautas, R., and Chan, C., Turbulent flame propagation in obstacle-filled tubes, *Proc. Combust. Inst.*, 20, 1663, 1984.
34. Peraldi, O., Knystautas, R., and Lee, J.H., Criteria for transition to detonation in tubes, *Proc. Combust. Inst.*, 21, 1629, 1986.

35. Markstein, G.H., *Non-Steady Flame Propagation*, McMillan, New York, 1964.
36. Wolański, P. and Wójcicki, S., On the mechanism of the influence of obstacles on the flame propagation, *Arch. Combust.*, 1, 69, 1981.
37. Tsuruda, T. and Hirano, T., Growth of flame front turbulence during flame propagation across and obstacle, *Combust. Sci. Technol.*, 51, 323, 1983.
38. Yatsufusa,T., Chao, J.C., and Lee, J.H.S., The effect of perturbation on the onset of detonation, *Proceedings of the 21st International Colloguium on the Dynamics of Explosions and Reactive Systems* on CD, Poitiers, 2007.
39. Guenoche, H. and Manson, N., Influence des conditions aux limits transversales sur la propagation des ondes de shock at de combustion, *Revie de l'Institut Francais du Petrole*, 2, 53, 1949.
40. Teodorczyk, A., Lee, J.H.S., and Knystautas, R., Propagation mechanism of quasi-detonations, *Proc. Combust. Inst.*, 22, 1723, 1988.
41. Teodorczyk, A., Lee, J.H.S., and Knystautas, R., The structure of fast turbulent flames in very rough, obstacle-filled channels. *Proc. Combust. Inst.*, 23, 735, 1990.
42. Teodorczyk, A., Lee, J.H.S., and Knystautas, R., Photographic study of the structure and propagation mechanisms of quasi-detonations in rough tubes, *Prog. Astr. Aeron.*, 138, 223, 1990.
43. Gamezo, V.N., Ogawa, T. and Oran, E.S., Numerical simulations of flame propagation and DDT in obstructed channels filled with hydrogen-air mixture, *Proc. Combust. Inst.*, 31, 2463, 2007.

8.5 Detonations

Bernard Veyssiere

8.5.1 Introduction

The detonation phenomenon in a gaseous mixture presents, at first sight, surprising features. It is possible to predict, with excellent accuracy, the average characteristics of the detonation front on the basis of the classical Chapman–Jouguet theory. This well-known model is an oversimplified description of detonation considering this supersonic combustion regime as a steady one-dimensional (1D) plane wave, where the reaction of gaseous components into burnt products occurs instantaneously, so that the detonation wave has a zero thickness. In reality, it has been established that the detonation wave has a finite thickness: the chemical energy is released in an extended zone behind a complex system of leading shock waves interacting with weak transverse shock waves, giving rise to triple points continuously moving perpendicularly to the direction of detonation propagation. Thus, the detonation wave has a multidimensional structure and is intrinsically unstable. To conciliate these two apparently antagonist points of view, it is necessary to have a better knowledge of the mechanisms through which the coupling between the shock front and the exothermic reaction zone is achieved, and understand how it is maintained so that the detonation is able to continue to propagate.

Investigation of the fine structure of the detonation front is a difficult problem. Because of high pressure effects generated by the detonation, experiments in the laboratory must be performed in pressure-resistant confinements, which is not a favorable situation for using in situ diagnostic methods. However, this difficulty can be partially avoided by conducting experiments at low initial pressure. Above all, real-time, high-resolution methods with fast characteristic response time (10^{-6} s or less) are required to explore the reaction zone. In the past, the schlieren technique was widely used. Currently, researchers have two main tools of very different nature at their disposal. One seems very rudimentary in its principle: it involves of registering the tracking of triple-point trajectory on soot plates. On the contrary, the second one is based on numerical simulations with high-performance computations using sophisticated numerical schemes. However, a limited number of studies have been performed using Rayleigh scattering or laser-induced fluorescence (LIF), but until now they have not permitted to provide results of a quality comparable with what has been obtained for turbulent flames. In this short survey, we attempt to illustrate a few problems where efforts are made at the present time to acquire a better understanding of the detonation wave structure.

8.5.2 The Cellular Structure

The instability and nonplanar character of the detonation front was first exhibited by Manson [1] and Fay [2] in certain particular cases of detonation propagation, such as the "spinning detonations." The multidimensional nature of the detonation front has been established by Voitsekhovskii [3], Denisov and Troshin [4],

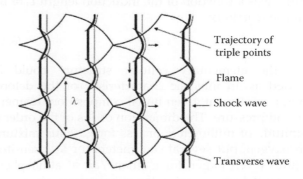

FIGURE 8.5.1
Schematic picture of the multidimensional structure of the detonation wave.

FIGURE 8.5.2
Typical soot patterns of the detonation in a $2H_2 + O_2 + 7Ar$ mixture at 90 torr initial pressure in a rectangular (3¼ in. × 1½ in.) tube. (Reprinted from Strehlow, R.A., *Astronaut. Acta*, 14, 539, 1969. With permission.)

White [5], and, then, by many other researchers. They displayed that transverse shock waves existed behind the main shock front. The triple points resulting from the interaction of this complex system of shock waves periodically oscillate perpendicularly to the direction of the leading front propagation.

The denomination "cellular structure," which is commonly used now in the literature, comes from the observation of the trajectory of these triple points by a special tracking technique. It is recorded on plates or foils disposed along the propagation direction of the detonation, over which a thin coat of soot had been preliminarily deposited. Because of high temperature and pressure conditions existing at the triple points, they draw a two-dimensional (2D) image of the history of their trajectory on the soot coating; the elementary picture of this drawing is called "cell" (see Figure 8.5.1). An example among many others is shown in Figure 8.5.2 taken out from the well-known experiments of Strehlow [6,7]. The typical shape of the elementary cell resembles a fish shell. Many parametric studies have been conducted subsequently to investigate the dimensions and the regularity of this cellular structure [8,9] (see also the database [10]). It has been shown, first by Schchelkin and Troshin [11] and other researchers [9,12], that the measure of the cell width λ was a function of the induction length L_i of the chemical reactions:

$$\lambda = BL_i \qquad (8.5.1)$$

Thus, the elementary cellular structure could be regarded as an intrinsic characteristic of the detonation in a mixture at given initial composition, temperature, and pressure. The dimension of λ is of the order of magnitude of millimeters or less for gaseous mixtures with oxygen, but several centimeters for less sensitive mixtures (even larger, for methane/air at atmospheric pressure). It decreases when the initial pressure increases. Its variation with the initial temperature is more complicated and depends on the value of the reduced activation energy of the chemical reactions. The value of

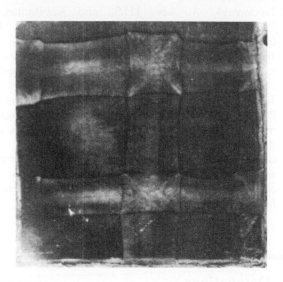

FIGURE 8.5.3
Soot patterns, recorded at the end plate of a 27 mm × 27 mm square tube, of the detonation of 25% $(2H_2 + O_2) + 70\%$ Ar at 400 torr. (Reprinted from Takai, R., Yoneda, K., and Hikita, T., *Proceedings 15th Symposium (International) on Combustion*, The Combustion Institute, Pittsburg, 1974, 69–78. With permission.)

λ also decreases exponentially with the increase in the detonation strength (case of overdriven detonations). The regularity of the cellular network is linked to the activation energy of the chemical reactions [9]: mixtures with low activation energy display a regular structure with cells having the same size, whereas high activation energy results in more irregular structure with an important dispersion of the cell-size values. Besides, the cellular structure is three-dimensional (3D) as attested by the few available frontal observations of the detonation front (see, e.g., Figure 8.5.3 obtained by Takai et al. [13]). However, the 3D cell-width is the same as in the 2D case.

8.5.3 Detonation Wave Structure

Many photographic records of the flow made using the schlieren technique have revealed its very high

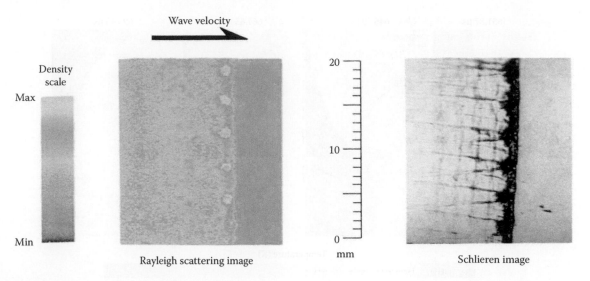

FIGURE 8.5.4
Example of Rayleigh and schlieren images acquired simultaneously from the same wave front. Mixture $H_2/O_2/Ar$ stoichiometric at 0.374 atm initial pressure. (Reprinted from Anderson, T.J. and Dabora, E.K., *Proceedings 24th Symposium (International) on Combustion*, The Combustion Institute, Pittsburg, 1992, 1853–1860. With permission.)

complexity. But, the density variations that they provide are integrated over a test section and it is not obvious to extract local information about the local structure behind the triple points. Using Rayleigh scattering, Dabora et al. [14] displayed (see Figure 8.5.4) that high-density regions at the wave front corresponded, spatially, to the points where the opposing transverse waves behind the front have just intersected and combustion has not yet occurred, as is revealed by schlieren observations.

Since the early works of Taki and Fujiwara [15], Oran et al. [16], or Markov [17], numerical simulations have been performed by several research teams to investigate the fine structure of the detonation front. They are based on the solution of Euler or Navier–Stokes equations of the reactive flow. The detonation model is the classical Zel'dovich–Von Neumann–Döring (ZND) model, which assumes the detonation as a wave of finite thickness, where the chemical reactions are thermally initiated by the leading shock. Arrhenius global (or simplified schemes) kinetics laws are considered for chemical reactions. In numerical simulations, the instability leading to the formation of the cellular detonation front from a planar detonation wave appears as a result of the nonlinearity of governing equations and may be triggered by numerical noise [18]. Stiffness of the problem owing to the nonlinear coupling between the fluid dynamics and the chemical kinetics (which is highly sensitive to temperature), together with the necessity of discretization of the reaction zone with a sufficiently good accuracy, require very large numbers of meshes in the numerical grid, which rapidly

leads to prohibitive computing times. Most of these computations have been made in 2D case, but several 3D-numerical simulations have also been performed [19]. These 2D simulations have displayed the existence of unburnt mixture pockets embedded in burnt gases behind the leading front, and also the role of turbulence attested by the generation of vortices at the interaction of transverse waves. Special attention has been devoted to these features in several numerical simulation works by Gamezo [20] and researchers of the Naval Research Laboratory [21]. An example is shown in Figure 8.5.5 in the case of marginal detonations in an acetylene/oxygen mixture. In this case, the induction time is quite long and one can clearly observe on the temperature fields how the transverse detonation formed behind the transverse wave interacts with the induction zone and burns the gaseous mixture.

Shepherd et al. [22] experimentally investigated the reaction-zone structure by combining the schlieren technique with the PLIF technique to follow the evolution of the OH radical, which is an intermediate species of the chemical reactions. In hydrogen/oxygen/argon mixtures, they have clearly shown (see Figure 8.5.6) that the changes in OH-concentration front location can be correlated with the density changes observed on schlieren images, thus, with local changes in the shock strength. However, no clear evidence of the existence of unreacted gas pockets detached behind the detonation front can be established from these experiments. Numerical simulations performed by Gamezo et al. [23] with very high accuracy (mesh size of $5\,\mu m$) after formation of the

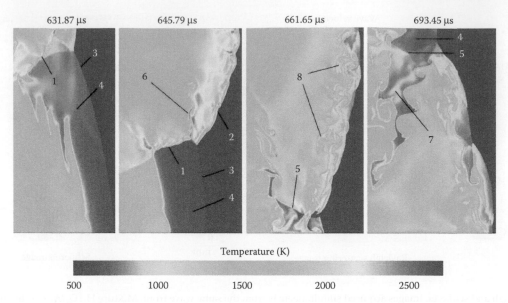

631.87 μs   645.79 μs   661.65 μs   693.45 μs

Temperature (K)

500   1000   1500   2000   2500

FIGURE 8.5.5
Temperature field behind the leading shock at different times obtained by the numerical simulations [21]. 1, transverse detonation; 2, strong part of the leading shock (overdriven detonation); 3, weak part of the leading shock (inert); 4, induction zone; 5, transverse shock; 6, unreacted tail; 7, primary unreacted pocket; and 8, secondary unreacted pockets. (Courtesy of V. Gamezo.)

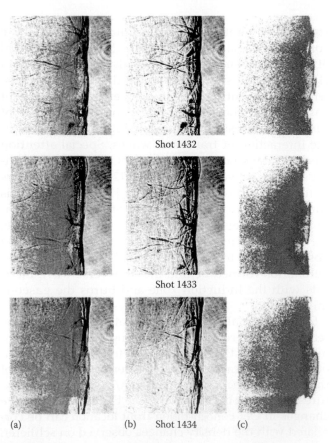

Shot 1432

Shot 1433

(a)   (b)   Shot 1434   (c)

FIGURE 8.5.6
Simultaneous schlieren and OH-fluorescence images behind detonation front in stoichiometric hydrogen/oxygen mixtures diluted with 85% argon (shots 1432 and 1433) and 87% argon (shot 1434) with initial pressure of 20 kpa, (a) overlay of PLIF and schlieren images, (b) schlieren, (c) PLIF. (Reprinted from Pintgen, F., Eckett, C.A., Austin, J.M., and Shepherd, J.E., *Combust. Flame*, 133, 211, 2003. With permission.)

detonation in hydrogen/air mixture at atmospheric pressure (see Figure 8.5.7) display the extreme complexity of the front structure even in the case of well-established detonations. Thus, comprehensive understanding of the detailed structure of the detonation wave remains an open issue of research in detonation.

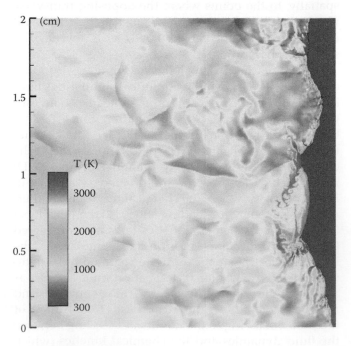

FIGURE 8.5.7
Temperature field obtained by numerical simulations [23] behind the front of a fully developed (0.3 ms after the detonation initiation) detonation in hydrogen/air at 1 atm. Minimum computational cell size is 5 μm. (Courtesy of V. Gamezo.)

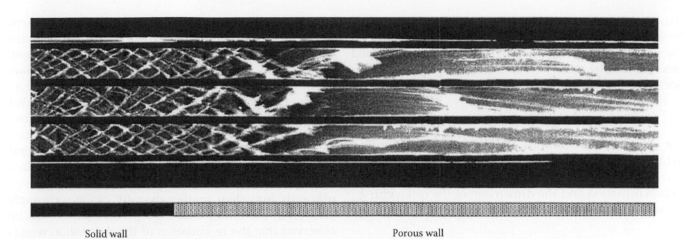

Solid wall   Porous wall

FIGURE 8.5.8
Example of the failure of the cellular structure at the passage from solid walls to porous walls in the detonation of $C_2H_2 + 2.5O_2$ mixture at 2.6 kPa initial pressure. (Reprinted from Radulescu, M.I. and Lee, J.H.S., *Combust. Flame*, 131, 29, 2002. With permission.)

8.5.4 Role of Transverse Waves

Transverse waves appear to play a fundamental role in the mechanism of detonation propagation. It might be speculated that the presence of walls in the case of detonations propagating in tubes is responsible for the formation of transverse waves owing to the interaction of triple points with the tube walls. This question was investigated mainly by Lee and his co-workers [24]. They performed experiments in tubes with sections equipped with porous walls to damp the transverse waves. An example is shown in Figure 8.5.8 for the detonation propagating in an acetylene/oxygen mixture, where the weakening of transverse waves is clearly observed at the entrance of the porous section when the triple points interact with the porous walls, leading to the failure of the entire wave structure. At the same time, the propagation velocity of the combustion front drops down to 40% of the CJ detonation velocity. This behavior attests the determining importance of transverse waves in the process of detonation propagation. However, this explanation has been established valid for undiluted hydrocarbon/oxygen detonations. On the contrary, for mixtures diluted by argon, experiments have not established that transverse waves could play a significant role in the mechanism of detonation propagation. Moreover, it is important to remember that the existence of the cellular structure has also been exhibited in the case of unconfined diverging detonations; since in this case, the number of cells increases as the detonation front proceeds, the problem arises in understanding how additional cells are generated. Thus, the precise mechanism of regeneration of new cells through the interaction of transverse waves and of the creation of extra cells in diverging detonation is far from being well understood.

8.5.5 Cellular Structure and Detonation Initiation

Since the cellular cell-width λ is related to the induction length of the chemical reactions, this parameter, intrinsic

for each reactive mixture, is very useful to evaluate its detonability. Direct initiation of an unconfined detonation by instantaneous energy release from a high power point source corresponds to the most severe conditions to achieve detonation initiation. It has been established [25] that at critical conditions for direct initiation, the detonation was formed abruptly behind the leading spherical shock. This can be clearly observed on soot tracks, as displayed in Figure 8.5.9: first, the soot coat deposited on the plate is not affected by the propagation of the spherical shock from the point source. But, suddenly, the cellular

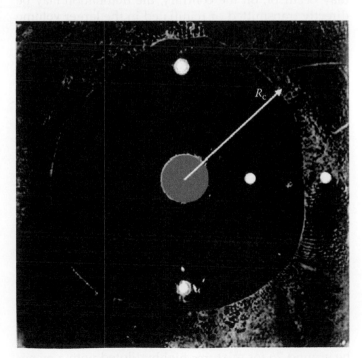

FIGURE 8.5.9
Soot patterns showing the predetonation radius in hemispherical critical detonation initiation of $C_2H_2 + 2.5O_2$ mixture at 30 torr initial pressure. (Courtesy of D. Desbordes.)

structure appears at the periphery of a circle of characteristic radius R_c, which delineates the frontier between the predetonation zone and the zone of fully developed detonation. The value of the critical radius can be connected to the cell width by the following relationship:

$$R_c = K \lambda_{cj} \qquad (8.5.2)$$

For classical hydrocarbons, the value of K is, according to Desbordes [25], approximately $K = 20$. In the same way, correlations between the critical energy for direct detonation initiation and the dimension of the cell width λ_{cj} have been proposed by different authors [25–27].

8.5.6 Detonation Transmission

Another important problem related to detonation formation is the transmission of a detonation propagating in a tube of constant cross section into a larger volume (either semi-infinite or in a confinement of larger size). The problem of critical conditions for detonation transmission is analogous to that of direct initiation. Besides, it is of great interest from the practical points of view: in the safety of industrial plants for preventing detonation transmission to the surrounding medium after an accident, as well as in the propulsion applications to reduce the predetonation distance needed to initiate the detonation regime in a combustion chamber. When the leading front exits from the tube, owing to the sudden lateral expansion of the flow, detonation failure may occur or, on the contrary, the detonation may be directly transmitted or reinitiated in the larger volume after an intermediate stage during which the leading shock and the reaction zone are decoupled. For the detonation to be transmitted, a minimum dimension d_c of the tube diameter is required to balance the lateral expansion effects due to abrupt change in the cross section and thus, to prevent detonation quenching. The well-known correlation proposed by Mitrofanov and Soloukhin [28]

$$d_c = 13\lambda \qquad (8.5.3)$$

is applicable for a large variety of mixtures in the case of transmission to a semi-infinite space. However, this correlation between the critical value of the characteristic cross section of the tube and the cell size should rather be expressed as:

$$d_c = k_c \lambda \qquad (8.5.4)$$

The value of k_c is not universal and depends on the nature of the reactive mixture (for example, the value of k_c may be around 26 for mixtures highly diluted with a monoatomic gas [29], or around 20 for hydrogen/air mixtures [30]), as well as on the diffraction process at the tube

exit. This diffraction process appears to play an essential role and illustrates the complex interaction between the detonation wave and the surrounding confinement. For example, it has been displayed that for tubes having a cross section different from the circular shape (e.g., square, polygonal, …) the value of k_c could be smaller than 13 [31]. Furthermore, for rectangular cross sections with one dimension much larger than the other, the critical height for detonation transmission through this kind of orifice is only 3λ [31]. When the diffraction takes place through a diverging cone, the value of k_c may drop significantly for values smaller than 40° of the cone angle [32]. From the example of Figure 8.5.10 [33], it can be observed that the re-initiation of the detonation results from the interaction of the leading wave with the walls, and a so-called super-detonation propagates along the front before giving rise to the self-sustained detonation once again. More complicated situations have been investigated, such as the case of transmission with an obstacle at the tube exit [34], as shown in Figure 8.5.11. In this example, the obstacle has a conic shape and the detonation is extinguished at the exit of the annular space. However, re-initiation occurs through a complicated process of interaction between the toroïdal shock transmitted and the rear wall of the obstacle. Numerical simulations performed for the same conditions as the experiments fit very well with the experimental results: mechanism of detonation transmission, characteristic shape of the flow, and distances of detonation re-formation, size of the different cellular structures, etc. With

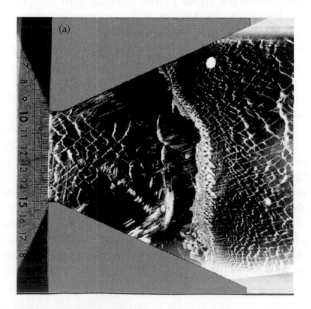

FIGURE 8.5.10
Soot patterns of detonation transmission from a tube through a cone ($\propto = 25°$) in a $C_2H_2 + 2.5\ O_2$ mixture at 30 mbar initial pressure. (Courtesy of V. Guilly.)

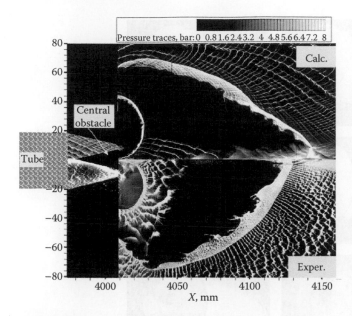

FIGURE 8.5.12
Soot patterns of the double-cellular structure of detonation in a H_2-NO_2/N_2O_4 mixture at equivalent ratio 1.1 and 0.5 bar initial pressure. (Courtesy of F. Joubert.)

FIGURE 8.5.11
Diffraction of detonation from a tube ($\phi=52$mm) to half-space through an annular orifice with central conical obstacle ($\propto - 15°$) in a $C_2H_2 + 2.5O_2$ mixture at 33 mbar initial pressure. Comparison of experimental soot patterns with numerical simulations. (Courtesy of B. A. Khasainov.)

such device at the tube exit, it is possible to divide the value of the transmission coefficient k_c by a factor of 2.

8.5.7 Heat Release Process

Until recently, most of the classical models used to describe the detonation wave (and to perform numerical simulations) assumed that the heat release in the detonation wave occurred in "one stage" (whether a global or a detailed chemical-kinetic scheme was considered). Since the so-called cellular structure is related to the induction length of chemical reactions, this results in a characteristic size λ of this structure typical for each mixture, as explained earlier. However, it appears that in certain mixtures with particular intermediate steps of chemical reactions, the heat-release process occurs non-monotonously in "two stages" (or even several), clearly separated in time. Registration of the cellular structure on soot foils in the detonation in this particular kind of mixture, reveals the existence of two different networks of cells [35,36], each with its characteristic size, as shown in Figure 8.5.12. A first network is composed of "large" cells, inside which a second network of "small" cells exists. The dimension of each cellular structure is related to the induction delay of the chemical reactions corresponding to the different stages of heat release, with the smallest cells corresponding to the first stage. From these observations, it is obvious that the problem of detonation transmission is even more complicated than what was believed previously, and more

sophisticated correlations between the cell size and a characteristic dimension of the confinement have to be considered.

8.5.8 Spinning Detonations

Spinning detonation is an interesting situation, because it corresponds to the limit case of detonation propagation with a unique triple point. When the propagation takes place in a tube of circular cross section, this triple point follows a helical trajectory along the wall and the leading front rotates around the tube axis as it proceeds in the longitudinal direction. Since its discovery at the end of the 1920s [37,38], the phenomenon of spinning detonations has been the subject of extensive experimental studies in various conditions to determine the characteristic parameters of this propagation regime, such as spin pitch, track angle, variations of the detonation velocity during one cycle, etc. Acoustic models were proposed by Manson [1] and Fay [2], explaining the periodic oscillations by the transverse vibration modes of the gases in the tube. However, the acoustic models cannot describe the detailed structure of the spinning detonations. Investigation of the detonation wave structure is a very difficult task on account of the large extent of the nonstationary-reaction zone behind the leading shock front. In recent years, 3D-numerical simulations of the spinning detonations [39] performed with reasonably good accuracy (see Figure 8.5.13) have permitted to progress in the understanding of how coupling between the leading shock front and the combustion zone is achieved. The difficulty to understand even this configuration with a unique triple point indicates how far we are still from giving a comprehensive model of the mechanisms of detonation wave propagation.

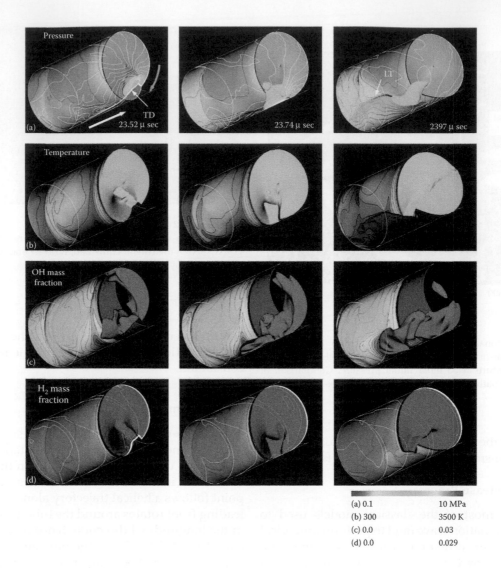

(a) 0.1	10 MPa
(b) 300	3500 K
(c) 0.0	0.03
(d) 0.0	0.029

FIGURE 8.5.13
Numerical simulation of a spinning detonation in H_2/air mixture in a circular tube at various times. Gray and green space isosurfaces in pressure are the detonation front and the pressure of 6 MPa. White arrow: propagating direction of the detonation front, pink arrow: rotating direction of the transverse detonation. TD—transverse detonation, and LT—long pressure trail. (Reprinted from Tsuboi, N., Eto, K., and Hayashi, A.K., *Combust. Flame*, 149, 144, 2007. With permission.)

References

1. N. Manson, Sur la structure de l'onde hélicoïdale dans les mélanges gazeux, *Comptes Rendus Acad. Sci. Paris*, 222, 46–51, 1946.
2. J.A. Fay, Mechanical theory of spinning detonations, *J. Chem. Phy.*, 20, 942–950, 1952.
3. B.V. Voitsekhovskii, *Doklady Akad. Nauk. SSSR*, 114(4), 717–720, 1957.
4. Yu.N. Denisov and Ya.K. Troshin, Pulsating and spinning detonation of gaseous mixtures in tubes. *Doklady Akad. Nauk. SSSR*, 125, 110–113, 1959.
5. D.R. White, Turbulent structure of gaseous detonation, *Phys. Fluids*, 4(4), 465–480, 1961.
6. R.A. Strehlow, The nature of transverse waves in detonations, *Astronautica Acta*, 14, 539–548, 1969.
7. R.A. Strehlow, Multi-dimensional detonation wave structure, *Astronautica Acta*, 15, 345–357, 1970.
8. R. Knystautas, C. Guirao, J.H.S. Lee, and A. Sulmistras, Measurements of cell size in hydrocarbon-air mixtures and prediction of critical tube diameter, critical initiation energy and detonability limits, *AIAA Prog. Astronautics Aeronautics*, 94, 23–37, 1984.
9. J.C. Libouton, A. Jacques, and P. Van Tiggelen, Cinétique, structure et entretien des ondes de détonation, *Actes du Colloque international Berthelot-Vieille-Mallard-Le Châtelier*, 2, 437–442, 1981.
10. J.E. Shepherd, http://www.galcit.caltech.edu/detn_db/html/db.html, 2005.

11. K.I. Schchelkin and Ya.K. Troshin, Non stationary phenomena in the gaseous detonation front, *Combust. Flame,* 7, 143–151, 1963.
12. I.O. Moen, J.W. Funk, S.A. Ward, G.M. Rude, and P.A. Thibault. Detonation length scales for fuel-air explosives. *Prog. Astron. Aeron.,* 94, 55–79, 1984.
13. R. Takai, K. Yoneda, and T. Hikita, Study of detonation wave structure. *Proceedings 15th Symposium (International) on Combustion,* The Combustion Institute, Pittsburg, pp. 69–78, 1974.
14. T.J. Anderson and E.K. Dabora, Measurements of normal detonation wave structure using Rayleigh imaging, *Proceedings 24th Symposium (International) on Combustion,* The Combustion Institute, Pittsburg, pp. 1853–1860, 1992.
15. S. Taki and T. Fujiwara, Numerical analysis of two-dimensional nonsteady detonations, *AIAA J.,* 16, 73–77, 1978.
16. E.S. Oran, J.P. Boris, T. Young, M. Flanigan, T. Burks, and M. Picone, Numerical simulations of detonations in hydrogen-air and methane-air mixtures. *Proceedings 18th Symposium (Int.) on Combustion,* The Combustion Institute, Pittsburgh, PA, pp. 1641–1649, 1981.
17. V.V. Markov, Numerical simulations of the formation of multifront structure of detonation wave. *Doklady Akademii Nauk SSSR,* 258, 314–317, 1981.
18. S.U. Schöffel and F. Ebert, Numerical analyses concerning the spatial dynamics of an initially plane gaseous ZND detonation. *AIAA Progr. Astron. Aeron.,* 114, 3–31, 1988.
19. D.N. Williams, L. Bauwens, and E.S. Oran, Detailed structure and propagation of three-dimensional detonations, *Proceedings 26th Symposium (International) on Combustion,* The Combustion Institute, Pittsburg, PA, pp. 2991–2998, 1996.
20. V.N. Gamezo, D. Desbordes, and E.S. Oran, Formation and evolution of two-dimensional cellular detonations, *Combust. Flame,* 116, 154–165, 1999.
21. V.N. Gamezo, A.A. Vasil'ev, A.M. Khokhlov, and E.S. Oran, Fine cellular structures produced by marginal detonations. *Proceedings 28th Symposium (International) on Combustion,* The Combustion Institute, Pittsburg, PA, 28, pp. 611–617, 2000.
22. F. Pintgen, C.A. Eckett, J.M. Austin, and J.E. Shepherd, Direct observations of reaction zone structure in propagating detonations, *Combust. Flame,* 133, 211–229, 2003.
23. V.N. Gamezo, T. Ogawa, and E.S. Oran, Flame acceleration and ddt in channels with obstacles: Effect of obstacle spacing. *Combust. Flame,* published online July 2008.
24. M.I. Radulescu and J.H.S. Lee, The failure mechanism of gaseous detonations: Experiments in porous wall tubes, *Combust. Flame,* 131, 29–46, 2002.
25. D. Desbordes, Correlation between shock wave predetonation zone size and cell spacing in critically initiated spherical detonations, *Prog. Astron. Aeron.,* 106, 166–180, 1986.
26. A.A. Vasiliev and V.V. Grigoriev, Critical conditions for gas detonations in sharply expanding channels, *Fizika Gorenyia i Vzryva,* 16, 117–125, 1980.
27. J.H.S. Lee, Dynamic parameters of gaseous detonations, *Ann. Rev. Fluid Mech.,* 16, 311–336, 1984.
28. V.V. Mitrofanov and R.I. Soloukhin, Diffraction of multifront detonation waves, *Doklady Akad. Nauk. SSSR,* 159(5), 1003–1006, 1964.
29. I.O. Moen, A. Sulmistras, G.O. Thomas, D.J. Bjerketvedt, and P.A. Thibault. Influence of regularity on the behaviour of gaseous detonations. *Prog. Astron. Aeron.,* 106, 220–243, 1986.
30. G. Cicarelli, Critical tube measurements at elevated initial mixture temperatures, *Combust. Sci. Tech.,* 174, 173–183, 2002.
31. Y.K. Liu, J.H.S. Lee, and R. Knystautas, Effect of geometry on the transmission of detonation through an orifice, *Combust. Flame,* 56(2), 215–225, 1984.
32. B.A. Khasainov, H.-N. Presles, D. Desbordes, P. Demontis, and P. Vidal, Detonation diffraction from circular tubes to cones, *Shock Waves,* 14(3), 187–192, 2005.
33. V. Guilly, Etude de la diffraction de la detonation des mélanges C_2H_2-O_2 stoechiométriques dilués par l'argon. Thèse de l'Université de Poitiers, 2007.
35. M.-O. Sturtzer, N. Lamoureux, C. Matignon, D. Desbordes et H.-N. Presles, On the origin of the double cellular structure of the detonation in gaseous nitromethane and its mixtures with oxygen, *Shock Waves,* 14(1–2), 45–51, 2005.
36. F. Joubert, D. Desbordes, and H.N. Presles, Structure cellulaire de la détonation des mélanges H_2-NO_2/ N_2O_4, *Comptes Rendus Acad. Sci. Mécanique,* 331, 365–372, 2003.
37. C. Campbell and D.W. Woodhead, *J. Chem. Soc.,* 3010–3021,1926.
38. W.A. Bone and R.P. Fraser, *Philos. Trans. R. Soc. Sect.,* A 228, 197–234, 1929.
39. N. Tsuboi, K. Eto, and A.K. Hayashi, Detailed structure of spinning detonation in a circular tube, *Combust. Flame,* 149, 144–161, 2007.

Index